BASIC WELDING TECHNIQUES
THREE BOOKS IN ONE

IVAN H. GRIFFIN,
EDWARD M. RODEN AND
CHARLES W. BRIGGS

ARC
OXYACETYLENE
TIG & MIG

VNR VAN NOSTRAND REINHOLD COMPANY
NEW YORK CINCINNATI ATLANTA DALLAS SAN FRANCISCO
LONDON TORONTO MELBOURNE

Van Nostrand Reinhold Company Regional Offices:
New York Cincinnati Chicago Millbrae Dallas

Van Nostrand Reinhold Company International Offices:
London Toronto Melbourne

Library of Congress Catalog Card Number: 76-53578
ISBN: 0-442-22861-9

Manufactured in the United States of America

Published by Van Nostrand Reinhold Company
450 West 33rd Street, New York, N.Y. 10001
Published simultaneously in Canada by Van Nostrand Reinhold Ltd.

15 14 13 12 11 10 9 8 7 6 5 4 3

Library of Congress Cataloging in Publication Data

Griffin, Ivan H
Basic welding techniques.

1. Electric welding. 2. Gas tungsten arc welding. 3. Gas metal arc welding.
4. Oxyacetylene welding and cutting. I. Roden, Edward M., joint author. II. Briggs, Charles W., joint author. III. Title.
TK4660.G744 1977 671.5'2 76-53578
ISBN 0-442-22861-9

BASIC ARC WELDING

PREFACE

The electric arc is one of the most common sources of the heat required to fuse metals. *Basic Arc Welding* emphasizes performance in electric arc welding.

The approach employed is to allow students to learn the fundamentals of electric arc welding, by welding pieces of steel plate. Each unit explores a new aspect of welding with an electric arc. The first few units of this book are introductory, and present the background knowledge the student will draw upon in completing the procedural units. After the introductory units each unit includes a procedure for a specific kind of weld or type of equipment. All of the units include review questions to check the student's progress. To provide an opportunity to see the effect of varying the procedure or manipulation of the equipment, the student is encouraged to experiment.

No previous skill, knowledge, or training in welding is required for the student to use this book. However, some experience with metals and basic metalworking tools will prove helpful.

Different welding courses for different trades or occupations can be constructed by selecting appropriate units from this book. However, care should be exercised not to omit essential prerequisite units.

CONTENTS

CHARTS

APPLICATION CHART – BASIC ARC WELDING

UNIT NUMBER	AUTO MECHANIC	BOILER MAKER	BRICKLAYER	CARPENTER	ELECTRICIAN	FARM EQUIPMENT REPAIR	GENERAL WELDING	IRON WORKER (ORNAMENTAL)	IRON WORKER (STRUCTURAL)	MACHINIST	PLUMBER	SHEET METAL WORKER	STEAMFITTER
1	■	■	■	■	■	■	■	■	■	■	■	■	■
2	■	■	■	■	■	■	■	■	■	■	■	■	■
3	■	■	■	■	■	■	■	■	■	■	■	■	■
4	■	■	■	■	■	■	■	■	■	■	■	■	■
5	■	■	■	■	■	■	■	■	■	■	■	■	■
6	■	■	■	■	■	■	■	■	■	■	■	■	■
7	■	■	■	■	■	■	■	■	■	■	■	■	■
8	■	■	■	■	■	■	■	■	■	■	■	■	■
9	■	■	■	■	■	■	■	■	■	■	■	■	■
10	■	■	■	■	■	■	■	■	■	■	■	■	■
11		■					■	■	■			■	■
12	■	■	■	■	■	■	■	■	■	■	■	■	■
13	■	■	■	■	■	■	■	■	■	■	■	■	■
14		■				■	■	■	■		■	■	■
15		■				■	■	■	■		■	■	■
16		■					■	■	■			■	■
17		■					■	■	■			■	■
18		■					■	■	■			■	■
19		■					■	■	■			■	■
20		■					■	■	■			■	■
21		■					■	■	■			■	■
22	■	■	■	■	■		■	■	■			■	■
23	■	■	■	■	■		■	■	■			■	■
24		■					■	■	■			■	■
25		■					■	■	■			■	■
26		■					■	■	■			■	■

Note: The above chart is in terms of suggested minimums only. The final choice of course content is a function of the individual instructor, often with the advice of an industry advisory committee.

■ – Required

UNIT 1 THE ARC WELDING PROCESS

One of the most important processes in industry is the fusion of metals by an electric arc. This is commonly called *arc welding.*

Briefly, the process takes place in the following manner. The work to be welded is connected to one side of an electric circuit, and a metal electrode is connected to the other side. These two parts of the circuit are brought together and then separated slightly. The electric current jumps the gap and causes a continuous spark called an *arc.* The high temperature of this arc melts the metal to be welded, forming a molten puddle. The electrode also melts and adds metal to the puddle, figure 1-1.

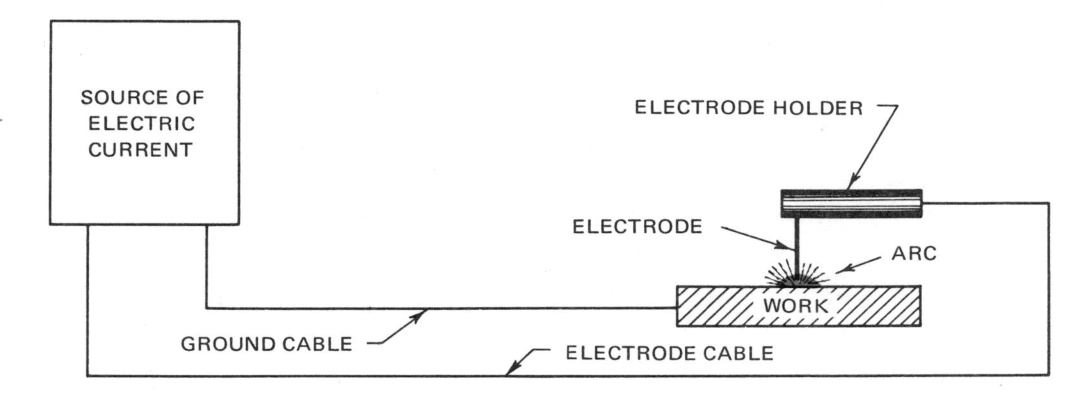

Fig. 1-1 Simple welding circuit

As the arc is moved, the metal solidifies. The metal fuses into one piece as it solidifies.

The melting action is controlled by changing the amount of electric current which flows across the arc and by changing the size of the electrode.

HAZARDS

Before arc welding is begun, the student should be fully aware of the personal dangers involved. Of course, the high-temperature arc and the hot metal can cause severe burns. However, the electric arc itself may be a hazard.

An electric arc gives off large amounts of ultraviolet and infrared rays. Infrared rays are also given off from the molten metal. Both types of rays given off from arc welding are invisible, just as they are when given off from the sun. They will cause sunburn, the same as they will from the sun, except that the rays given off from the electric arc, burn much more rapidly and deeply. Since these rays are produced very close to the operator, they can cause severe damage to the eyes in a very short period of time.

During arc welding there is a danger that small droplets of molten metal may leave the arc and fly in all directions. These so-called sparks range in temperature from 2,000 degrees F.

to 3,000 degrees F., and in size from very small to as large as 1/4 inch in diameter. They may cause burns plus they are a fire hazard when they fall on flammable material.

PROTECTIVE DEVICES

For protection from the rays of the arc and the flying sparks, the welding operator must use a helmet, figure 1-2, and other protective devices. The welding helmet is fitted with filter plates that screen out over 99% of the harmful rays. The helmet must be in place before attempting to do any arc welding. The arc is harmful up to a distance of 50 feet and all persons within this range must be careful that the rays do not reach their eyes.

Most welding helmets are made of pressed composition material or molded plastic. If they are dropped or if material is dropped on them they may be unfit for use. Each helmet is equipped with an adjustable headband. Any attempt to use wrenches or pliers to force the adjusting device may destroy the helmet.

The filter plate in each helmet is a special, costly glass which should be handled with great care. All filter plates should be protected from the flying globules of molten metal in the manner indicated in figure 1-3.

All welding stations should be equipped with curtains or other devices which keep the arc rays confined to the welding area. For the protection of others, the welder should make sure that these curtains are in place before starting any welding.

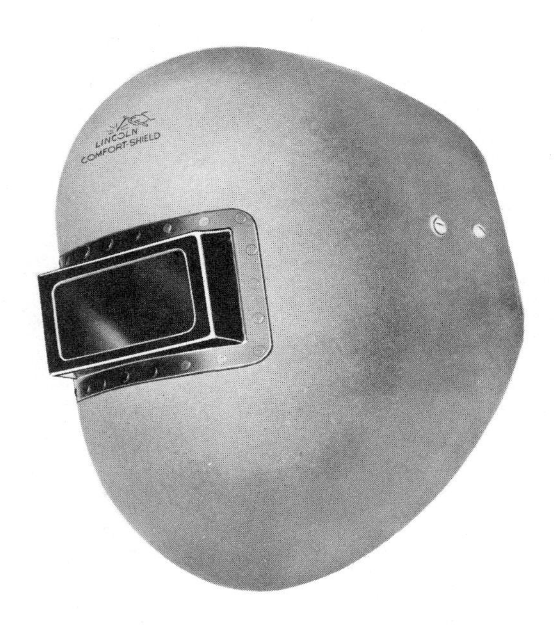

Fig. 1-2 A modern welding helmet

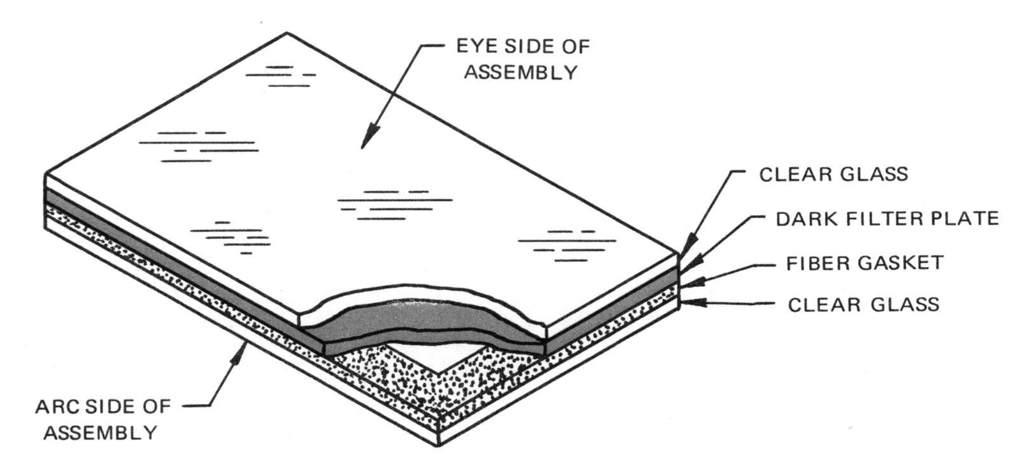

Fig. 1-3 Filter plate protective assembly

Fig. 1-4 Note protective devices

The arc rays will penetrate one thickness of cloth and cause sunburn. Therefore, the operator must protect himself with fire-resistant aprons, sleeves, and gloves to eliminate this hazard plus the hazard of fire, figure 1-4. In fact, all clothing worn by the operator should be reasonably flame resistant. Clothing which has a fuzzy surface can be a serious fire hazard, particularly if it is cotton.

Naturally, all other types of flammable material such as oil, wood, paper, and waste should be removed from the welding area before any welding is attempted. Each welder should be acquainted with the location and operating characteristics of all fire extinguishers.

REVIEW QUESTIONS

1. Why is clear glass used on the eye side of the filter plate assembly in figure 1-2?

2. What determines how fast the weld metal melts?

3. Who is responsible for the protection of workers in the area of the welding operation?

4. What type of rays are given off by the electric arc?

5. Can the rays given off by the electric arc be seen?

UNIT 2 SOURCES OF ELECTRICITY FOR WELDING

TYPES OF WELDING MACHINES

Electric current for the welding arc is generally provided by one of two methods. A transformer which reduces the line voltage can provide *alternating current* (AC). This current reverses direction 120 times per second. The transformer has no moving parts.

Direct current (DC) for the welding arc may be produced by a direct-current generator connected by a shaft to an AC motor, figure 2-1. A gasoline engine or other type of power may also be used to turn the generator. Direct current flows in the same direction at all times. In any case, the welding machine must have the ability to respond to the need for rapid changes in the welding voltage and current.

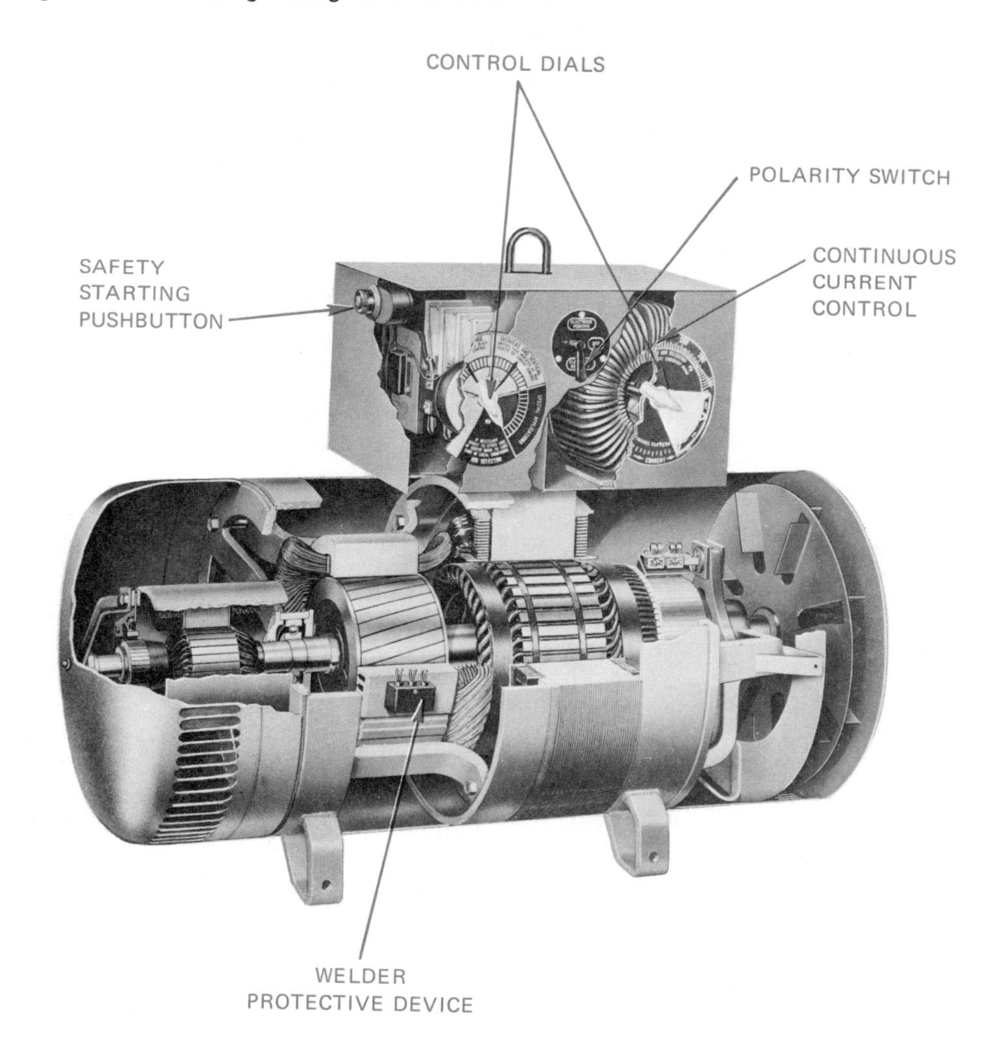

Fig. 2-1 A motor generator-type of DC welding machine

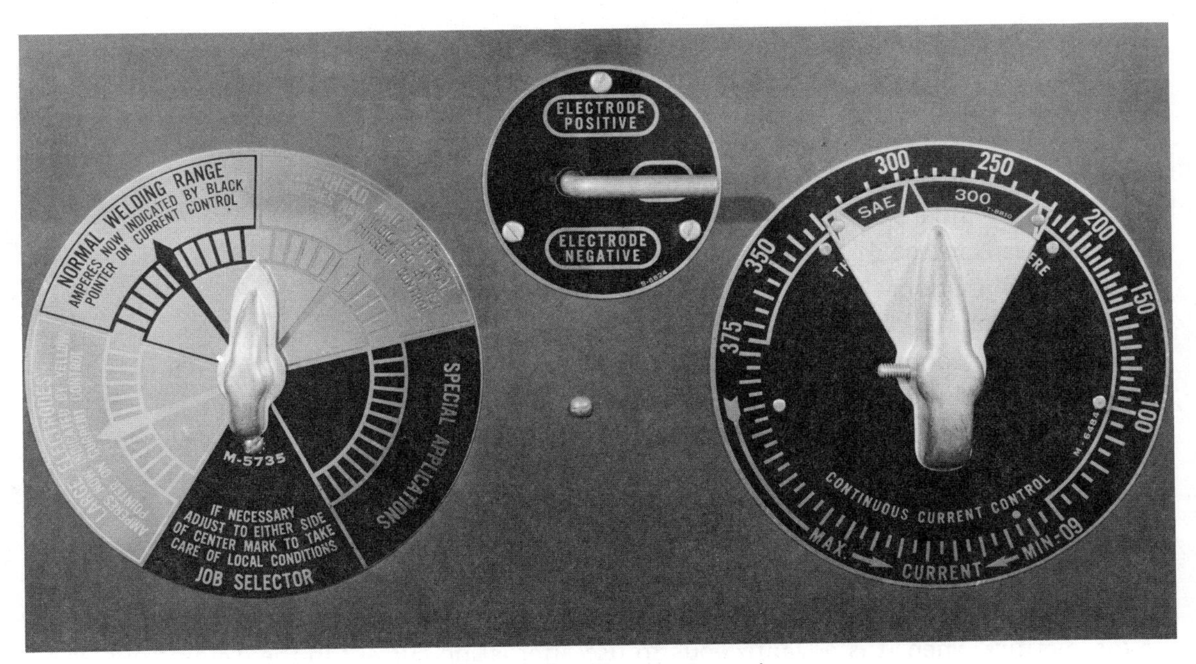

Fig. 2-2 Dual-current control

Both types of machines are widely used in industry, but the DC type is slightly more popular. Both are supplied in various sizes, depending on the use to which they are to be put, and are designated by the maximum continuous current in amperes which they can supply, (e.g., 150a or 300a).

CURRENT CONTROL

Different types of welding operations require different amounts of current (amperes). Therefore, arc welding machines must have a way of changing the amount of current flowing to the arc.

The DC generator may have either a *dual-current control,* figure 2-2, (the more popular type) or a *single-current control.* In the dual-current control type, two handwheels or knobs adjust the electrical circuits to provide the proper current to the arc. In the single-control

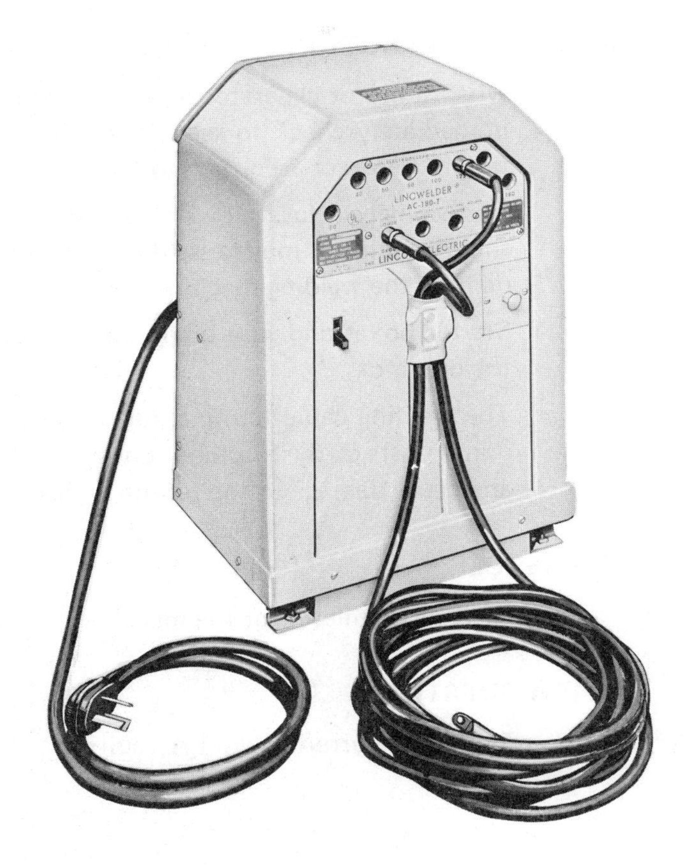

Fig. 2-3 AC machine with taps for current control

type, a single wheel adjusts the current. The other controls on a DC machine are an on-off switch and a polarity switch.

In one type of current control for the AC machine, the output cables are moved from one tap to another on the secondary side of the transformer winding, figure 2-3. The disadvantage of this type is that the number of current combinations is limited by the number of taps in the machine.

The other popular type of AC machine has a movable core in the transformer. On this type of machine the operator selects the desired current rating by turning a handwheel.

Another type of machine which has gained popularity is the transformer-type with a built-in rectifier. The rectifier converts alternating current to direct current for welding. In some types of machines, the alternating current can be taken ahead of the rectifier when it is advantageous to use alternating current in the welding operation, figure 2-4.

Fig. 2-4 AC/DC machine – single-current control

CARE AND PRECAUTIONS

- When a motor-generator type of welding machine is turned on, the operator should immediately check to see if the armature is rotating and that the direction of rotation is correct according to the arrow on the unit.
- Occasionally a fuse blows or a starter contact becomes burned or worn. Either of these conditions may cause the machine to overheat if it is left on. This can rapidly damage the welding machine.
- Starter boxes and fuse boxes carrying 220 or 440 volts should not be opened by the operator.
- The welding cable terminal lugs should be clean and securely fastened to the terminal posts of the machine. Loose or dirty electrical connections tend to overheat and cause damage to the terminal posts.

REFERENCE

Manufacturer's bulletin for the machine to be used.

REVIEW QUESTIONS

1. What is a direct-current welding circuit?

2. What does the term current control mean?

3. What effect do loose connections have on the welding circuit?

4. What are the three types of welding machines?

5. How many times per second does alternating current reverse direction?

UNIT 3 THE WELDING CIRCUIT

PARTS OF THE WELDING CIRCUIT

In addition to the source of current, the welding circuit consists of:

- The work
- The welding cables
- The electrode holder
- The electrode

The *work* with which the welder is concerned may be steel plate, pipe, and structural shapes of varying sizes and thicknesses. It should be suitably positioned for the job being done. The work is a conductor of electricity and thus, is a part of the circuit.

The *welding cables,* figure 3-1, are flexible, rubber-covered copper cables of a large enough size to carry the necessary current to the work and to the electrode holder without overheating. The size of the cable depends on the capacity of the machine, and the distance from the work to the machine. A *ground clamp,* figure 3-2, is attached to the end of one of the cables, so that it may be connected to the work.

The *electrode holder,* figure 3-3, is a mechanical device on the end of the welding cable which clamps the welding rod or electrode in the desired position. It also provides

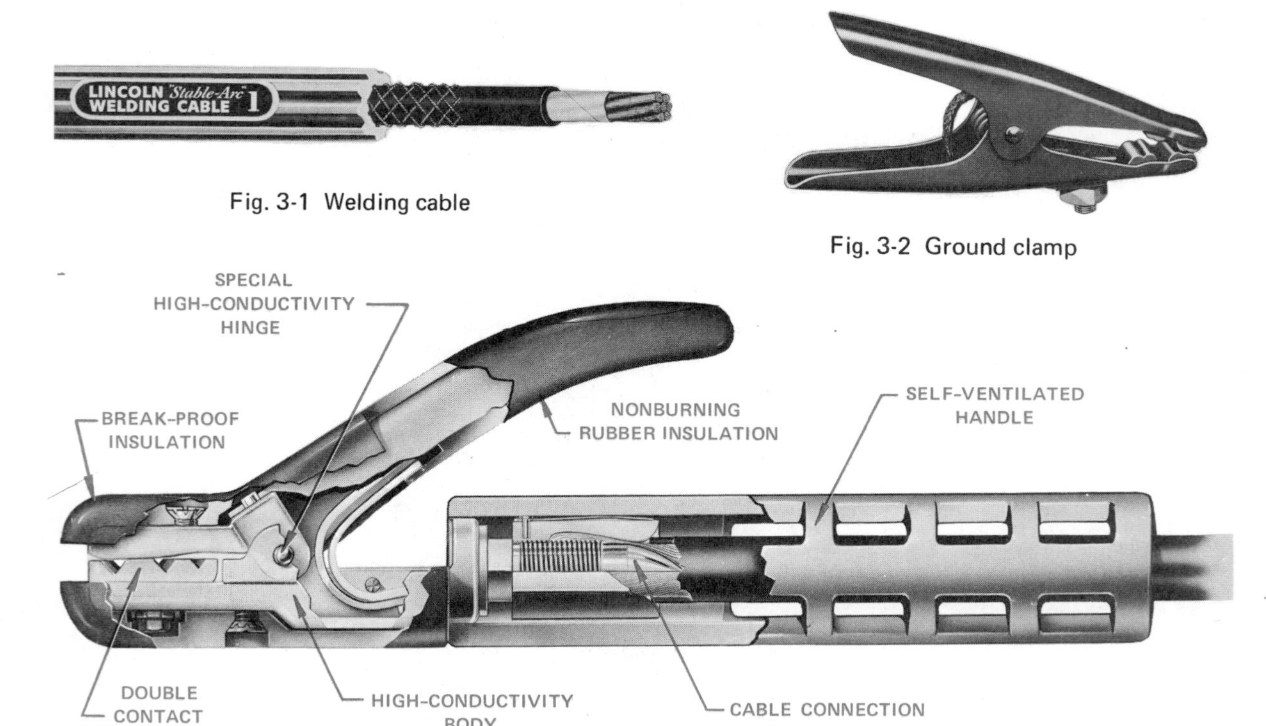

Fig. 3-1 Welding cable

Fig. 3-2 Ground clamp

Fig. 3-3 A cutaway view of an electrode holder

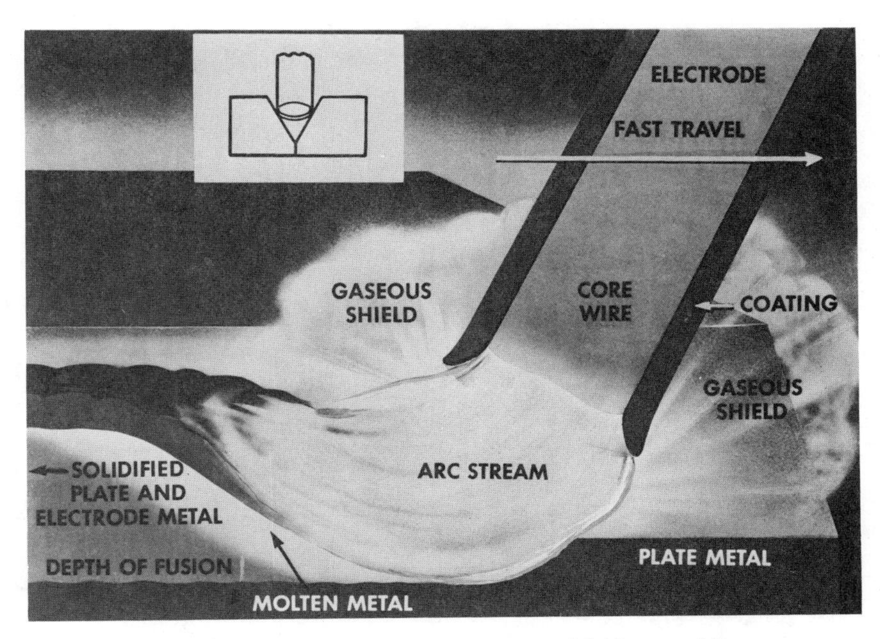

Fig. 3-4 Coated electrodes form a gaseous shield around the arc

an insulated handle, with which the operator can direct the electrode and arc. These holders come in various sizes depending on the amperage which they are required to carry to the electrode.

THE ELECTRODE

The electrode usually has a steel core. This core is covered with a coating containing several elements, some of which burn under the heat of the arc to form a gaseous shield around the arc, figure 3-4. This shield keeps the harmful oxygen and nitrogen in the atmosphere away from the welding area.

Other elements in the coating melt and form a protective slag over the finished weld. This slag promotes slower cooling and also protects the finished weld or bead from the atmosphere. Some coated electrodes are designed with alloying elements in the coating which change the chemical and physical characteristics of the deposited weld metal.

The result of using properly designed coated electrodes is a weld metal which has the same characteristics as the work, or base metal, being joined.

Electrodes are supplied commercially in a variety of lengths and diameters. In addition, they are supplied in a wide variety of coatings for specific job applications. These applications are discussed in other units.

The American Welding Society and The National Electrical Manufacturer's Association classify electrodes according to the type of coating, operating characteristics, and chemical composition of the weld metal produced. Chart 3-1 indicates some of these electrodes with their color code markings. These represent the commonly used rods. There are many more. Most electrode manufacturers supply, free of charge, a chart of all the electrodes they make.

A growing number of electrode manufacturers mark the grip end of the electrodes with their classification, either EXXXX, as indicated in Chart 3-1, or simply XXXX so the welder does not have to memorize the color code.

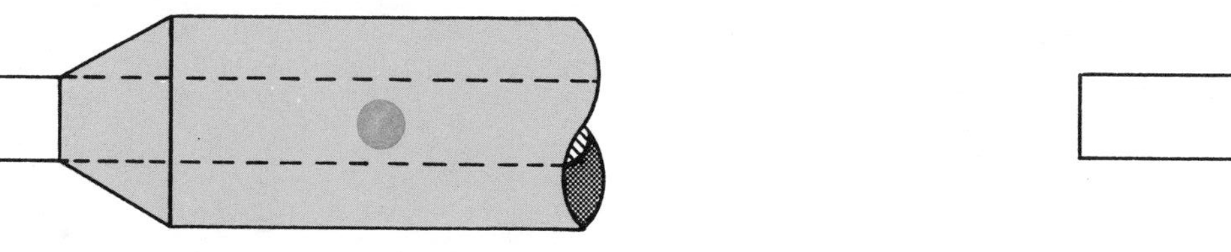

COLOR CODE LOCATION

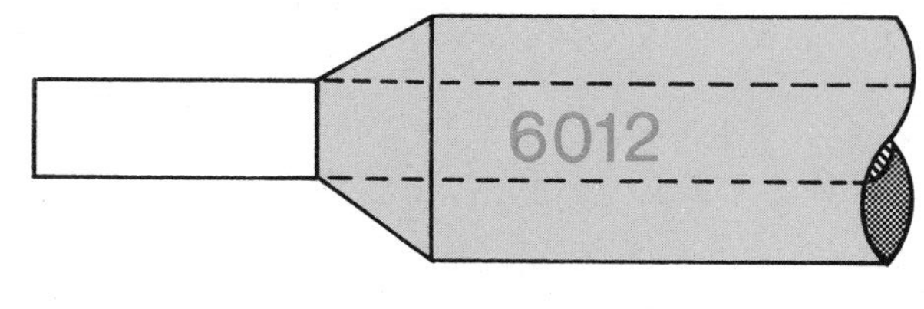

A W S NUMBER LOCATION

AWS Classification	Type	Current and Polarity	Tensile Strength P.S.I.	Yield Point P.S.I.	% Elongation in 2 inches	Color Code			Applications
						End	Spot	Group	
E-4510	Bare	d.c. Negative	45-50,000	40-45,000	5-7%	None	None	None	Building up worn surfaces and training operators.
E-6010	Coated	d.c. Positive	65-72,000	53-58,000	27-30%	None	None	None	All-position, all-purpose high-impact, high-ductility code welding
E-6011	Coated	a.c. or d.c. Positive	65-72,000	53-58,000	27-30%	None	Blue	None	All-position, primarily for producing welds with a.c. current equal to E-6010
E-6012	Coated	a.c. or d.c. Negative	68-78,000	58-68,000	20%	None	White	None	All-position, for poor "fitup" work and work where resistance to impact and low ductility are not too important.
E-6013	Coated	a.c. or d.c. Negative	75,000	62,000	20%	None	Brown	None	For all-position work primarily with a.c. current. Compares to E-6012 series in general applications
E-6020	Coated	a.c. or d.c. Negative	68,000	56,000	32%	None	Green	None	For flat-position welding, used in code welding. Being replaced slowly by E-6027 and E-7024
E-7014	Coated Iron Powder	a.c. or d.c. Negative	72-82,000	62-72,000	23-32%	Black	Brown	None	High-speed production work, faster deposition rate than E-6012 or E-6013
E-7024	Coated Heavy Iron Powder	a.c. or d.c.	75-83,000	63-75,000	17-25%	Black	Yellow	None	Fast deposition rate, excellent appearance — for joints not requiring deep penetration
E-6027	Coated Iron Powder	a.c. or d.c. Negative	62-69,000	52-60,000	25-35%	None	Silver	None	Iron powder version of E-6020 — faster deposition for flat and horizontal positions

Chart 3-1 Mild Steel Electrode Chart

Only the code markings are important when determining the type of electrode. The coating colors should not be depended on for recognition as they vary with manufacturers. Some producers also place dots and other markings on the coating. These are only trademarks, and are not to be confused with the code markings, which appear only on the grip end of the electrodes.

Any of the electrodes shown in Chart 3-1 may be purchased in a wide variety of sizes and lengths.

AWS CLASSIFICATION NUMBERS

All AWS (American Welding Society) numbers consist of three parts. For example, E-6010.

1. The E in all cases indicates an electric arc welding electrode or rod.
2. The number following the E (in this case, 60) indicates the minimum tensile strength of the weld metal in thousands of pounds per square inch, (in this case, 60,000 p.s.i.). This number could be 80, 100, or 120 which would indicate minimum tensile strengths respectively of 80,000 p.s.i., 100,000 p.s.i., or 120,000 p.s.i.
3. In a four-digit number the third digit indicates the positions in which the electrode may be used: 1—indicates all positions; 2—flat or horizontal; 3—deep groove.
4. The fourth digit indicates the operating characteristics, such as polarity, type of covering, bead contour, etc.

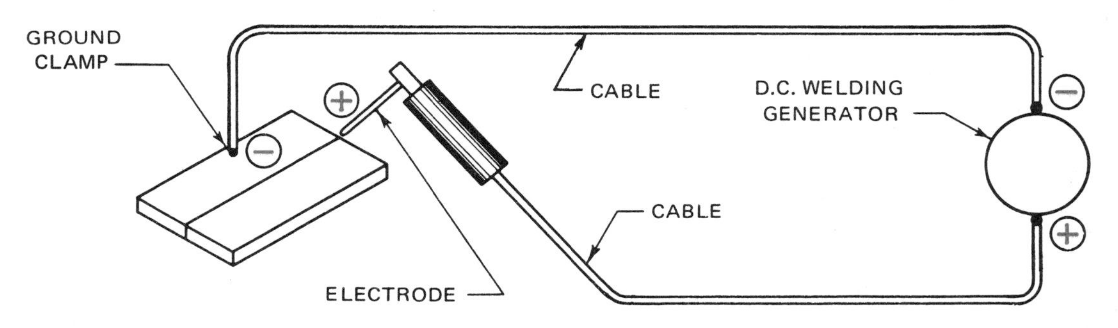

POSITIVE OR REVERSE POLARITY

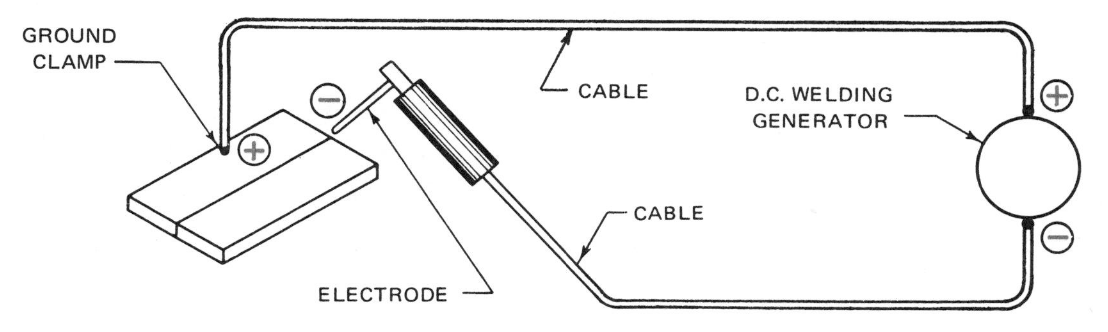

NEGATIVE OR STRAIGHT POLARITY

Fig. 3-5 The welding circuit

POLARITY

In welding with direct current, the electrode must be connected to the correct terminal of the welding machine. This polarity may be changed by a switch on the welding machine. The polarity to be used is determined by the type of electrode and is indicated in the electrode chart, Chart 3-1.

When the electrode is connected to the negative terminal (–), the polarity is called *negative or straight.* When connected to the positive terminal (+), it is called *positive or reverse,* figure 3-5. The use of incorrect polarity produces a poor weld. When welding with alternating current, polarity is not considered.

A simple test for checking the polarity of an electric welding machine is as follows:

1. Place a carbon electrode in the electrode holder.
2. Strike an arc. Maintain the puddle and weld for 5 or 6 inches.
3. Check the plate for smears or black smudges. If these are present, the machine is in reverse polarity.

REFERENCE

Manufacturer's chart of electrodes

REVIEW QUESTIONS

1. What is the primary purpose of the AWS code markings on welding rods?

2. What are the effects of oxides and nitrides in the weld metal?

3. Referring to Chart 3-1, what AWS type of electrode is used if the strength of the weld is the only characteristic of importance for a given job?

4. If it is necessary to weld pressure pipe in the overhead position and the only machine available is an alternating-current type (AC), what AWS classification rod is used?

UNIT 4 FUNDAMENTALS OF ARC WELDING

VARIABLES

Four things greatly affect the results obtained in electric arc welding. To make good welds, each one must be adjusted to fit the type of work done and the equipment being used.

They are:

- Current setting or amperage
- Length of arc or arc voltage
- Rate of travel
- Angle of the electrode

CURRENT SETTING

The current which the welding machine supplies to the arc must change with the size of the electrode being used. Large electrodes use more current than smaller sizes. A good general rule to follow is: when welding with standard coated electrodes, the current setting should be equal to the diameter of the electrode in thousandths of an inch.

Thus, a 1/8-inch electrode measures .125 inch and operates well at 125 ± a few amperes. Similarly, a 5/32-inch rod measures .156 inch and operates well at 150 ± a few amperes. The ± indicates that these electrodes will operate well in a range of current values either below or above the indicated amperage. For example, a value of 125 ± 10 amperes indicates a range of values with a low of 115 amperes and a high of 135 amperes.

When indicating the diameter of the electrodes, reference is made only to the steel or alloy core of the rod, figure 4-1. The overall diameter including the rod coating is not the indicated electrode size.

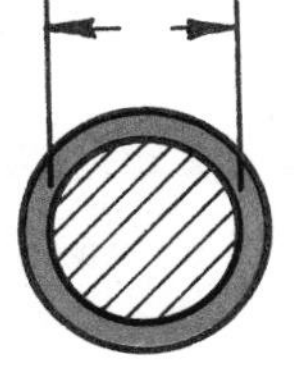

Fig. 4-1 Measure core of rod

LENGTH OF ARC

The arc length is one of the most important considerations in arc welding. Variations in arc length produce varying results.

The arc length increases as the arc voltage increases. For example, an arc 3/16 inch long requires three times the voltage of a 1/16-inch arc, figure 4-2.

The general rule on arc length states: The arc length shall be slightly less than the diameter of the electrode being used.

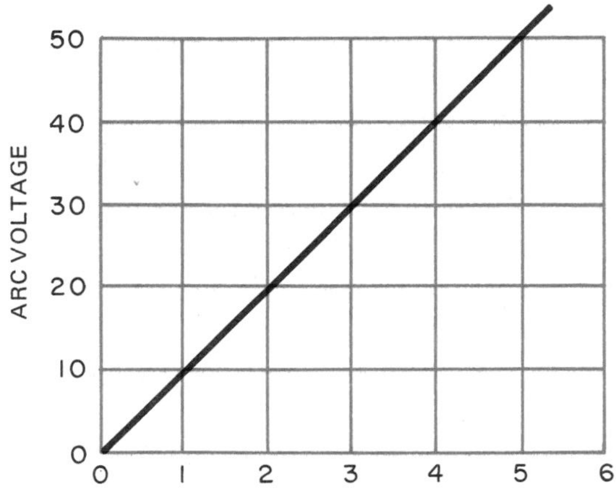

Fig. 4-2 Arc length in 1/16-inch increments

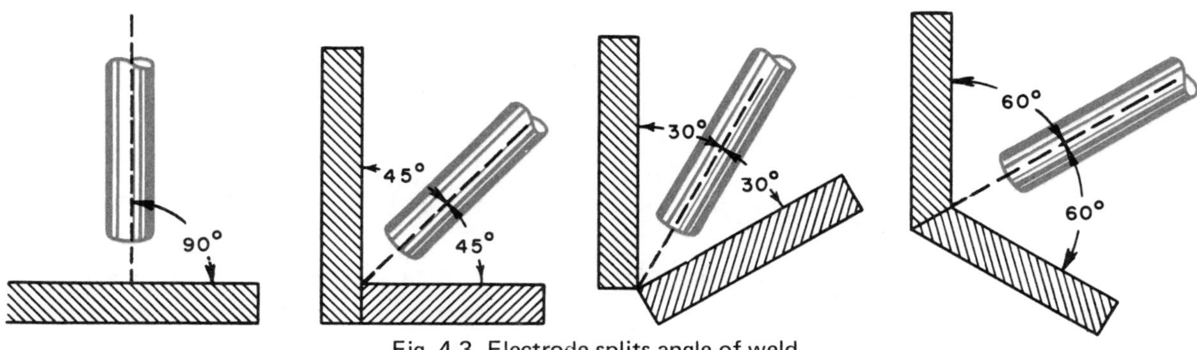

Fig. 4-3 Electrode splits angle of weld

Thus, a 5/32-inch diameter electrode operates well between 1/8 inch and 5/32 inch of arc gap, or 20-22 arc volts according to the chart, figure 4-2.

It is almost impossible for the operator to measure the arc length accurately when welding. However, the welder can be guided by the sound of the arc. At the proper arc length, the sound is a sharp, energetic crackle. Proper arc length is determined by noting the difference in the sound of the arc when it is set too far, and at just about the right distance from the work. By practicing this, the operator will be able to judge good arc length by the distinctive sound.

RATE OF TRAVEL

The *rate of travel* of the arc changes with the thickness of the metal being welded, the amount of current, and the size and shape of the weld, or bead, desired.

The welding student should begin by making welds known as single-pass stringer beads. The arc length and arc travel should be such that the puddle of molten metal is about twice the diameter of the rod used.

ANGLE OF ELECTRODES

When welding on plates in a flat position, the electrode should make an angle of 90 degrees with the work. In other than flat work, good results are obtained if the rod splits whatever angle is being welded, figure 4-3. In general practice it is found that this angle may vary as much as 15 degrees in any direction without affecting the appearance and quality of the weld. The electrode angle should be no greater than 20 degrees toward the direction of travel.

REVIEW QUESTIONS

1. Using the general rule for current setting, what is the proper setting, to the nearest round figure for electrodes with the following diameters: 3/32 inch, 3/16 inch, and 1/4 inch?

2. From figure 4-2 and the general rule for arc length, what is the voltage across the arc when welding with a 3/16-inch electrode?

3. What is the arc voltage for a 1/8-inch diameter electrode?

4. If the first attempt at making a stringer bead produces a weld that is too narrow, what adjustment must be made in the rate of travel to produce a bead of the proper width?

5. What is the proper angle of the electrode in the following sketch?

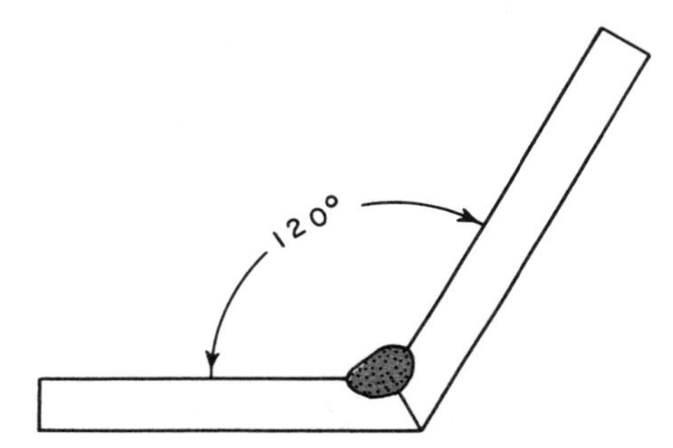

6. What indication does the operator have that the arc is the correct length for the diameter rod being used?

UNIT 5 WELDING SYMBOLS

DESCRIBING WELDS ON DRAWINGS

Welding symbols form a shorthand method of conveying information from the draftsman to the fabricator and welder. A few good symbols give more information than several paragraphs.

The American Welding Society has prepared a pamphlet, *Standard Welding Symbols* (AWS A2.0-68). This publication provides the draftsman with the exact procedures and standards to be followed so fabricators and welding operators are given all the information necessary to produce the correct weld.

The standard AWS symbols for arc and gas welding are shown in figure 5-1.

EXAMPLES OF THE USE OF SYMBOLS

Each of the symbols in this unit should be studied and compared with the drawing which shows its significance. They should also be compared with the symbols shown in figure 5-1.

In each of the succeeding units a symbol related to the particular job is shown together with its meaning. A study of these examples will clarify the meaning of the welding symbols.

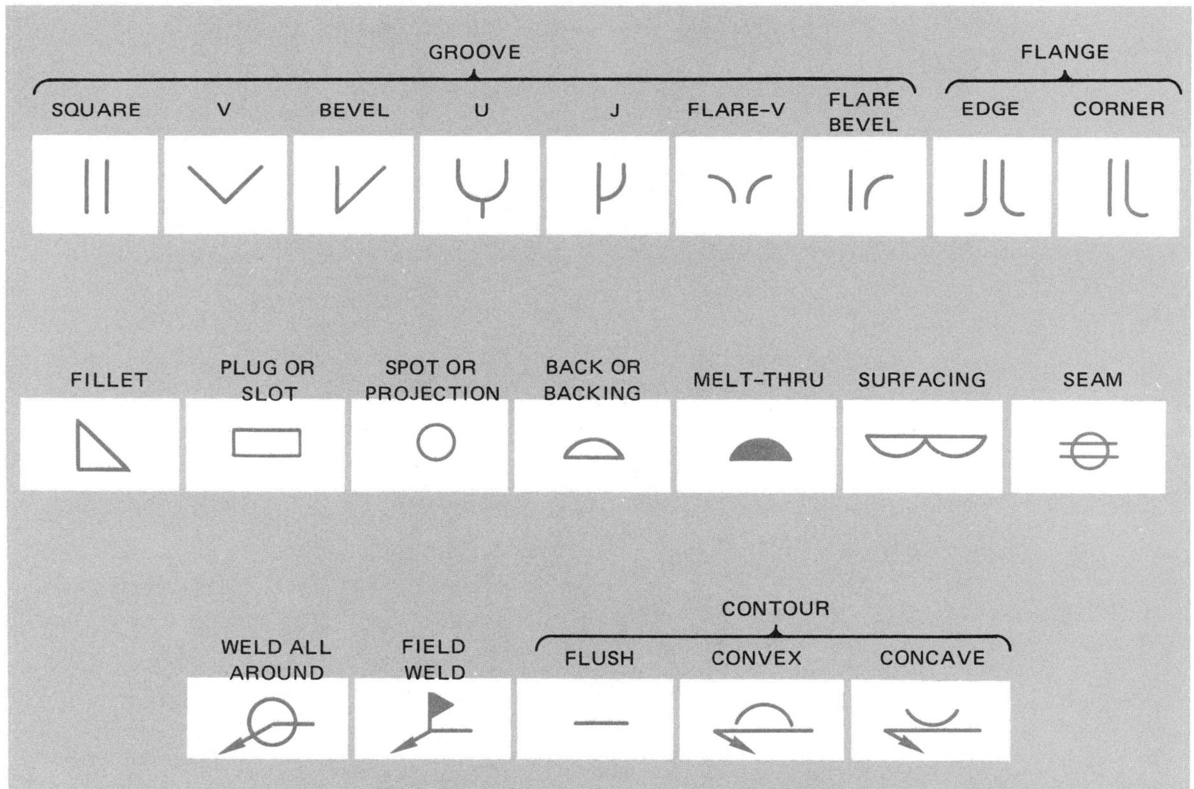

Fig. 5-1 Standard welding symbols

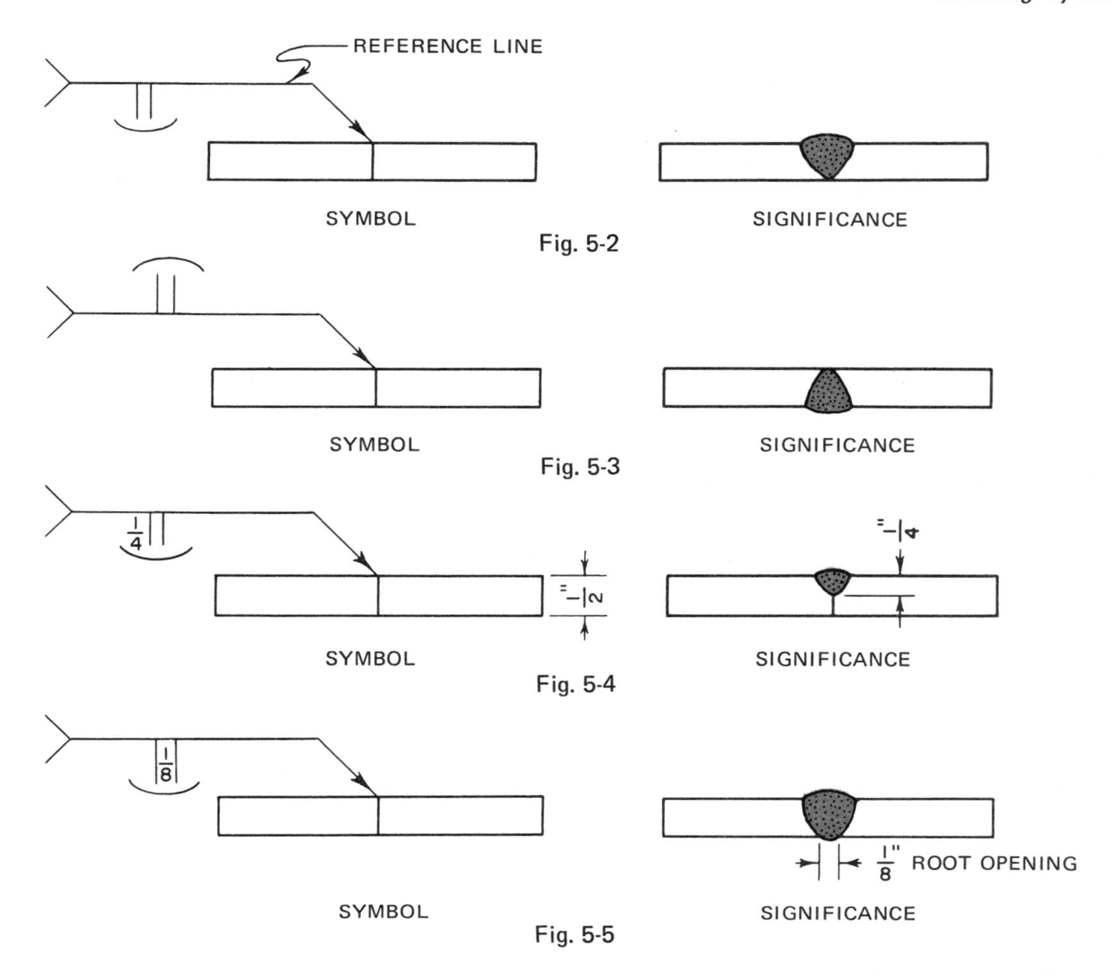

Fig. 5-2

Fig. 5-3

Fig. 5-4

Fig. 5-5

The symbols from the chart are placed at the midpoint of a reference line. When the symbol is on the near side of the reference line, the weld should be made on the arrow side of the joint as in figure 5-2.

If the symbol is on the other side of the reference line, as in figure 5-3, the weld should be made on the far side of the joint or the side opposite the arrowhead.

All penetration and fusion should be complete unless otherwise indicated by a dimension positioned as shown by the 1/4 in figure 5-4.

To distinguish between root opening and depth of penetration, the amount of root opening for an open square butt joint is indicated by placing the dimension within the symbol, figure 5-5, instead of at one side of the symbol, as in the preceding drawing.

The included angle of beveled joints and root opening is indicated in figure 5-6, page 20. If no root opening is indicated on the symbol, it is assumed that the plates are butted tight unless the manufacturer has set up a standard for all butt joints.

The tail of the arrow on reference lines is often provided so that a draftsman may indicate a particular specification not otherwise shown by the symbol. Such specifications are usually prepared by individual manufacturers in booklet or looseleaf form for their engineering and fabricating departments. The specifications cover such items as the welding process to be used (i.e. arc or gas), the size and type of rod or electrode, and the preparation for welding, such as preheating.

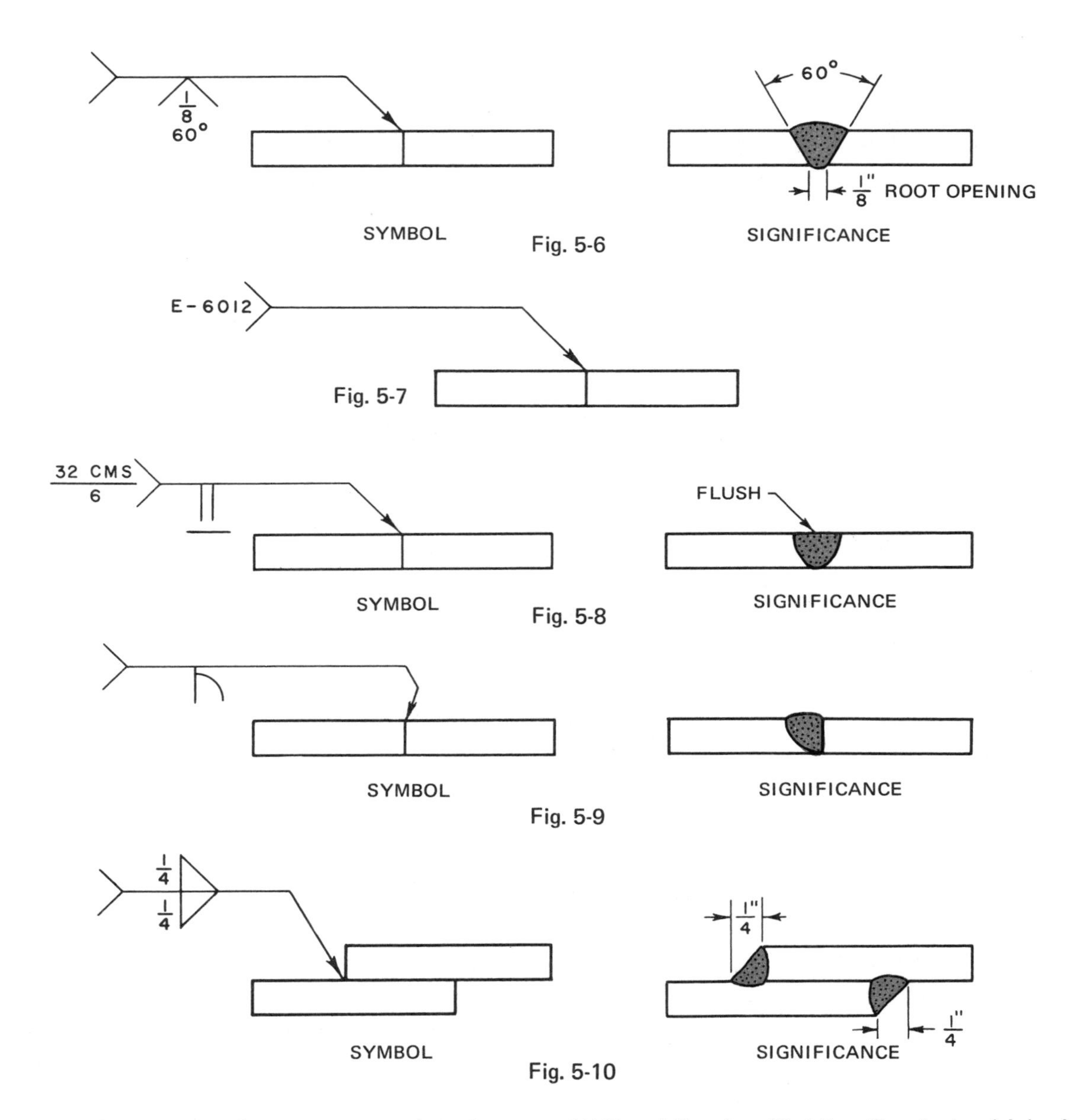

Fig. 5-6

Fig. 5-7

Fig. 5-8

Fig. 5-9

Fig. 5-10

Many manufacturers are using the new AWS publication *Welding Symbols* which gives very complete rules and examples for welding symbols, as well as a complete set of specifications with letters and numbers to indicate the process.

One method of indicating the type of rod to be used is shown in figure 5-7. This figure indicates that the butt weld is to be made with an AWS classification E-6012 electrode.

In figure 5-8, the rod to be used is indicated as a number 32 CMS (carbon mild steel) type and the 6 indicates the size of the rod in 32nds of an inch. In this case it is 3/16-inch diameter rod. In addition the symbol indicates that the finished weld is to be flat or flush with the surface of the parent metal. This may be accomplished by G = grinding, C = chipping, and M = milling.

When only one member of a joint is to be beveled, the arrow makes a definite break back toward the member to be beveled, figure 5-9.

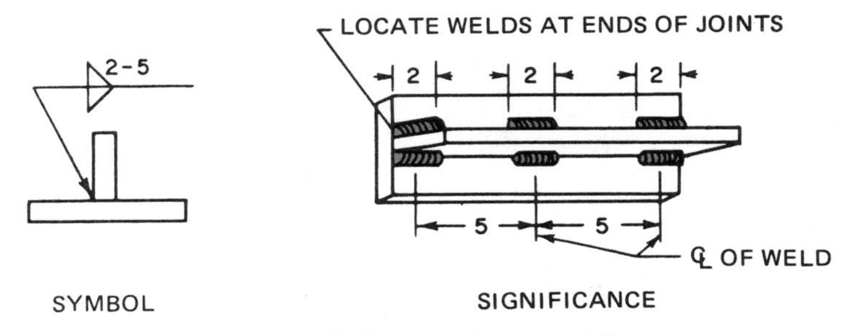

Fig. 5-11 Chain intermittent welding

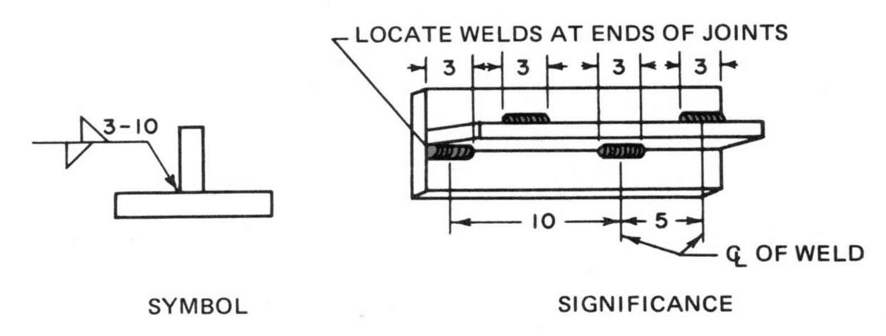

Fig. 5-12 Staggered intermittent welding

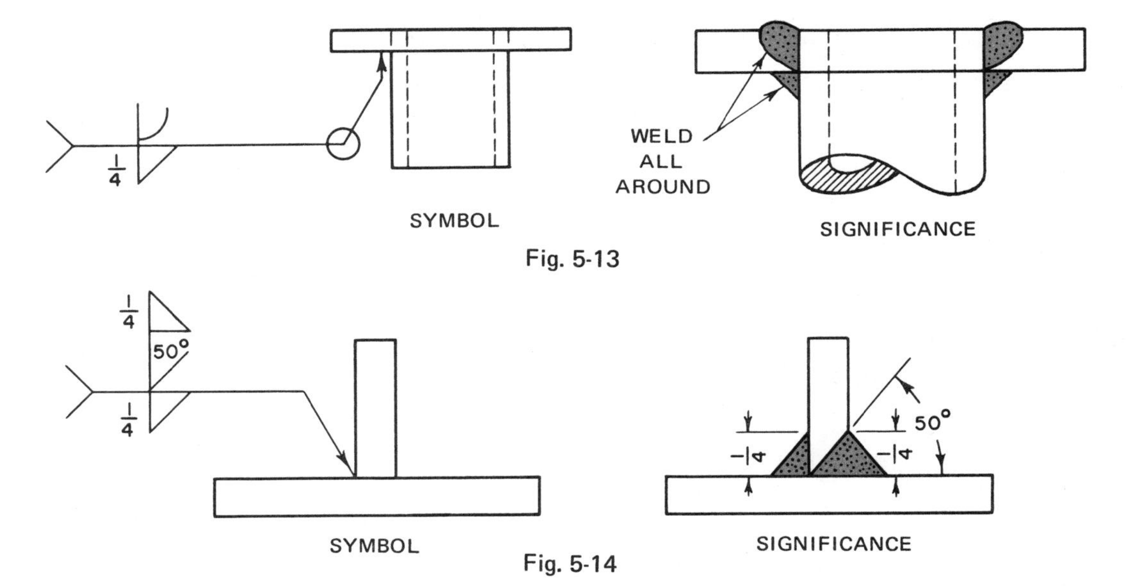

Fig. 5-14

The size of fillet and lap beads is indicated in figure 5-10. In all lap and fillet welds, the two legs of the weld are equal unless otherwise specified.

If the welds are to be chain intermittent, the length of the welds and the center-to-center spacing is indicated, as in figure 5-11.

When the weld is to be staggered intermittent, the symbols and desired weld is made as in figure 5-12.

An indication that the joint is to be welded all around is shown by placing the weld all around symbol, as in figure 5-13.

Several symbols may be used together when necessary, figure 5-14.

Field welds (any welds not made in the shop) are indicated by placing the field weld symbol at the break in the reference line, as in figure 5-15.

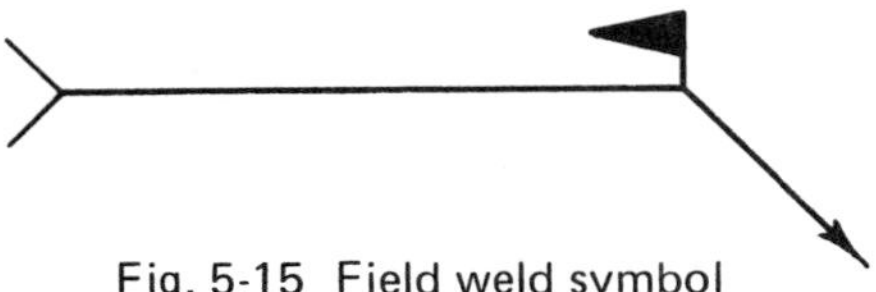

Fig. 5-15 Field weld symbol

REVIEW QUESTIONS

1. What is the symbol for a 60-degree closed butt weld on pipe?

2. What is the symbol for a U-groove weld with a 3/32-inch root opening?

3. What is the symbol for a double V, closed butt joint in plate?

4. What is the symbol for a 1/2-inch fillet weld in which a column base is welded to an H-beam all around?

5. What is the symbol for a J-groove weld on the opposite side of a plate joint?

UNIT 6 STARTING AN ARC AND RUNNING STRINGER BEADS

The quality and appearance of an electric arc weld depend almost entirely on the following:

- Length of the arc
- Rate of Travel
- Angle of the electrode
- Amount of current

Experimentation with each of these variables is helpful in learning correct welding procedures. This unit provides an opportunity to experiment with these variables and observe the results.

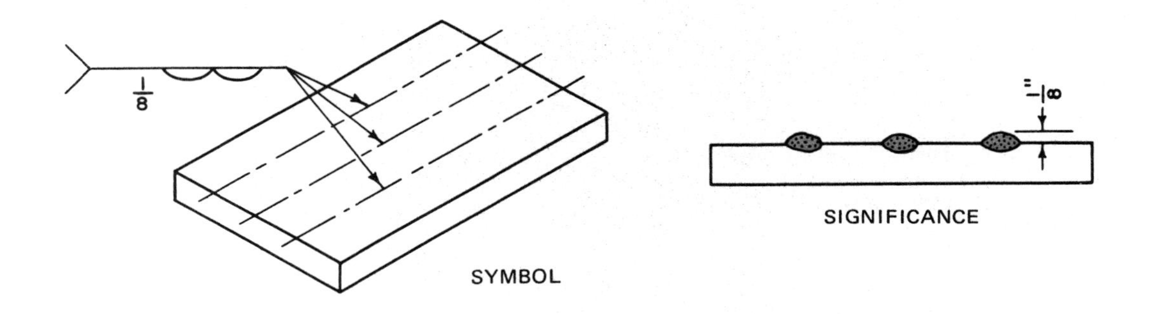

Fig. 6-1 Stringer beads

Materials

Steel plate 3/16" or thicker 6 in. x 9 to 12 in.

DC or AC welding machine

1/8-inch or 5/32-inch diameter E-6012 or E-6013 electrodes

Procedure

1. Start the machine, check the polarity and adjust the current setting as described in unit 4.

   CAUTION: Make sure all protective devices are in place. Use all recommended safety devices to protect the body, especially the eyes, from the arc rays. Failure to do so results in severe and painful radiation burns. Wear safety glasses.

2. Start the arc on the plate according to the method indicated in figure 6-2.

1. HOLD ELECTRODE 1" OFF PLATE, BRING HELMET OVER EYES
2. TOUCH PLATE WITH ELECTRODE
3. RAISE ELECTRODE 1/4"
4. LOWER TO NORMAL ARC LENGTH (UNIT 4)

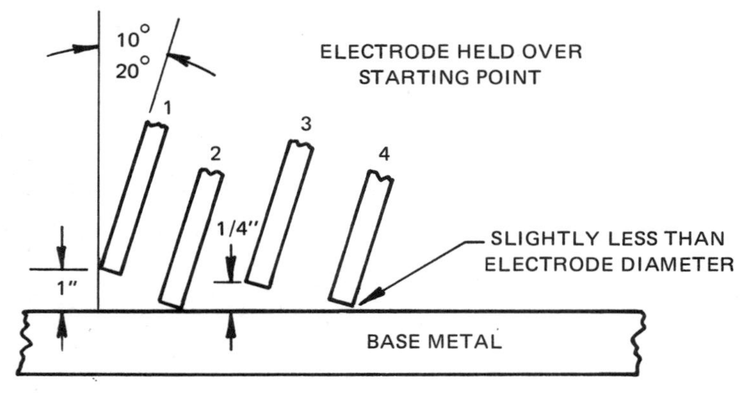

Fig. 6-2 Establishing an arc

3. Listen for the sound indicating the correct arc length, and observe the behavior of the arc.
4. Make straight beads or welds. Note that the electrode must be fed downward at a constant rate to keep the right arc length. Move the arc forward at a constant rate to form the bead.

   Note: Right-handed welders will see better welding from left to right. Left-handed welders should weld from right to left.

5. Remove the slag and examine the bead for uniformity of height and width.

   CAUTION: When removing slag from a weld with a chipping hammer, eye protection is very important. Safety glasses should always be worn.

6. Continue to make stringer beads until each weld is smooth and uniform.
7. Make a series of beads similar to those in figure 6-3. Note the difference of each bead as the variables are changed.

A. NORMAL BEAD
B. ARC TOO LONG
C. ARC TOO SHORT
D. RATE OF TRAVEL TOO HIGH
E. RATE OF TRAVEL TOO LOW
F. CURRENT TOO LOW
G. CURRENT TOO HIGH
H. ROD ANGLE TOO LOW

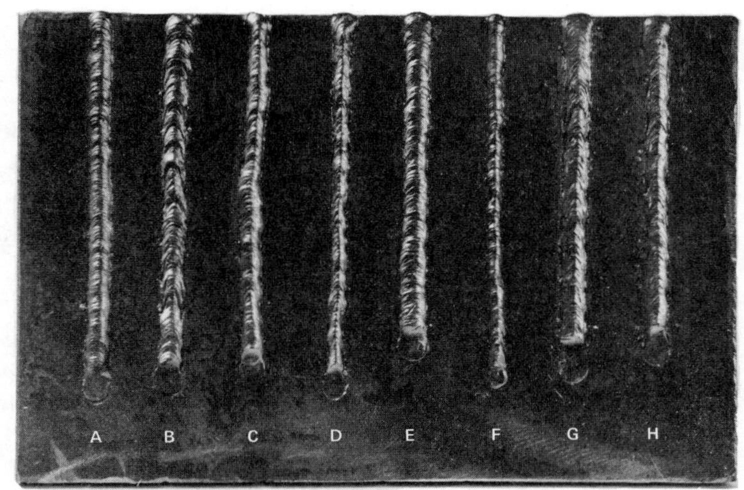

Fig. 6-3 Bead variables using E-6012 electrode and DC straight polarity.

AWS-ASTM Class	Hobart	Air Products	Airco	Canadian Liquid Air Ltd.	Canadian Rockwell	P & H	Lincoln	Marquette	McKay Co.	M & T (Murex)	N.C.G. (Sureweld)	Reid-Avery Co. (Raco)	Shober	Canadian Liquid Carbonic	Westinghouse
E-4510 **E-4520**	Sulkote		41 63	LA-SC-15		Washcoat	Stable Arc	101	21 3	Sulcoat Thincoat	4510	Raco Type D & M			Sulcoat 18
E-6010	10 10-IP	6010 6010-IP	6010	LA-6010 6010-P	R60	SW-610 AP-100	Fleetweld 5 Fleetweld 5P	105	6010 6010-IP	Speedex 610	6010-Y 6010-X	6010	32	P&H-704D 610-P	XL-610 XL-610A ZIP-10
E-6011	335-A	6011 6011-C	6011 6011-C	LA-6011-P 6011-F	R61	SW-14 SW-14-IMP	Fleetweld 35, 180 35-LS	130	6011 6011-IP	Type A 611-C	6011-Z	6011 6011-IP	11 13	P&H-504D 611-P	ACP-611
E-6012	12 212-A 12-A	6012-GP 6012-SF 6012-IP	6012 6012-C	LA-6012 6012-P	R62	SW-612 PFA SW-17 SW-29	Fleetweld 7, 7MP 77	120	6012	Genex-M Type N-13	6012W	6012 6012-F	34	604	FP-612 FP-2-612 ZIP-12 ZIP-AF
E-6013	13A 413 447-A	6013-GP 6013-SF	6013 6013-C	LA-6013 6013-P	R63	AC-3 SW-16 SW-15	Fleetweld 37	140	6013	Murex U, U-13	6013-Y	6013	13 35	613	SW-613 SW-2M-613
E-7014	714 14-A	7014-IP	Easyarc 7014	LA-7014	R74	DH-6 SW-15-IP	Fleetweld 47	146	7014	Speedex U		7014	14	P&H-714	ZIP-14
E-6020	111	6020	6020	LA-6020FS	R620	DH-3			6020	Murex FHP,D	6020-X				DH-620
E-7024	24 24-A	7024-IP	7024	6024	R724	SW-44 SW-624	Jetweld 1 3	24	7024	Speedex 24			36	P&H-624	ZIP-24
E-6027	27	6027-IP	6027	LA-6027	R627	DH-27	Jetweld 2			Speedex 27				P&H-727	ZIP-27
E-7010-A1	710	7010-A1	7010-A1	LA-7010-P		SW-75	Shieldarc 85			710-Mo		7010-A1		P&H-710	
E-7020-A1	111-HT	7020-A1	7020-A1				Jetweld 2HT			Murex DM		7020-A1			

Chart 6-1 Comparative index of mild steel and alloy electrodes

REVIEW QUESTIONS

1. What must be controlled to make good arc welds?

2. What is a stringer bead?

3. What are the points to look for in a good weld?

4. What is the current value for a 5/32-inch diameter electrode?

5. What two factors best determine a correct arc?

UNIT 7 RUNNING CONTINUOUS STRINGER BEADS

Running long stringer beads demands good control of the welding electrode if the beads are to be straight and uniform in appearance and size. Much practice is required to develop a high degree of skill.

Changing electrodes in the middle of the bead, or starting an arc which has been accidentally stopped is a basic and important skill. This unit provides experience in restarting a bead.

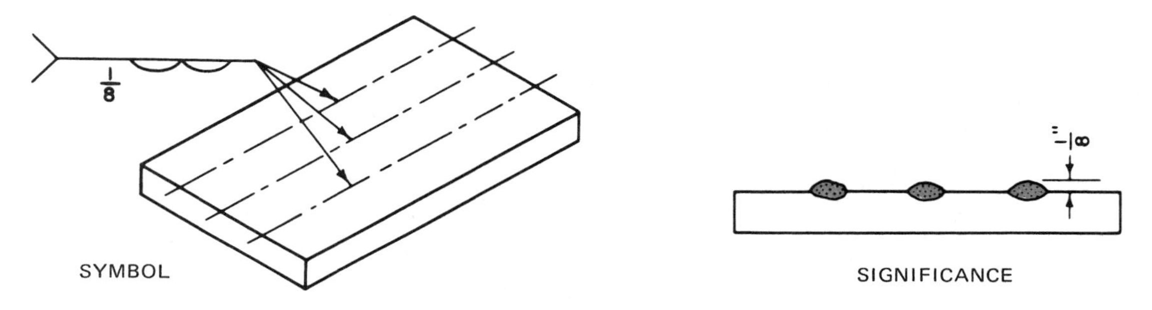

Fig. 7-1 Stringer beads

Materials

Steel plate, 3/16 inch thick, 6 in. x 9 to 12 in.

DC or AC welding machine

1/8- or 5/32-inch diameter E-6012 or E-6013 electrodes

Procedure

1. Start the machine, check the polarity, and adjust the current for the size of electrode being used.
2. Run a continuous stringer bead on the plate, using the full length of the electrode before stopping. Make this bead parallel to and about 1/2 inch from the edge of the plate.
3. Run additional beads at 1/2-inch intervals, being sure to keep each bead straight. Check the arc length and rate of travel constantly to produce smooth, uniform beads.
4. Continue to make this type of weld until each one is of uniform appearance for its entire length.
5. Make a bead 2 or 3 inches long and stop the arc. Start the arc again ahead of the crater. Move the electrode back to the crater, using an extra long arc. Bring the rod down rapidly to the proper arc length and make sure that the new puddle just fuses into the last ripple of the crater. Proceed with the weld for another 2 or 3 inches and stop.

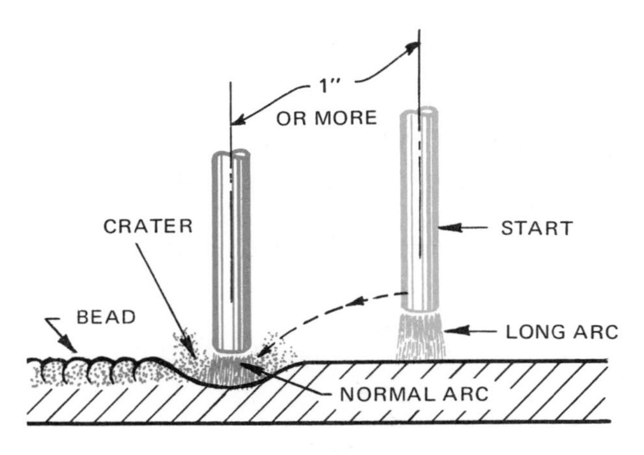

Fig. 7-2 Establishing a continuous bead

6. Continue this procedure until there is very little difference in appearance at the point where the arc was restarted. Figure 7-2 shows the right procedure.
7. Start an arc directly in the crater and notice the difference in the appearance of the connection.

   Note: The ideal length at which electrode stubs should be discarded is 1-1/2".

8. Try to eliminate the crater at the end of the finished bead by moving the arc back over the crater and finished bead slowly. As the arc is moved back gradually increase the arc length until the crater is filled as in figure 7-3.
9. Try to eliminate the crater by using a very short arc in the crater and pausing until there is enough buildup.

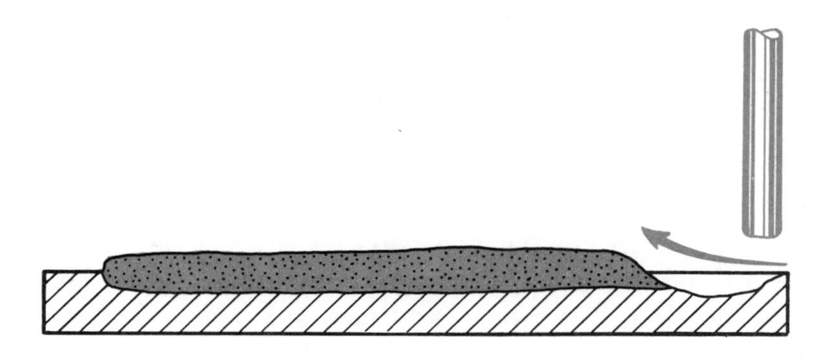

Fig. 7-3 Rod motion for filling crater

REVIEW QUESTIONS

1. What is the difficulty in trying to restart an arc directly in the crater?

2. What are the advantages of restarting a long arc ahead of the crater then backing up to the crater and shortening the arc?

3. Why is it necessary to gain skill in connecting the beads?

4. Why is it important for a welding student to learn to run a bead in a straight line?

5. What does the term polarity mean?

UNIT 8 RUNNING WEAVE BEADS

It is often necessary when welding large joints or making cover passes to produce beads wider than stringer beads. These are called *weave beads.* Weaving is done with a back and forth sidewise motion of the electrode and a slow forward movement, figure 8-1.

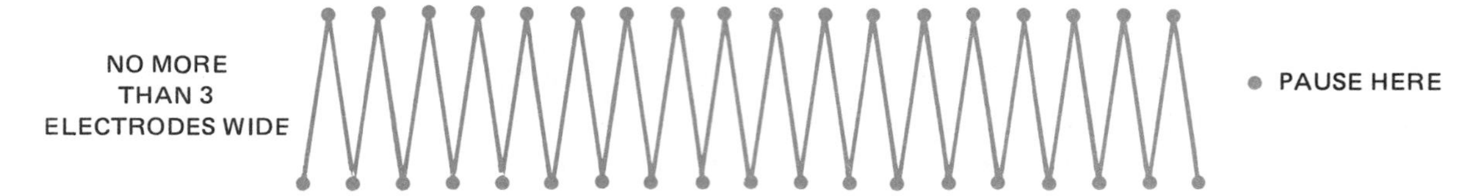

Fig. 8-1 Straight weave

The height of the bead depends on the amount the electrode is advanced from one weave to the next. The number of ripples depends on the speed and frequency of the weaving motions.

The pause at each side of the weave is important for puddle flow and penetration. Failure to pause causes an undercut along the sides of the weld, figure 8-2.

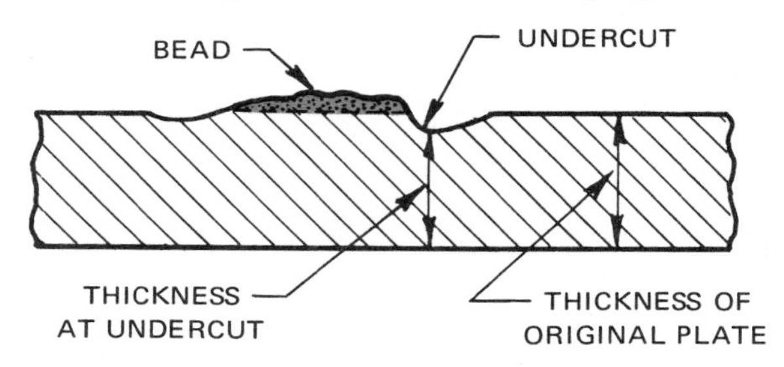

Fig. 8-2 Undercut bead

Materials

Steel plate

1/8- or 5/32-inch diameter E-6012 or E-6013 electrodes

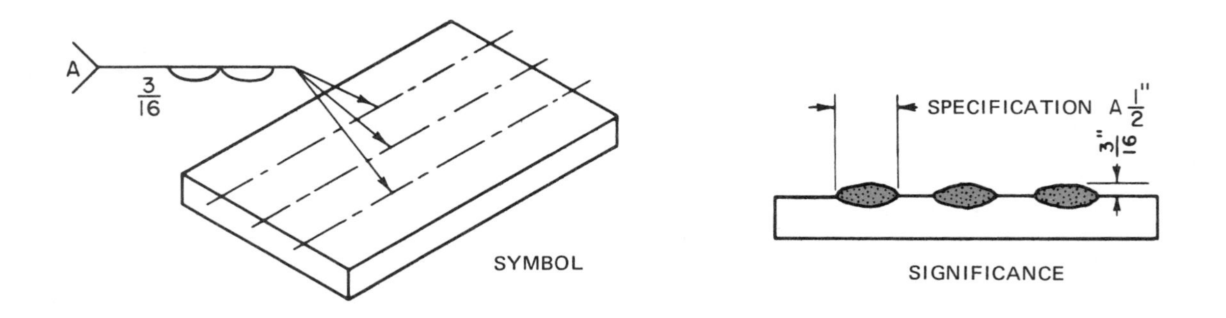

Fig. 8-3 Weave beads

Procedure

1. Start the machine, check the polarity, and adjust the current for the size of the electrode being used.
2. Start an arc and make a bead approximately 3 times wider than the diameter of the electrode being used.
3. Continue to make weave beads until they have a uniform height and width for their entire length.

   Note: Beginners have a tendency to let the weld become progressively wider with each pass of the arc. In an attempt to correct this, there is a tendency to decrease each weave motion.

4. Try stopping the arc and fusing the new bead to the original. Be sure to *slag* (remove the slag with a wire brush or hammer) the crater before starting each bead. Practice this until the starting and stopping points blend in smoothly.
5. Change the length of pause at each side. Change the travel speed and the amount of advance.
6. Clean the beads and compare the results.

REVIEW QUESTIONS

1. Why should there be a definite pause at each side of the weave?

2. What can be done to produce a weave bead with many fine ripples rather than a few coarse ones?

3. When restarting the bead, how does the rate of travel for the first two or three passes compare with the normal rate of travel?

4. What steps are taken to prevent undercutting?

5. How should the size of the molten puddle compare with the diameter of the electrode?

UNIT 9 PADDING A PLATE

Experience in padding helps the welding student develop an eye for following a joint. It helps the student compare beads for uniform appearance. Padding is used for building up pieces prior to machining and for depositing hardfacing metal on construction equipment.

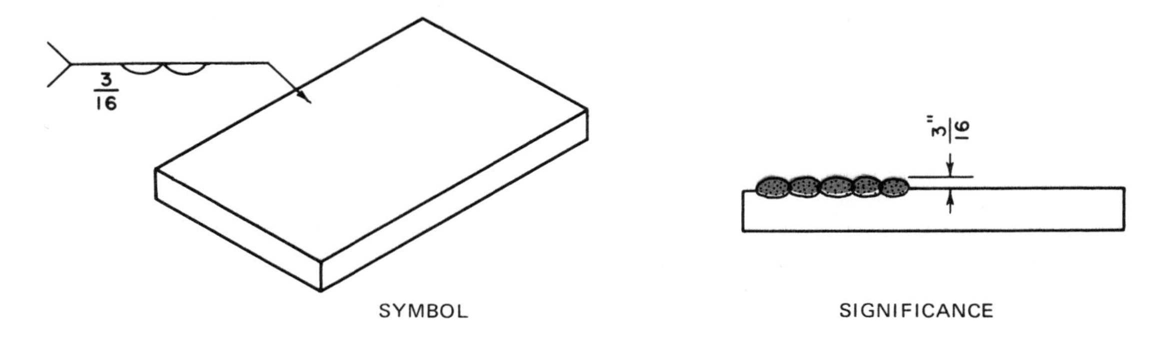

Fig. 9-1 The pad

Materials

2 Pieces of plate 6 in. x 6 in.

1/8- or 5/32-inch E-6012 or E-6013 electrodes

Procedure

1. Establish an arc and run a stringer bead close to and parallel with the far edge of the practice plate.

   Note: Observe that a crater is left at the end of this weld. Prevent this crater in the following manner: Upon reaching the end of the plate, pull the electrode out of the crater, letting the heat die down. When the color has disappeared restart the arc in the crater, depositing a small amount of weld. This can be done several times to fill the crater.

2. Run more beads alongside of the previous bead, figure 9-2. Make sure that the far edge of bead being deposited is in the center of the previous bead.

Fig. 9-2 A partially completed pad using E-6012 electrode (right-handed operator welding left to right).

Note: The electrode must be directed at the point where the previous bead meets the base metal.

Always chip the slag from each bead before welding.

3. Continue to cover the plate with weld using this technique. Keep the weld as straight as possible.
4. On another plate follow the same procedure and make a pad using the weave technique.

   Note: As the newly padded surface cools, the plate may bow upward. This can be corrected by welding a pad on the opposite side of the plate.

5. For additional practice, the plate can be turned 90 degrees for a second layer.
6. After the padding operation is finished, cut the material by saw or torch and examine the weld. There should be no holes or bits of slag imbedded in the weld.

REVIEW QUESTIONS

1. What is weld padding?
2. What characteristics of a weld are most easily examined through padding?
3. Where can the padding operation be used?
4. Why is it important to slag every bead before running the next one?
5. How is a washed-out area at the edge of a plate prevented?

UNIT 10 SINGLE-PASS, CLOSED SQUARE BUTT JOINT

In making a single-pass, closed square butt joint, penetration is extremely important. Welding from one side not only makes complete penetration difficult, but the joint strength depends directly on the depth of penetration.

By experimenting with this type of joint and testing the results, one can determine the best procedures.

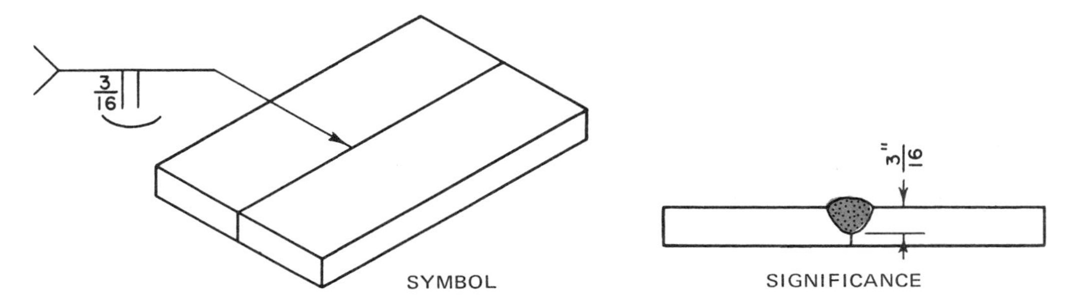

Fig. 10-1 Single-pass, closed square butt joint

Materials

Two steel plates, 1/4 inch thick, 2 in. x 9 in. each

1/8- or 5/32-inch diameter E-6012 or E-6013 electrodes

Procedure

1. Place the plates on the worktable so that the two 9-inch edges are in close contact.
2. Tack the two plates together using *tack welds* about 1/2 inch long. Start the tacks 1/2 inch to 1 inch from the ends of the plate to avoid having excessive metal and poor penetration at the start of the weld.
3. Proceed with the weld as in making stringer beads, but be very careful to keep the centerline of the arc exactly centered on the joint. Half the weld should be deposited on each plate.
4. Cool the finished assembly. Clean the bead and examine it for uniformity.
5. Check the depth of penetration of the weld by placing the assembly in a vise with the center of the weld slightly above and parallel to the jaws. Bend the plate toward the face of the weld so that the joint opens, figure 10-2. Examine the original plate edges.
6. Continue the bend until the weld breaks. Notice that the broken weld metal has a bright, shiny appearance, and that the metal that was not welded is much darker. This bright weld metal indicates the depth of penetration.
7. Make more joints of this type. Start with a setting of 150 amps and increase the amperage with each joint.

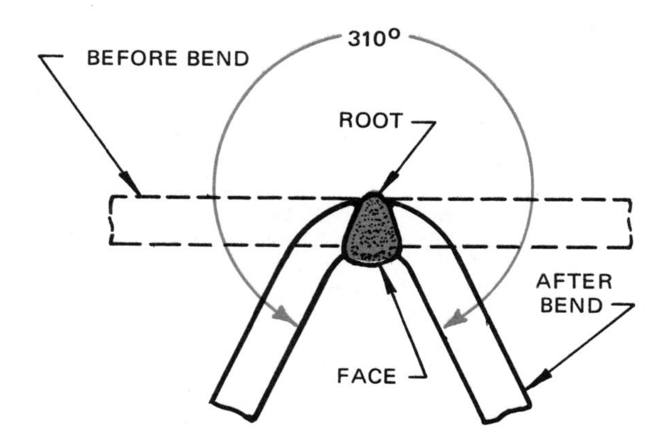

Fig. 10-2 Bend test for butt weld

8. Cool, break, and examine these test plates and compare the amount of penetration with that in the first weld made.
9. Set up another test plate and weld, using a weaving figure-8 motion, figure 10-3.
10. Cool, break, and examine this plate and compare the penetration with the other welds that have been made.

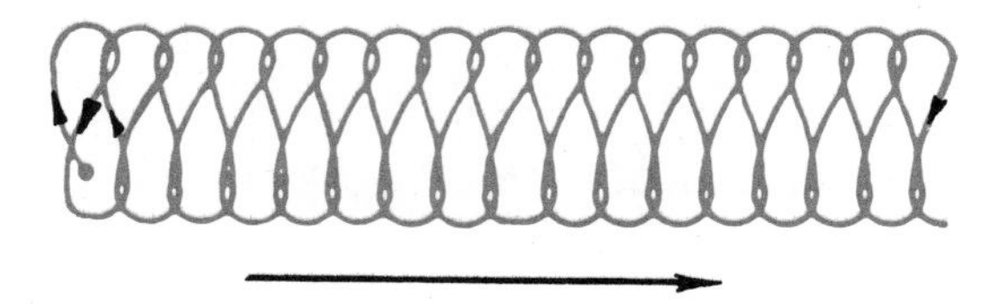

Fig. 10-3 Figure-8 weave

REVIEW QUESTIONS

1. How does increasing the amperage affect the depth of penetration?

2. How does the figure-8 weave bead affect the depth of penetration? Why?

3. Is there any advantage if the figure-8 weave is used in buildup operations?

4. What is penetration?

5. What is the purpose of a tack weld?

UNIT 11 OPEN SQUARE BUTT JOINT

An open butt joint presents some additional problems in penetration. By changing the space between the plates, and by welding one set of plates from one side only and another set from both sides, it is possible to compare the quality of the welds and, particularly, the penetration. The open square butt joint differs from the closed butt joint in that the open-type joint has some spacing between the plate edges.

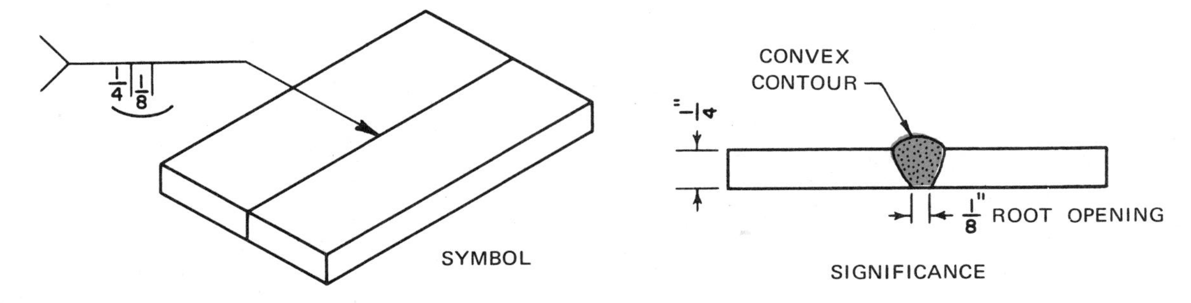

Fig. 11-1 Open square butt joint

Materials

Two steel plates, 1/4 inch thick, 1-1/2 to 2 in. x 9 in. each

5/32-inch diameter E-6012 electrodes

Procedure

1. Place the two plates on the welding bench, align and space them, and tack them as shown in figure 11-2. Tacks should be long enough to withstand the strain of the expanding metal being welded without cracking.
2. Make the weld in the same manner as a closed butt joint, but use a slight amount of weaving to allow for the additional width of the joint.
3. Cool, clean, and inspect the finished weld for uniform appearance. Examine the root side of the weld for penetration.

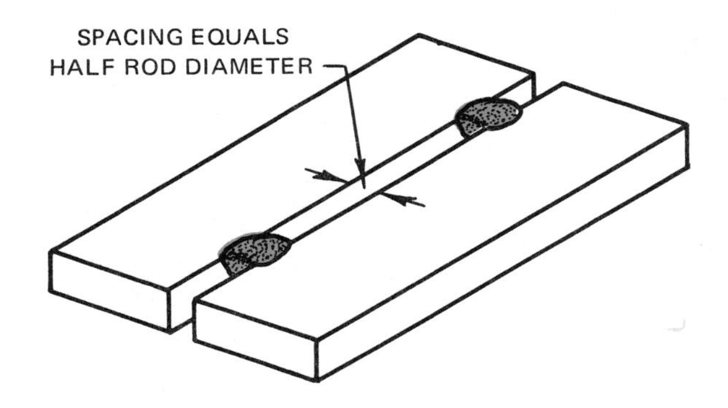

Fig. 11-2 Setup for open butt joint

4. Break the welded joint in the same manner as the closed butt joint. Check the amount of penetration. It should be a little more than one-half the thickness of the metal being welded.
5. Make additional open butt joints, using plate spacings narrower and wider than the first. After welding, check the plates for bead appearance and penetration.

   Note: If the spacing between the plates is too great it may be necessary to run another bead over the first one to build the weld to desired dimensions. The first bead is referred to as the burning-in or root-pass bead, and the second bead is called the finish bead.

6. Once beads of good appearance can be made consistently, make additional joints, but weld from both sides. Check these welds by cutting the plates in two and examining the cross section for holes and *slag inclusions* (nonmetallic particles trapped in the weld).

REVIEW QUESTIONS

1. What can be done to prevent holes or slag inclusions in the weld?

2. What advantages does the open butt joint have over the closed butt joint?

3. Sketch a cross section of the weld made in step 2 showing penetration and fusion.

4. Sketch a cross section of the weld made in step 5, with the spacing too wide. Show what is wrong with this weld.

5. What is the space between two pieces of plate being welded called?

UNIT 12 SINGLE-PASS LAP JOINT

The welded lap joint has many applications and is economical to make since it requires very little preparation. For maximum strength it should be welded on both sides. A single pass or bead is enough for the plate used for this job. For heavier plate, several passes must be made.

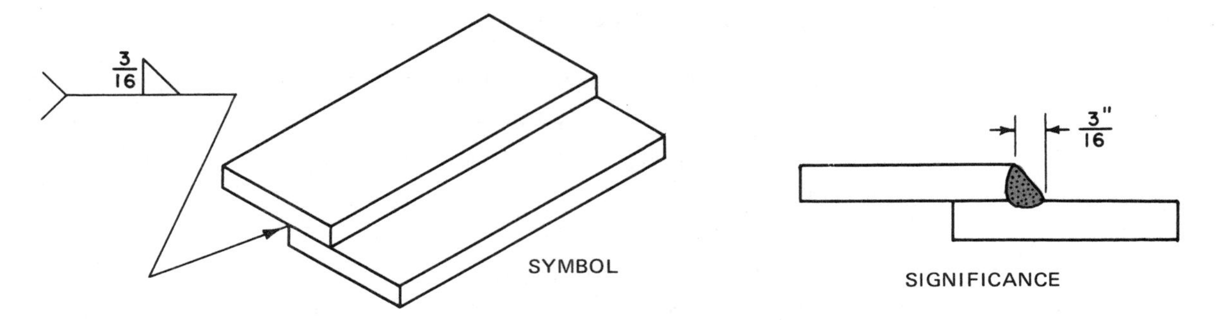

Fig. 12-1 Single-pass lap joint

Materials

Two steel plates, 3/16 inch thick, 2 in. x 9 in. to 12 in. each

1/8- or 5/32-inch diameter, E-6012 or E-6013 electrodes

Procedure

1. Set up the two plates as shown in figure 12-2. Make sure that the plates are reasonably clean and free of rust and oil. Be sure that the plates are flat and in close contact with each other.
2. Weld this joint with the electrode at an angle of 45 degrees from horizontal. Make sure that the weld metal penetrates the root of the joint, and that the weld metal builds up to the top of the lapping plate. Figure 12-3 shows the cross section, or end view, of a lap-welded joint, indicating the electrode angle and the size and

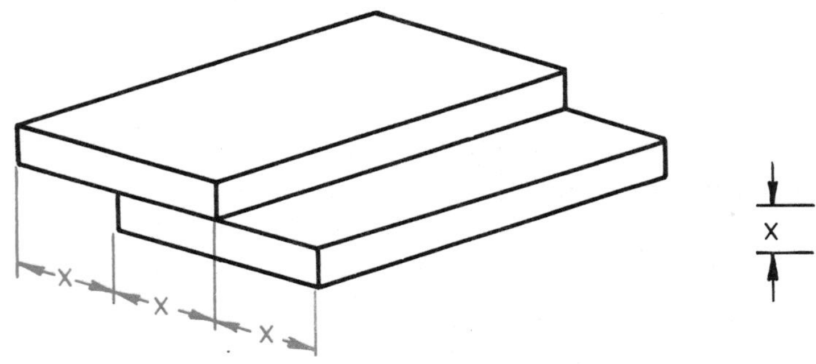

Fig. 12-2 Setup for lap joint

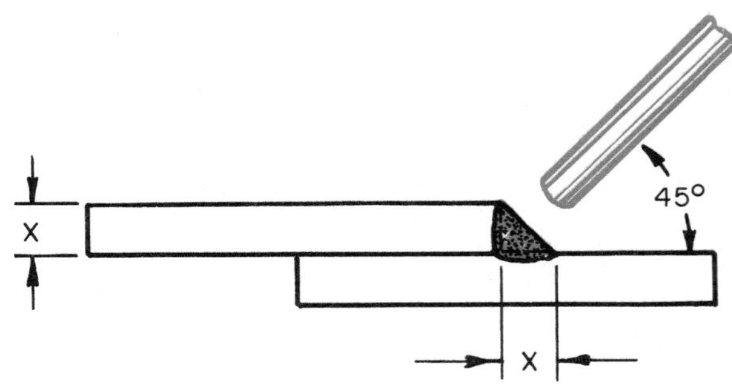

Fig. 12-3 Lap joint weld

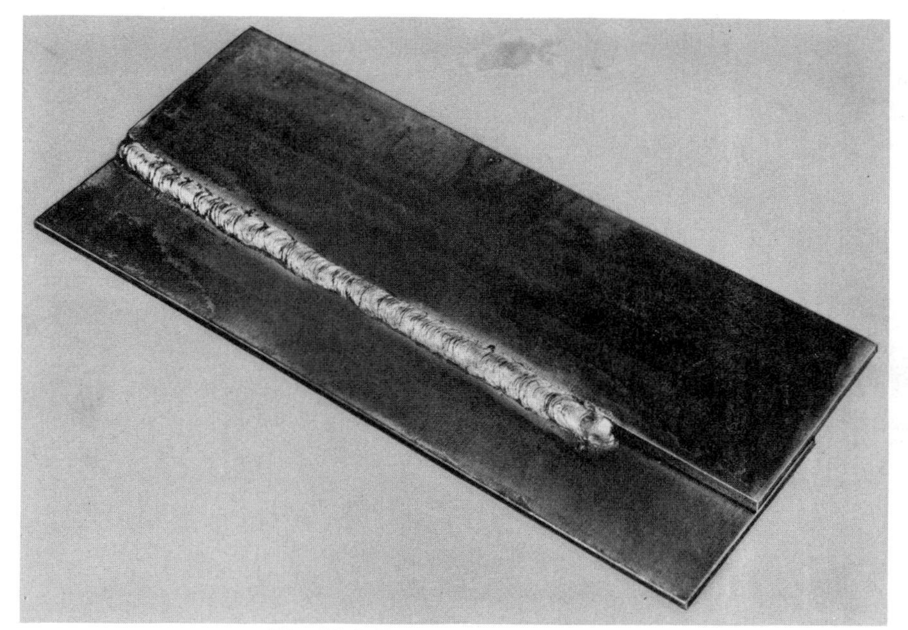

Fig. 12-4 Properly welded lap joint

shape of the weld. Notice that the weld makes a triangle with each side equal to the thickness of the plate. (See "X" in figure 12-3.)

3. Cool and clean this bead and examine it for uniformity. Pay particular attention to the line of fusion with the top and bottom plates. This should be a straight line with the weld blending into the plate, figure 12-4.

   Note: Too slow a rate of travel deposits too much metal and causes the weld to roll over onto the bottom plate. This forms a sudden change in the shape. The extra metal is a waste of material, and actually weakens the joint by causing stresses, figure 12-5.

4. Continue to make this type of joint until beads of uniform appearance can be made each time. Be sure to weld both sides of the assembly.
5. Make a test plate in the same manner as the other lap joints but weld it on only one side.
6. Place this plate in a vise so that the top plate can be bent or peeled from the bottom plate, figure 12-6. Bend this top plate until the joint breaks. Examine the break for penetration and uniformity. Another test may be made by sawing a lap-welded specimen in two and examining the cross section for penetration.

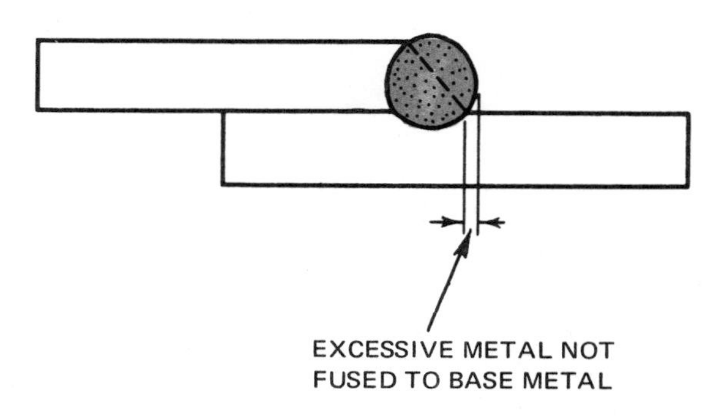

Fig. 12-5 Lap joint with excessive weld metal

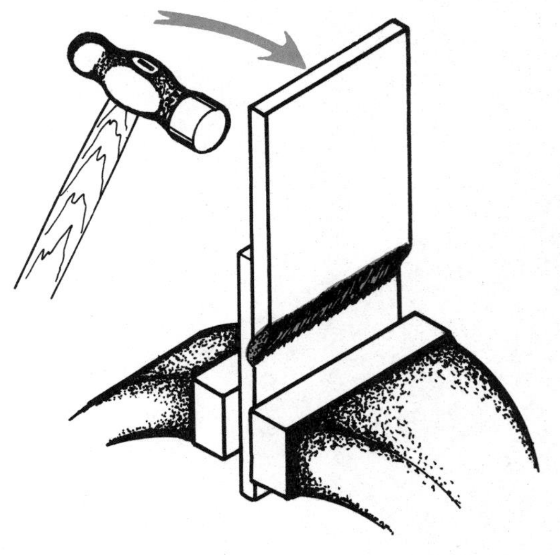

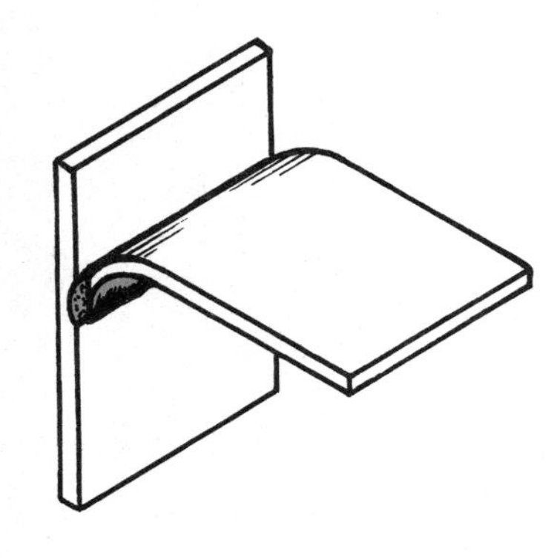

Fig. 12-6 Testing a lap weld

REVIEW QUESTIONS

1. Beginning students usually produce beads with an irregular line of fusion along the top edge of the overlapping plate. How is this corrected?

2. If the test shows lack of fusion at the root, how is this corrected?

3. What effect does too great a rod angle have on this type of joint?

4. What factor is most important in determining the location of the bead on a lap weld?

5. What is the result of depositing too much weld metal in a lap weld?

UNIT 13 SINGLE-PASS FILLET WELD

The fillet or T-weld is similar to the lap weld but the heat distribution is different. It has many industrial applications.

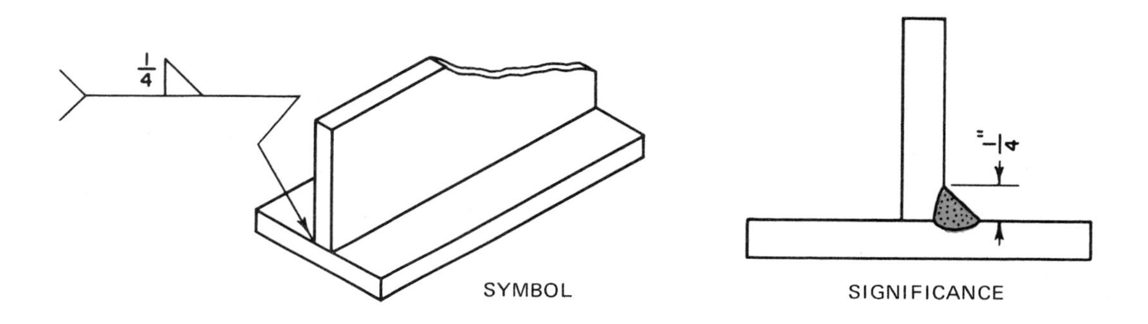

Fig. 13-1 Single-pass fillet weld

Materials

Two steel plates, 3/16 inch or 1/4 inch thick, 3 in. x 9 to 12 in. each

1/8- or 5/32-inch diameter, E-6012 electrodes

Procedure

1. Set up the two plates, figure 13-1. Be sure that the tack welds used to hold the plates in place are strong enough to resist cracking during welding, but not large enough to affect the appearance of the finished weld. This can be done by using a higher amperage and a higher rate of arc travel when tacking.
2. Make the fillet weld in much the same manner as the lap weld was made. The electrode angle is essentially the same. Both legs of the 45-degree triangle made by the weld must be equal to the thickness of the work for the full length of the joint.
3. Clean each completed weld and examine the surface for appearance. Look for poor fusion along both edges of the weld. Examine it for undercutting on the upstanding leg. If undercutting does exist, it is probably being caused by either too long an arc or too high a rate of travel.
4. When good fillet welds can be made every time, make a test weld on only one side of the joint. Then bend the top plate against the joint until it breaks, figure 13-2. Examine the break for root penetration and uniform fusion.

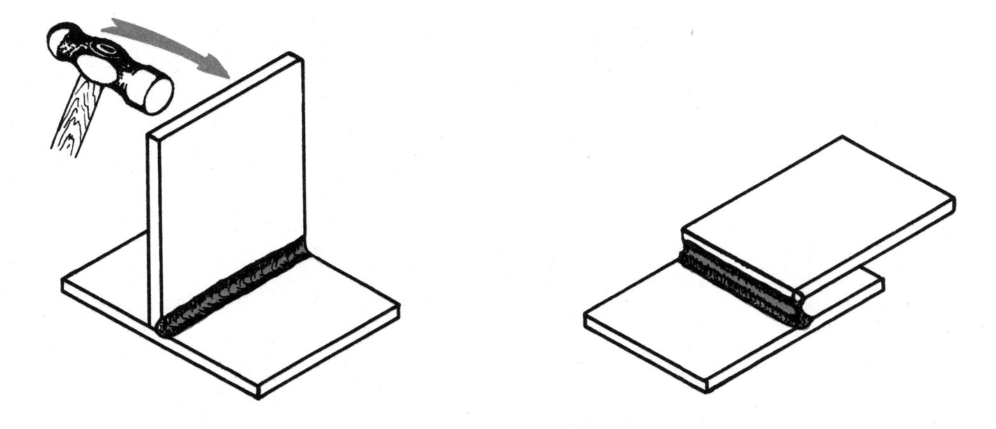

Fig. 13-2 Testing a fillet weld

REVIEW QUESTIONS

1. What can be done to prevent a hump at the start of the weld?

2. What causes undercutting?

3. What can be done to correct poor penetration and fusion in the root of the weld?

4. How is the size of a fillet weld measured?

5. If a fillet weld appears to be more on the flat plate than on the vertical plate how can it be corrected?

UNIT 14 MULTIPLE-PASS LAP JOINT

When heavier plate is used it is impossible to cover the joint with one bead. Joints of this type require three or more passes or beads. The additional beads require that the rod angle be changed for each bead.

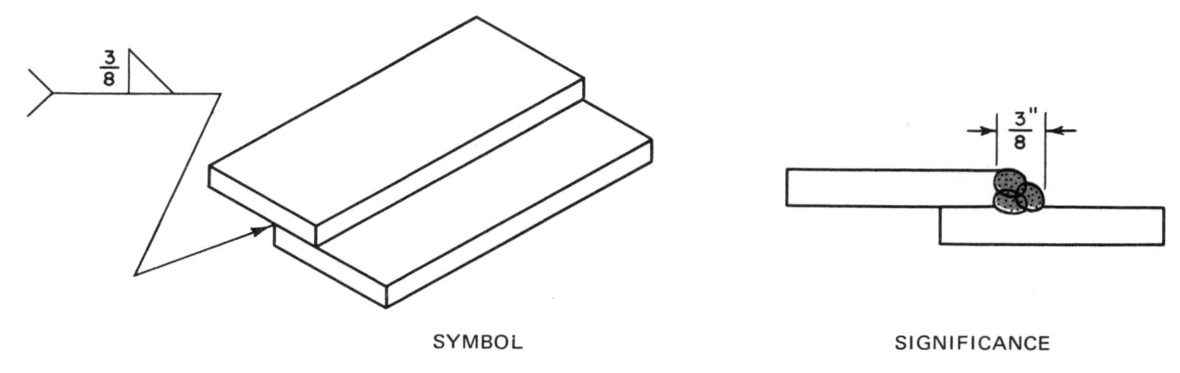

Fig. 14-1 Multiple-pass lap joint

Materials

Two steel plates, 3/8 inch thick, 2 in. x 9 to 12 in. each

5/32-inch diameter E-6012 electrodes

Procedure

1. Set up the plates as for a single-pass lap joint.
2. Make the weld in three passes, being sure to clean each bead before going on to the next.
3. Check the angle of the electrode as each bead is deposited. Adjust this angle to suit the bead being welded. Each bead must be deposited at a different angle. Figure 14-2 indicates the rod angle and position of the arc for each bead.

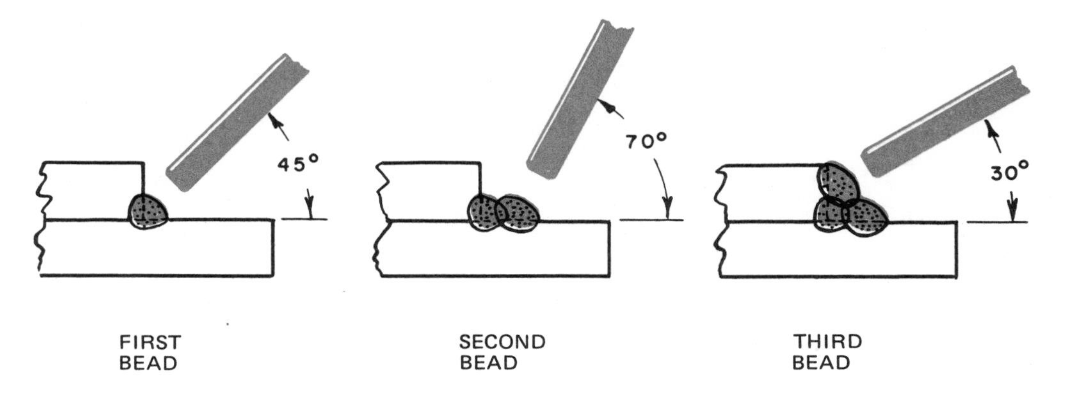

Fig. 14-2 Multiple-pass weld

Note: Observe that the second bead covers all but a very small portion of the first bead. This is necessary if the finished joint is to have a 45-degree angle face. The most common fault of beginners is not covering the bead enough, and then trying to stretch the third bead to correct the original fault. This results in a finished weld of low quality.

4. Continue to make this type of joint until beads of acceptable appearance are made each time. Check a test plate by cutting it in two and examining the section for penetration and slag inclusions.

REVIEW QUESTIONS

1. Figure 14-2 indicates a different electrode angle for each pass. Why is this necessary?

2. If the finished bead has a tendency to flatten out and become too wide, what steps are taken to insure that the face of the weld is at an angle of 45 degrees?

3. Is it necessary to slag the weld between beads? Why?

4. How wide should one bead be for a multiple-pass lap weld?

5. How much of the second bead should be visible when the third bead has been applied?

UNIT 15 MULTIPLE-PASS FILLET WELD

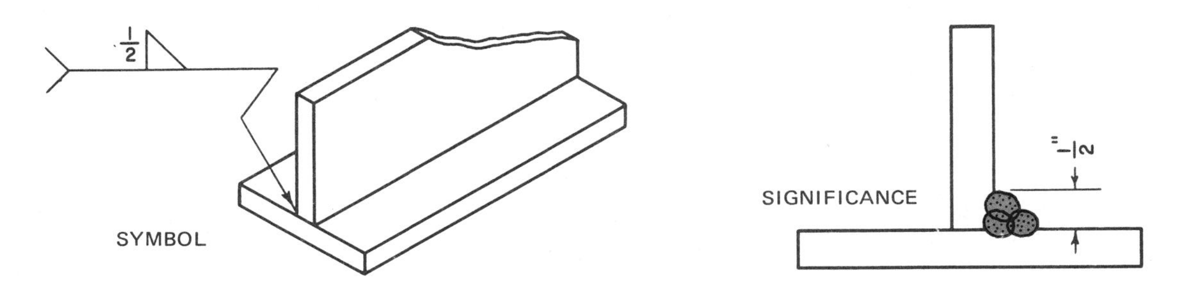

Fig. 15-1 Multiple-pass fillet weld

The heavy plate used in this unit requires 3 beads to complete the fillet.

Materials

Two steel plates, 1/2 inch thick, 2 in. x 9 to 12 in. each

5/32-inch diameter, E-6012 electrodes

Procedure

1. Set up and tack the plates in the manner used for the single-pass fillet welds.
2. Make a three-pass fillet weld, following the procedures shown in figures 15-1 and 15-2. Pay close attention to the angle of the electrode. When making the third pass, check the arc length frequently to make sure that an undercut does not develop along the upstanding leg of the weld.
3. Make additional fillet welds, welding both sides of the joint. Do not make all beads on one side of the joint first, but rather alternate the sequence.

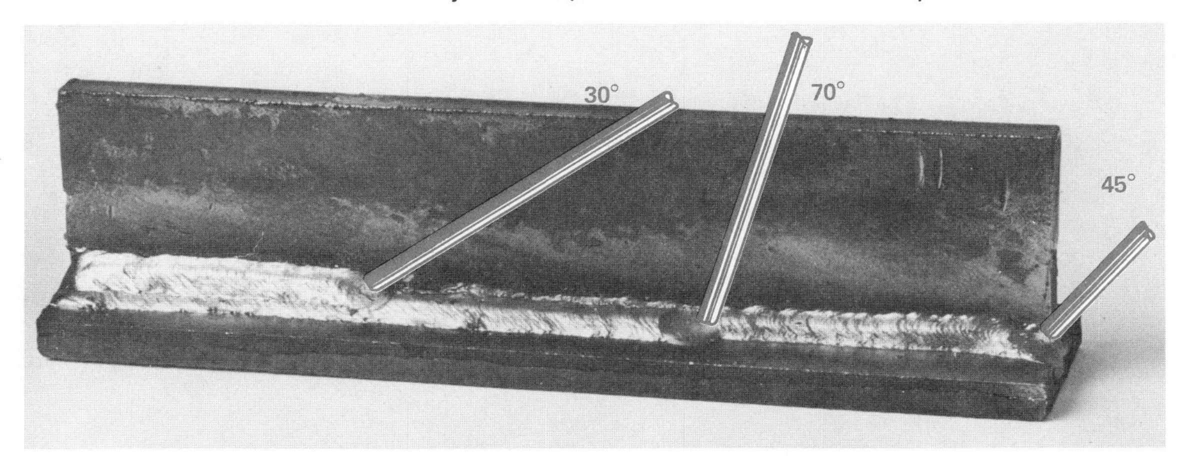

Fig. 15-2 Three-pass fillet weld using E-6012 electrode (right-handed operator).

4. Compare the distortion made by alternating sides with that made when only one side is welded.
5. Check the finished test plates in the same manner as the single-pass fillet welds were checked.

REVIEW QUESTIONS

1. What is the effect if the first bead on one side is followed by all three beads on the opposite side before completing the other two beads on the initial side? Why?

2. What should the height of the bead be for a fillet weld on 1/2-inch plate?

3. This unit indicates a weld made with three beads in two layers. How is a third layer applied?

4. How does the temperature of the base metal affect the overall appearance of the weld?

5. How does a hole in the first bead affect the second and third beads?

UNIT 16 WEAVING A LAP WELD

To lap-weld heavy plate with a single pass, a weaving motion of the rod is necessary. This requires a special technique to avoid an irregular, defective weld.

The weaving motion produces an oval pool of molten metal. The weld makes an angle of about 15 degrees with the edge of the plate.

The angle of the puddle varies with the amount of current, length of arc, and speed of welding as well as with the thickness of the welded metal. In general, larger welds at higher amperages need an angle slightly greater than 15 degrees.

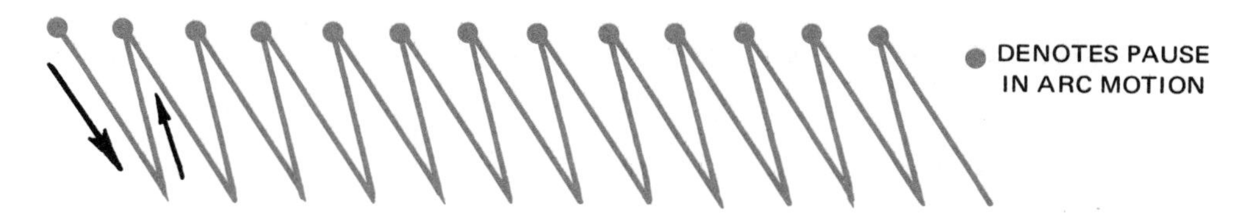

Fig. 16-1 Weaving a lap weld in heavy plate

Materials

Two steel plates, 3/8 inch thick, 2 in. x 9 to 12 in. each

5/32-inch diameter, E-6012 electrodes

Procedure

1. Position the plates for a lap weld.
2. Weld the joint using a weaving motion to keep the bottom edge of the molten puddle ahead of the top of the puddle.
3. When making this type of weld, pause with the arc at the top of each weave motion but not at the bottom of each weave, figure 16-1. This method prevents burning away or undercutting of the top of the weld.
4. Cool, clean and examine the face of the finished weld. Look for undercutting of the top leg and overlapping of the bottom leg of the weld.

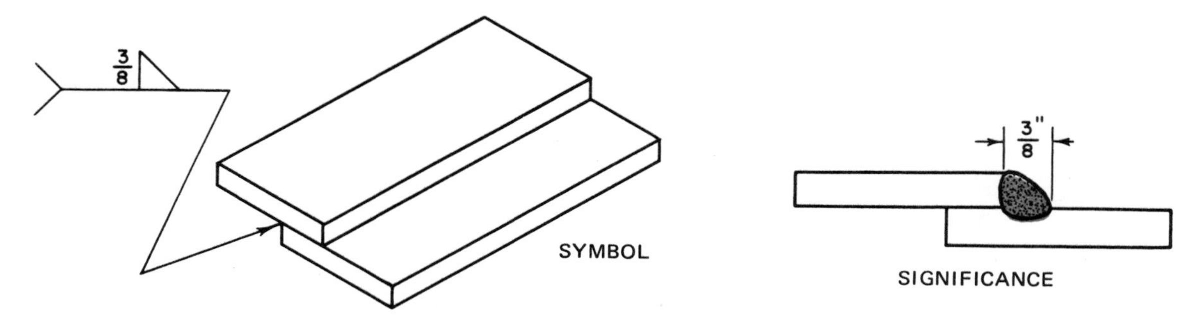

Fig. 16-2 Weaving a lap weld

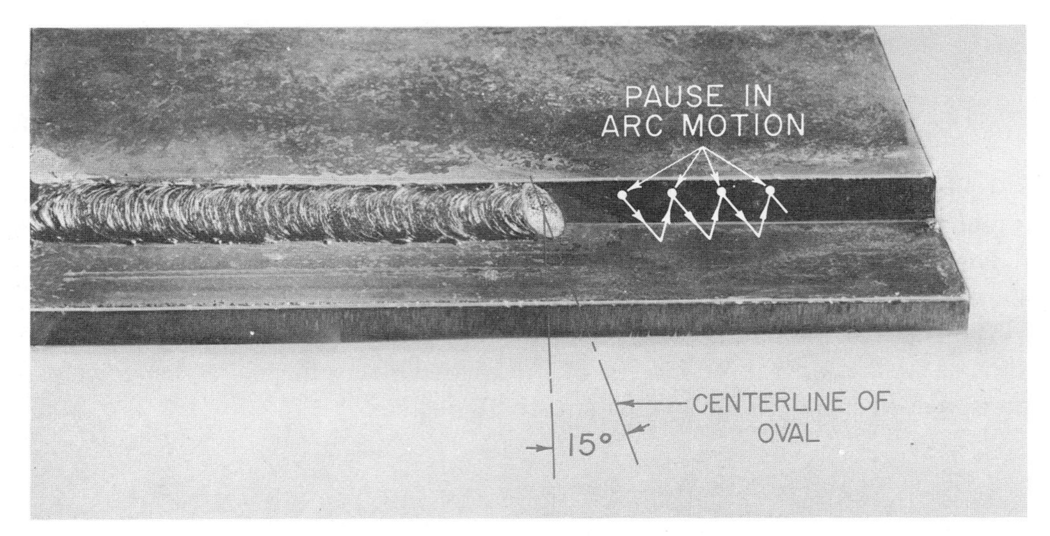

Fig. 16-3 Weaving a lap weld

5. Break this test plate and examine the root of the weld for good penetration.
6. Make another lap weld with the centerline of the molten puddle at a 90-degree angle to the line of the weld. Compare the line of fusion of the bottom leg of the weld with that of the previous bead.
7. Make more welds and examine them for appearance. All differences in ripple shape and spacing are caused by differences in arc control.

REVIEW QUESTIONS

1. Other than pausing at the top of each weave, how can irregularities and undercutting be controlled?

2. Does the undercutting referred to in this unit have the same appearance as the undercutting of a fillet weld?

3. Step 6 indicates an electrode motion with no lead at the bottom of the weave cycle. How does this affect the appearance of the finished bead and the line of fusion with the bottom plate?

4. Make a sketch of the cross section of the weld made in step 6. Show the correct shape for a lap weld with a dotted line.

5. How does the time for weaving a weld compare with that for a multiple-pass weld?

UNIT 17 WEAVING A FILLET WELD

This unit provides more practice in weaving beads to produce a multiple-pass weld of large size.

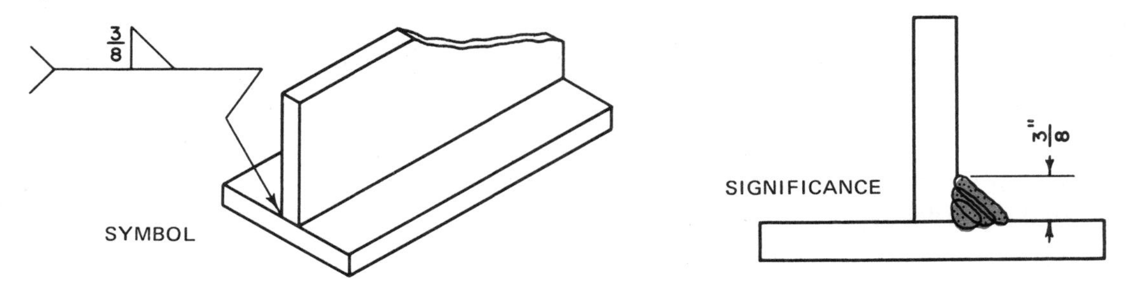

Fig. 17-1 Weaving a fillet weld

Materials

Two steel plates 3/8 inch thick, 2 in. x 9 to 12 in. each

5/32-inch diameter E-6012 electrodes

Procedure

1. Set up the plates as in figure 17-1.
2. Weld the joint using the electrode angle and weave motion described in unit 16. Weld a 3/8-inch fillet (i.e., each leg of the triangle formed by the weld should measure 3/8 inch).

   Note: The difference between a weave lap weld and a weave fillet weld is that the pause at the top of the puddle must be slightly longer for the fillet weld. Also, the length of the arc striking the upstanding leg must be kept very short to prevent undercutting.

3. Continue to make this type of weld. Examine each bead as it is made to determine what corrections are necessary to produce welds with uniform ripples and fusion.
4. Weave another bead over the original bead, using a very short arc and a definite pause as the arc is brought against the upstanding leg. Then bring the arc toward the bottom plate at a normal speed so that the bottom of the puddle leads the top of the puddle by 15 to 20 degrees. Return the arc to the top of the fillet rapidly with a rotary motion, figure 17-2.
5. Clean and inspect this bead for undercutting along the top leg of the weld and for poor fusion along the bottom edge. Also check for uniformity of the bead ripples.
6. Weave a third bead over the first two so that the bottom of the bead leads the top slightly more than 20 degrees. As the arc reaches the bottom of the fillet, try hooking the bead by moving the arc along the bottom before returning to the top of the weld.

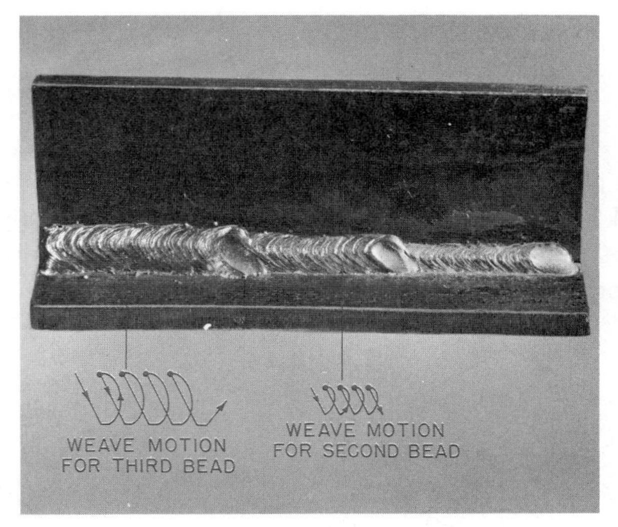

Fig. 17-2 Weave motions for a fillet weld

7. Clean and inspect weld as in step 5.

REVIEW QUESTIONS

1. What effect does hooking the bead have on the finished bead?

2. In making the second and third beads, is welding accomplished on the return stroke of the weave?

3. A common mistake in welding this type of fillet is depositing too much metal on the bottom plate. How is this fault corrected?

4. Is the bead made at step 6 good on 3/8-inch plate? Why?

5. Is undercutting more or less of a problem on the T joint than it is on the lap joint? Why?

UNIT 18 BEVELED BUTT WELD

This important joint can be very strong if it is well made. The multiple-pass procedure used in this unit is likely to produce a better joint than a thick single-pass method.

The beveled butt joint requires skill in weaving beads of two different widths, cleaning the preceding bead, and controlling the width of the bead.

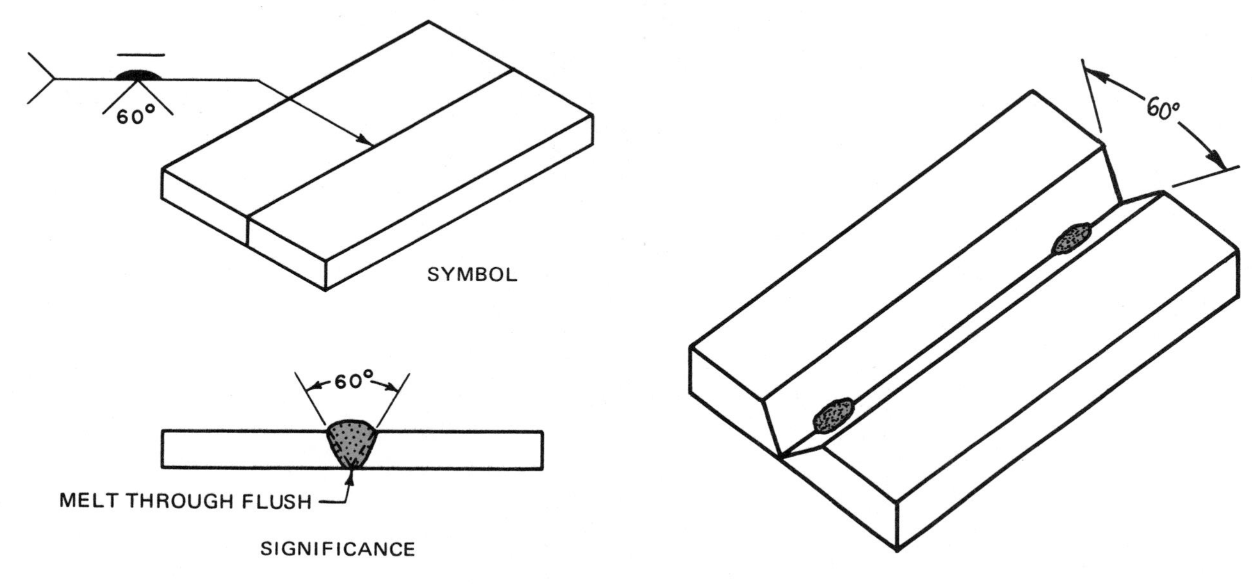

Fig. 18-1 Beveled butt weld

Fig. 18-2 Setup for beveled butt weld

Materials

Two steel plates, 3/8 inch thick, 4 in. x 9 to 12 in. each with one long edge beveled at 30 degrees

5/32-inch diameter E-6012 electrodes

Procedure

1. Align the plates on the welding bench and tack them as shown in figure 18-2.
2. Run a single-pass stringer bead in the root of the V formed by the two plates.
3. Clean the first bead and deposit a second bead by using a slight weaving motion. Allow the arc to sweep up the sides of the bevel in order to give this bead a slightly concave surface.
4. Clean the bead and run the third bead using a wide weaving motion. Do not allow the weld to become too wide. The actual width of the face of the weld should be slightly wider than the distance between the top edges of the V. Figure 18-3 is a cross-section view showing the size and contour of each bead.

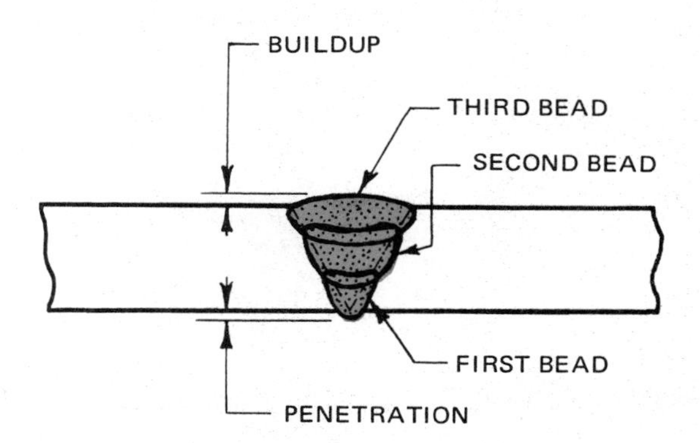

Fig. 18-3 Multiple-pass beveled butt weld

5. Clean and examine the finished weld for root penetration, evenness of fusion lines, and equal spacing of the ripples. Any variation in fusion, penetration, or ripple is caused by variations in arc manipulation. Uniform results can only be obtained by following uniform procedure.

   Note: Usually only three beads are required to make a bevel butt weld in 3/8-inch plate. However, if the first two beads are thin, do not attempt to make up for this by building the third bead much heavier. Instead, apply a normal third bead and a fourth if necessary. Heavy or thick buildups in one pass tend to produce holes and slag inclusions.

6. Align and tack a second set of plates as before, but leave an opening at the root of the V about one-half the rod diameter in size.
7. Weld these plates in the same manner as the first set and inspect visually.

REVIEW QUESTIONS

1. Why should the second bead have a slightly concave surface?

2. What effect does leaving a slight gap at the root of the V have on the finished bead?

3. If it is difficult to make the root pass with the plates gapped because of burn-through, what step is taken to correct this difficulty? Why?

4. How should the width of the finished bead compare with the width of the joint?

UNIT 19 OUTSIDE CORNER WELD

Outside corner welds are frequently used as finished corners after they have been smoothed by grinding or other means, figure 19-1. In this case, the shape of the bead and the smoothness of the ripples are very important. Roughness, caused by too much or not enough weld metal and uneven ripples, requires a lot of smoothing. This results in higher cost.

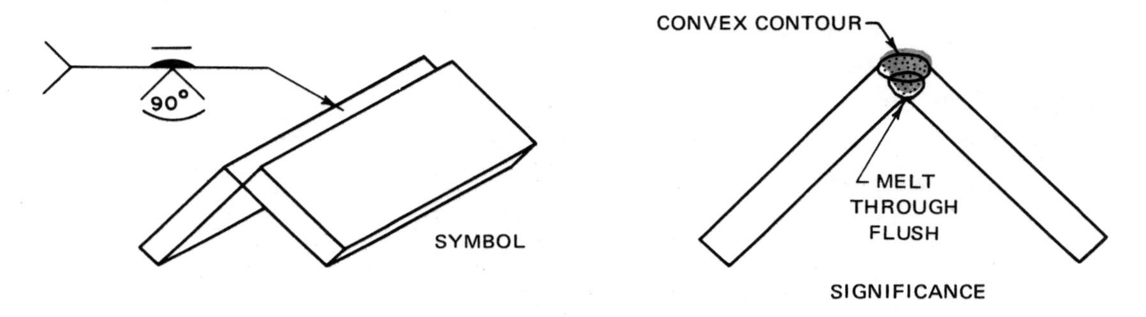

Fig. 19-1 Outside corner fillet

Materials

Two steel plates, 3/8 inch thick, 2 in. x 9 to 12 in. each

5/32-inch diameter E-6012 electrodes

Procedure

1. Set up the plates and tack them as shown in figure 19-2.
2. Make the weld in three or more passes as for a beveled butt joint. When making the final pass, observe all the precautions for weave welding to avoid any possibility of the finished bead overhanging the plate edges, figure 19-4. Place the assembly on the welding bench so that the weld can be made in the flat position.

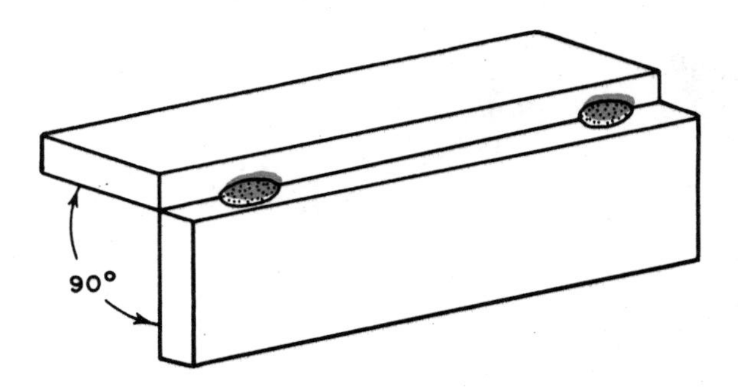

Fig. 19-2 Setup for outside corner weld

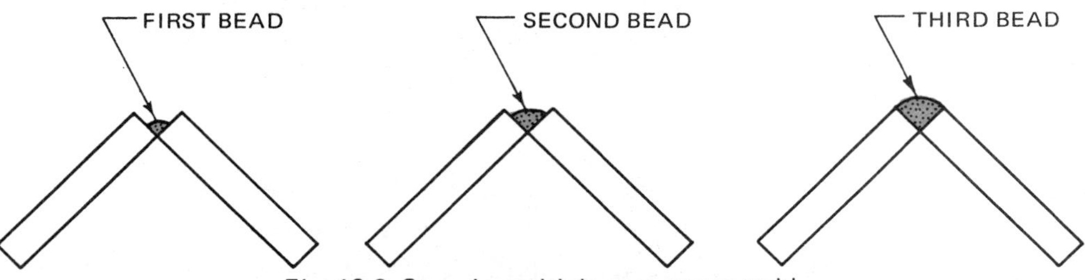

Fig. 19-3 Steps in multiple-pass corner weld

3. Check the finished weld for uniform appearance and to see if the angle of the plates is 90 degrees. Make any necessary corrections when setting up the next set of plates.
4. Test the weld by placing the assembly on an anvil and hammering it flat. Examine the root fusion and penetration, figure 19-5.
5. Make additional joints of this type, checking each weld for uniform appearance and for the shape shown in figure 19-4.

Fig. 19-4 Multiple-pass outside corner weld

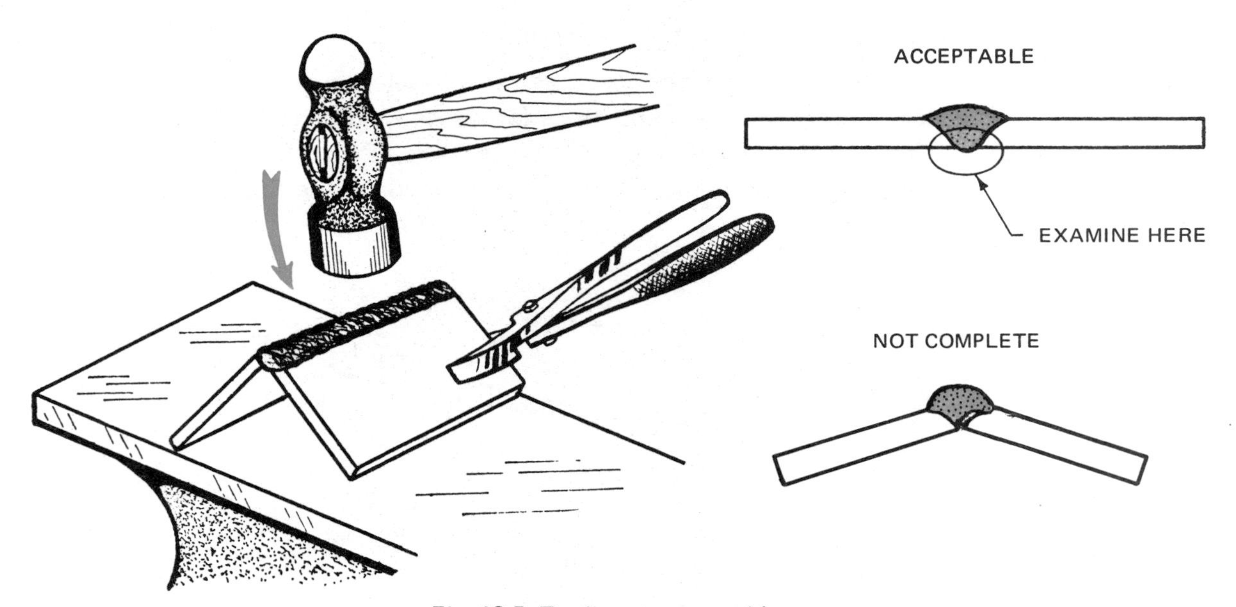

Fig. 19-5 Testing a corner weld

REVIEW QUESTIONS

1. How are poor fusion and uneven penetration corrected?

2. How are deep holes between the ripples in a finished weld corrected?

3. What can be done to prevent the final pass from building up too much, causing a hump along the line of fusion?

4. Is undercutting possible on an outside corner weld? How?

UNIT 20 OUTSIDE CORNER AND FILLET WELDS (HEAVY-COATED ROD)

This unit provides an opportunity to gain skill and knowledge in the use of heavy-coated electrodes. These electrodes are recommended for use in the downhand or flat position only. This unit also provides experience in adjusting the electrode angle, arc length, and current setting.

Research and development have produced an electrode of this type with a large amount of iron powder added to the coating. This type of electrode makes a weld with good physical characteristics, and very good appearance and contour. The slag is so easily removed that it is described as self-cleaning.

The current values and arc voltages required to make this weld are quite high. As a result, the weld metal deposits rapidly. As much as 17 pounds of weld metal per hour may be deposited when using a heavy-coated 1/4-inch electrode.

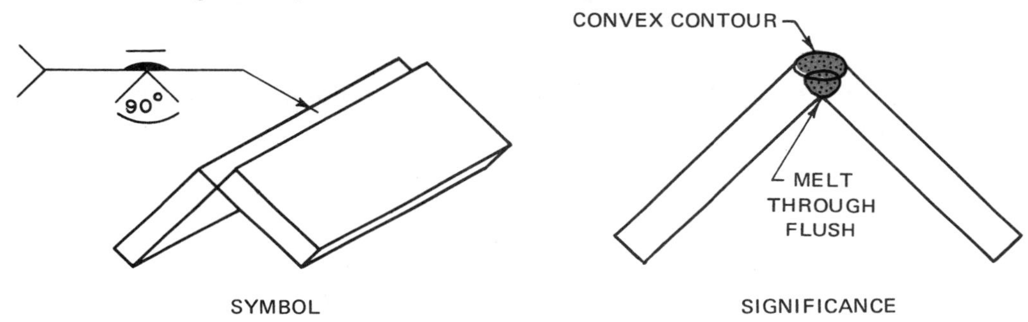

Fig. 20-1 Outside corner weld

Materials

Steel plate, 3/8 inch thick, 2 in. x 9 to 12 in.

5/32-inch diameter x 14-inch E-7024 electrode

Procedure

1. Set up and tack the plates as indicated.
2. Adjust the current according to the manufacturer's recommendations, or use the formula for current settings. For a heavy-coated rod, 20 percent should be added to the current value, so that the minimum setting is about 190 amperes.
3. Make a root pass with little or no weaving motion.
4. Clean and examine this bead for appearance and note the ease of slag removal. Also examine the end of the electrode and note that the metal core has melted back into the coating to form a deep cup.

Note: The cupping characteristic makes this rod good for *contact welding.* In contact welding the coating is allowed to touch the work. In this manner the operator does not have to maintain the arc length. However, this limits the possibility of controlling

Fig. 20-2 Outside corner weld with heavy-coated electrode

the rod, so the width of the weld is limited to stringer beads. For this reason most experienced operators prefer to maintain a free arc which can be controlled.

5. Apply a second bead with a weaving motion and a rate of travel that builds up a bead as thick as the plates being joined.
6. Clean and inspect the bead. Especially notice the line of fusion, appearance of ripples, and the shape of the finished joint. Figure 20-2 shows both beads.
7. Make a series of joints of this type, but increase the current for each joint by 10 to 15 amperes until a final joint is made at 250 amperes. Also vary the amount of *lead* (the angle at which the electrode points back toward the finished bead).
8. Clean and inspect the joints. Hammer each joint flat on an anvil. Examine the weld metal for holes and slag inclusions, and compare the grain sizes.

   Note: Valuable practice in making fillet welds with the E-7024 electrode is gained by using the plates welded into outside corner welds, steps 1 through 7.

9. Set up the assembly to weld the inside corner as shown in figure 20-3. Make sure that any slag, which may have penetrated this inside corner from the original welds, is thoroughly cleaned from the joint.
10. Adjust the current values as in step 2, and proceed to make a single-pass 3/8-inch fillet weld. The electrode should point back toward the finished weld at an angle of 15 to 20 degrees.
11. Clean and inspect this bead for the shape of the weld, straight, even fusion line between the weld and plate, and any evidence of holes or slag inclusions in the

Fig. 20-3 Fillet weld with heavy-coated electrode

finished bead. Slag inclusions are usually caused by either poor cleaning of the original assembly, or a rod angle that caused the slag to flow ahead of the arc and molten pool.

12. Run a second and third bead on top of the first, using a uniform, slightly weaving motion to produce enough width. Weld other joints with current values up to 250 amperes.
13. Clean and inspect, as in step 11.
14. Set up plates and make a three-pass fillet weld as in steps 10 through 12, figure 20-3.
15. Break this weld by hammering the two plates together. Check for penetration and grain structure.
16. Weld more joints with one leg of the final assembly flat and the other vertical. Try varying the electrode angle from more than 45 degrees to less than 45 degrees from horizontal, and check the results for fusion and bead shape.
17. Make additional joints as in steps 10 through 12, but incline the plates so that the welding proceeds in a slightly downhill direction. Then make some joints with the weld proceeding in a slightly uphill direction.
18. Clean and inspect these joints and compare the results in each case with the welds made when the joint was in a perfectly flat position.

REVIEW QUESTIONS

1. What polarity does the manufacturer recommed for this rod?

2. It is suggested that various rod angles be tried while making these joints. What does this experiment show?

3. Too much current applied to a standard-coated rod causes the coating to break down and burn some distance up the rod. Is this true with the iron-powder type of electrode?

4. How do the higher amperages used in step 7 affect the surface appearance of the joint?

5. How does the speed of making these fillet welds compare with the speed of similar welds made with an E-6012 electrode?

6. What can be done to correct the tendency of the slag to run ahead of the arc and make holes in the bead?

7. How does the cleaning time for welds made in this unit compare with that for similar welds made with E-6010 and E-6012 electrodes?

UNIT 21 DEEP-GROOVE WELD (HEAVY-COATED ROD)

Although deep groove welds can be made with a wide variety of electrode types, this unit provides practice and experience in using an electrode designed specifically for this type of joint.

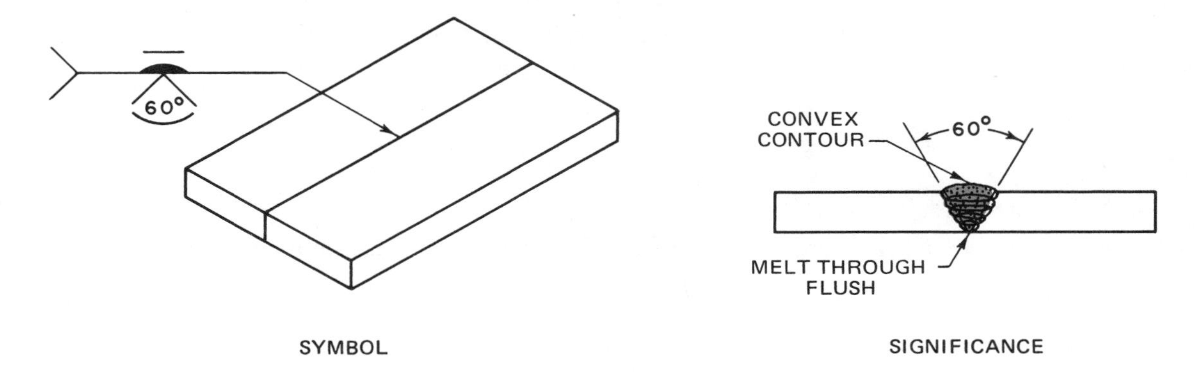

Fig. 21-1 Deep groove weld

Materials

Steel plates, 3/4 inch thick, 3 in. x 9 to 12 in. each

5/32-inch diameter, E-6027 electrodes

Procedure

1. Align and tack the beveled plates as shown.
2. Adjust the current for heavy-coated rod and make the root pass. Use a free arc rather than the contact method.
3. Clean and inspect this bead, especially for shape, and observe the ease of slag removal.
4. Apply additional beads in layers 1/8 to 3/16 inch thick until the joint is completed. Use a weaving motion as necessary to provide a good surface appearance.

   Note: This electrode is designed to cause the weld metal to wash up the sides of the groove and form a bead with a slightly concave surface from which the slag can be easily removed between passes. Also notice the absence of undercutting with this electrode. Figure 21-2 shows a joint made in a series of steps to show the bead shape.

5. Clean and inspect the finished job for fusion at the root of the joint as well as along the edges of the weld. Cut a cross section from the finished joint and inspect for holes and slag inclusions. If the equipment is available, break the test section and inspect the grain structure.

Fig. 21-2 Deep groove weld with E-6027 electrode

Figure 21-2 shows the plates provided with a *run-out tab.* This is a device used to provide for continuing the weld beyond the ends of the work. It is then cut or broken from the work. It eliminates the necessity of filling the crater at the extreme edge of the job. This tab can be applied to any type of joint which ends suddenly. It is time-saving when the weld must have the same shape and size for its entire length. Run-out tabs are especially helpful in some automatic welding processes, in which there is no opportunity to manipulate the arc to fill the crater.

REVIEW QUESTIONS

1. How does slag removal for this type electrode compare with that for previous welds?

2. What are the advantages of heavy-coated electrodes?

3. What surface contour should intermediate passes have on deep-groove welds?

4. Can the E-6027 electrode be used for weave beads?

5. Why is a run-out tab used on some joints?

UNIT 22 REVERSE-POLARITY WELDING

Direct current straight-polarity type electrodes are applied by a straight-line, steady motion. These electrodes produce a smooth, crowned bead when properly applied.

This unit covers the basic knowledge needed for the correct manipulation of E-6010 and E-6011 type arc welding electrodes. The E-6010 and E-6011 electrodes must be applied with a definite whipping motion. This makes a rougher surface appearance, but the penetration and bead shape is uniform in all positions.

DOWNHAND BEAD WITH WHIPPING MOTION AND HOT CRATER IN THE FLAT POSITION

Materials

Two steel plates, 3/8 inch thick, 6 in. x 6 in. each

5/32-inch diameter, E-6010 or E-6011 electrodes

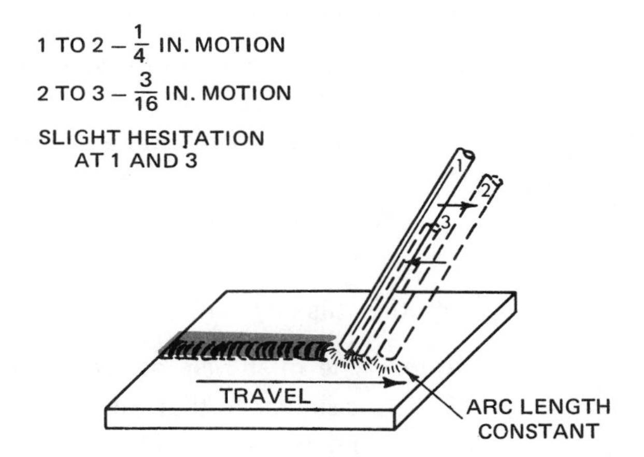

Fig. 22-1 Hot-crater method

Procedure

1. Set up a 3/8-in. thick, 6 in. x 6 in. plate in the flat position.
2. Set the amperage at 130 to 150 for the 5/32-inch electrode.
3. The welder should have good visibility of the arc area.
4. Hold the electrode perpendicular and pointed 10 to 20 degrees in the direction of travel.
5. Strike the arc and maintain standard arc length using a whipping motion.

   Note: The whipping motion shown by figure 22-1 is a forward and backward motion in the direction of travel. This motion may vary in length, but could be 1/4-inch motion ahead and 3/16-inch back. At the end of the back motion, a slight pause deposits the bead. The forward motion controls the amount of penetration. Various bead results can be attained by varying the pause, travel motion length, and speed of travel.

6. Practice this whipping motion in the flat position until a uniform, closely rippled bead is produced.

BEAD WITH WHIPPING MOTION AND COOL CRATER IN THE FLAT POSITION

Materials

Two steel plates, 3/8 inch thick, 6 in. x 6 in. each

5/32-inch diameter, E-6010 or E-6011 electrodes

Procedure

1. Set up a 3/8-in. thick, 6 in. x 6 in. plate in the flat position.
2. Set the amperage, position, and angles as for a downhand bead.

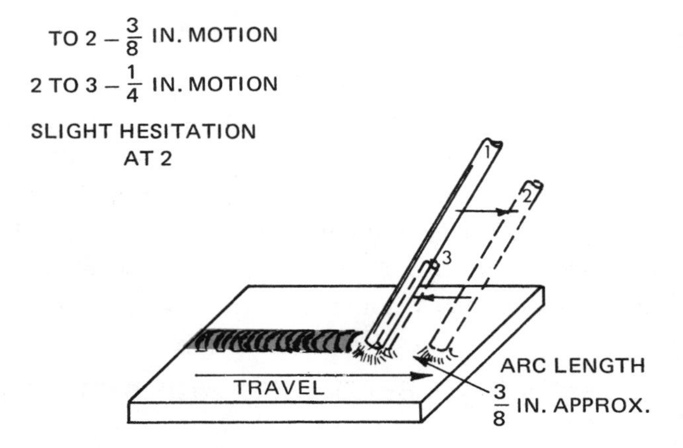

Fig. 22-2 Cool-crater method

3. Strike an arc and, using a whipping motion, proceed as in figure 22-2.

   Note: When welding *out of position* (any position other than flat) a slight change in the whipping action is needed. To overcome the pull of gravity on a molten mass, a cooler puddle is desirable. This is done by pausing at the end of the forward motion, and holding a long arc. On the backward motion, a normal arc length is used with no pause in the puddle.

4. Practice this until uniform results are attained.

   Note: The fillet welding technique for E-6010 and E-6011 electrodes is similar to that for other electrodes. The first bead, however, is a downhand bead.

REVIEW QUESTIONS

1. What is the difference in the way the E-6010 and E-6012 electrodes are handled?

2. When is the hot-crater technique used in welding with the E-6010 electrode?

3. What advantages does the cool-crater method have over the hot-crater method in the vertical-up welding position?

4. Is the electrode positive or negative in reverse-polarity welding?

5. What positions can the E-6010 and the E-6011 electrodes be used in?

UNIT 23 LOW-HYDROGEN ELECTRODES

The low-hydrogen type of electrode replaces the E-6010 and E-6012 electrodes in many industrial and construction applications. The welding student should have knowledge of the characteristics and uses of this electrode.

AWS-ASTM Class	Hobart	Air Products	Airco	Alloys Rods	Arcos	Canadian Liquid Air Ltd.	Canadian Rockwell	P&H	Lincoln
E-7016	16	7016	7016 7016-M		Tensilend 70		Tensiarc 76	70LA-2	
E-7018	LH-718	7018 7018-IP	Easyarc 7018 7018-C 7018-M	Atom-Arc 7018	Ductilend 70	7018-S LA-7018 Atom-Arc 7018	Hyloarc 78	170-LA SW-47	Jetweld LH-70
E-7028	LH-728	7028	7028					DH-170	LH-3800
E-7018-A1	LH-718-MO	7018-A1		Atom-Arc 7018-MO	Ductilend 70-MO	Atom-Arc 7018-MO		718-A1	
E-8018-C2	LH-818-N2	8018-C2		Atom-Arc 8018-N		Atom-Arc 8018-N		SW-818-C2	

AWS-ASTM Class	Marquette	McKay Co.	M&T (Murex)	N.C.G.	Reid-Avery	Shober	Canadian Liquid Carbonic	Westinghouse
E-7016		7016 Pluralloy 70-AC	HTS HTS-18 HTS-180		Raco 7016		P&H-716	LOH2-716
E-7018	7018	7018	Speedex HTS HTS-M-718	Atom-Arc 7018	Raco 7018	718	P&H-718	Wiz-18
E-7028			28					
E-7018-A1	7018-MO	7018-A1		Atom-Arc 7018-MO	Raco 7018-A1		710-P	
E-8018-C2	8018-N	8018-C2		Atom-Arc 8018-N	Raco 8018-C2			Wiz 818-C2

Chart 23-1 Low-hydrogen electrode comparison chart

CHARACTERISTICS AND TYPES

The gaseous shield formed by the electrode coating normally uses hydrogen as a shielding element. The gaseous shield formed by low-hydrogen electrodes contains a small amount of hydrogen.

There are several types of low-hydrogen electrodes and it is important that the correct one be used. Chart 23-1 compares electrodes made for different applications.

REASONS FOR USING LOW-HYDROGEN ELECTRODES

1. To provide better physical and mechanical properties.
2. To reduce bead and under-bead cracking in certain kinds of steel in a molten state. This cracking may be caused by hydrogen.
3. To improve *ductility* (to make the metal easier to work).
4. To reduce the temperature needed to preheat the weld.
5. To provide better low-temperature impact properties.

PROCEDURE GUIDELINES

- E-XX15 electrodes are used with DC reverse polarity. E-XX16 and E-XX18 are used with AC or DC reverse polarity. These electrodes are all-position type electrodes.
- E-XX28 are iron-powder electrodes. They are used in the flat position, with AC or DC reverse polarity.
- Do not use a whipping motion with low-hydrogen electrodes.
- The arc must be held very close.
- Stringer beads should be used rather than weave beads in all positions. The weave bead is likely to trap slag and gas in the bead.
- When making multiple-pass beads, all slag must be removed from each pass.
- Pinholes, usually found at the start or the end of a weld, may be caused by:
    1. Incorrect arc striking or stopping
    2. Moisture in the electrode coating
    3. Chipped coating on the electrode
    4. Moisture in the weld
    5. Arc length too great
- When striking an arc, always strike ahead of the point where the weld is to begin. Shorten the arc immediately, pushing the weld metal back to the starting point.
- When stopping the arc at the end of the joint, keep a short arc. If washout occurs, chip the slag, clean, and restrike as before.
- Low-hydrogen electrode coatings pick up moisture easily. Therefore, they should be stored in a thermostatically controlled oven or dry box. A temperature of 300 to 400 degrees F. is required.

- The operator should always examine every electrode he places in his electrode holder. A chip in the coating of a low-hydrogen electrode makes a hard spot and pinholes in the weld. A chipped rod can also cause the arc to be unstable.
- In some structural applications, the base metal may contain moisture. Preheating the weld area immediately before welding insures successful results.
- When welding out of position, undercutting may occur at the edges of the weld. To overcome this hold a closer arc and, in the case of vertical-up welding, use a slight U-shaped weave and hesitate on each side. Better control can be obtained by using an electrode one size smaller. The welding current should not be too high.
- *Arc blow* (the tendency of the arc to wander off its path) is sometimes a problem in out-of-position welding with low-hydrogen electrodes. This is a greater problem with large electrodes (5/32- to 1/4-inch size). To overcome this:
    1. Insure a good ground.
    2. Clean the base metal as well as possible.
    3. Keep the arc short.

REVIEW QUESTIONS

1. How is a low-hydrogen electrode different from others?

2. When welding with a low-hydrogen electrode, what is the most important thing for the operator to keep in mind?

3. Why is moisture bad for low-hydrogen electrodes and base material?

4. What is the tensile strength of a bead made with E-7018 electrodes?

5. Why is it difficult to restart an arc with a used E-7018 electrode?

UNIT 24 HORIZONTAL WELDING

When the ability to make good welds in the flat position has been developed, it is important that out-of-position welding be learned. The easiest out-of-position welding is horizontal welding. It is different from flat welding because of the effect of gravity on the molten puddle.

Many jobs must be welded in the horizontal position. This is because the welded part cannot be moved, due to its size or location.

WELDING A HORIZONTAL PAD

Materials

Steel plate 1/4-inch or heavier, 4 inches x 8 inches.

1/8-inch E-6010 and E-7018 electrodes with a DC welding machine or 1/8-inch E-6011 electrode with an AC welding machine.

Procedure

1. Set the material up with the surface in a vertical position.
2. Set the current at 110-125 amps for 1/8-inch E-6010 electrode.
3. Right-handed operators weld from left to right.

   Note: Hold the electrode parallel with the floor no more than 10 degrees below horizontal. The electrode should also be angled toward the direction of travel 15 to 20 degrees.

   The E-6010 and E-6011 electrodes are applied with a whipping motion, but low-hydrogen electrodes are not.

4. Start the arc and make the whipping motion.

   Note: The molten metal will have a tendency to drop down to the bottom of the puddle if the electrode is not handled correctly. To overcome this sag, change the amount of pause at the back of the puddle and shorten the arc length.

   The low-hydrogen electrode is applied with a very short arc along a straight line.

   The same angles are used for both electrodes.

5. Continue making the horizontal pad by welding each bead above the preceding bead, figures 24-1 and 24-2.

   Note: The ability to do out-of-position welding requires considerable practice.

Forward progress is approximately 1/16" with each motion. Motion is straight back and forth with the pause just right of the crater center.

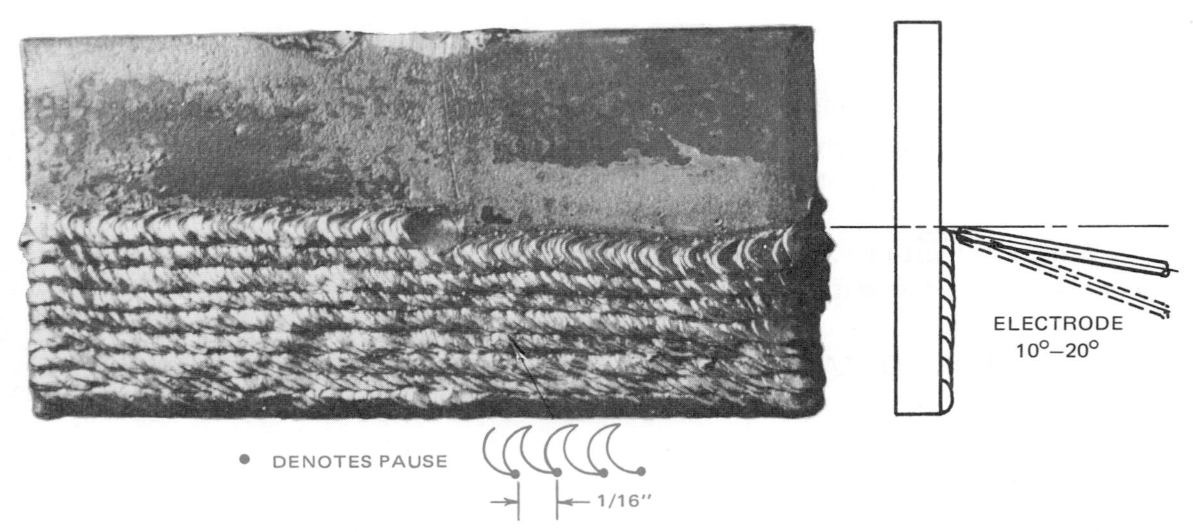

Fig. 24-1 Horizontal pad using E-6010 electrode (right-handed operator travelling left to right).

HORIZONTAL V-GROOVE BUTT JOINT

Materials

Two pieces of plate 1/4-inch or heavier, 2 inches x 8 inches, with a 30-degree bevel on one edge of each

1/8-inch E-6010 and E-7018 electrodes

DC welding machine

Procedure

1. Tack weld the material and position it for a horizontal weld. It may be necessary to experiment with the root opening to insure 100% penetration of the weld.

No motion with this electrode. Hold very close arc length.

Fig. 24-2 Horizontal pad using E-7018 electrode.

Motion is straight back and forth on all stringer beads, with slight hesitation in crater. Progress is approximately 1/16" with each motion.

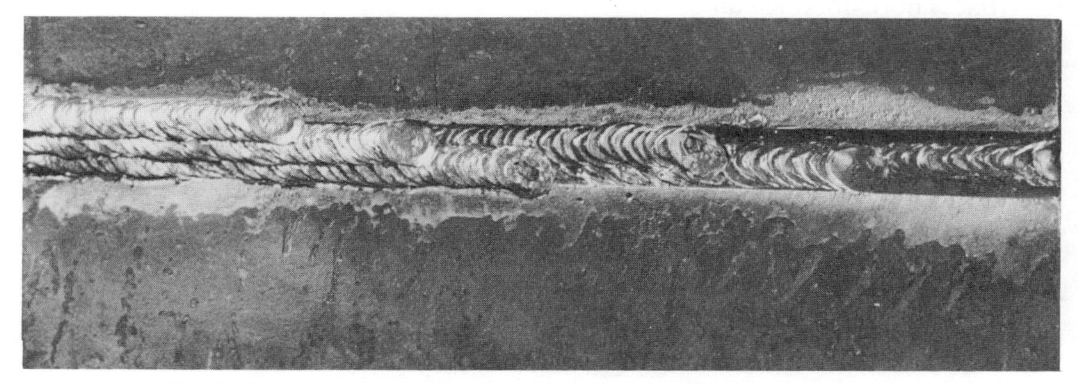

Fig. 24-3 Horizontal V-groove butt joint using E-6010 electrode.

2. Weld the root pass with a stringer bead.

   Note: Make sure the slag is removed from all passes. Run the beads in the order shown in figures 24-3 and 24-4.

3. The number of beads for a joint may vary, but they are always applied from the bottom up.
4. The surface appearance should be smooth, with no undercutting or overlapping.
5. When a good weld can be made in this manner, try welding heavier plate using 5/32-inch electrodes.
6. The V-groove should be tested by grinding the root side and face side flush with the surface. Cut 1-inch strips across the weld and test by bending the welded area in a vise. The cut will show any pinholes in the weld.

CAUTION: Personal safety precautions must be observed in out-of-position welding. Falling spatter and sparks are always present.

No motion is required. Use close arc length.

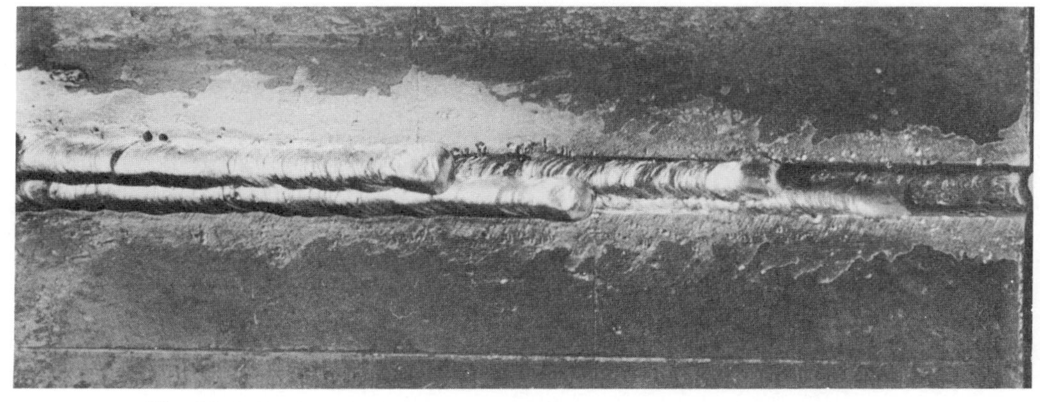

Fig. 24-4 Horizontal V-groove butt joint using E-7018 electrode.

REVIEW QUESTIONS

1. What is horizontal welding?

2. What additional factor has to be considered in controlling the size and shape of the bead in horizontal welding?

3. What does whipping mean in welding and why is it used?

4. Is whipping necessary with the low-hydrogen electrode?

5. Is it practical to make weave beads in the horizontal position? Why?

UNIT 25 OVERHEAD WELDING

When the ability to do horizontal welding is developed, the overhead fillet weld should not be difficult. The procedure is very similar. It is often necessary to use the overhead position in industry.

As in all welding processes, the welder must be comfortable to do a good job. Overhead welding can be tiring when it is done for a long period of time. Anything the operator can do to help keep the hand steady is an advantage.

OVERHEAD FILLET WELD

Materials

1/4-inch or heavier plate, 2 inches x 8 inches

1/8-inch E-6010 and E-7018 electrodes

DC welding machine

Procedure

1. Tack weld two pieces of plate together for a T joint.
2. Place the material in the overhead position. A holding tab can be tacked to the top of the plate.

   Note: For best results the joint line should be only slightly above eye level.

3. The root pass should be centered in the joint. A slight whipping motion is used with E-6010. E-7018 is not used with a whipping motion.
4. If it is difficult to start the arc, hold a longer arc and hesitate more.
5. The order in which the beads are run with each type of electrode is shown in figures 25-1 and 25-2.

First pass motion straight back and forth with slight hesitation in crater. Second pass oscillated with pauses at dot (•).

Fig. 25-1 Two-pass overhead fillet weld using E-6010 electrode (right-handed welder travelling left to right).

No motion required. Use close arc length.

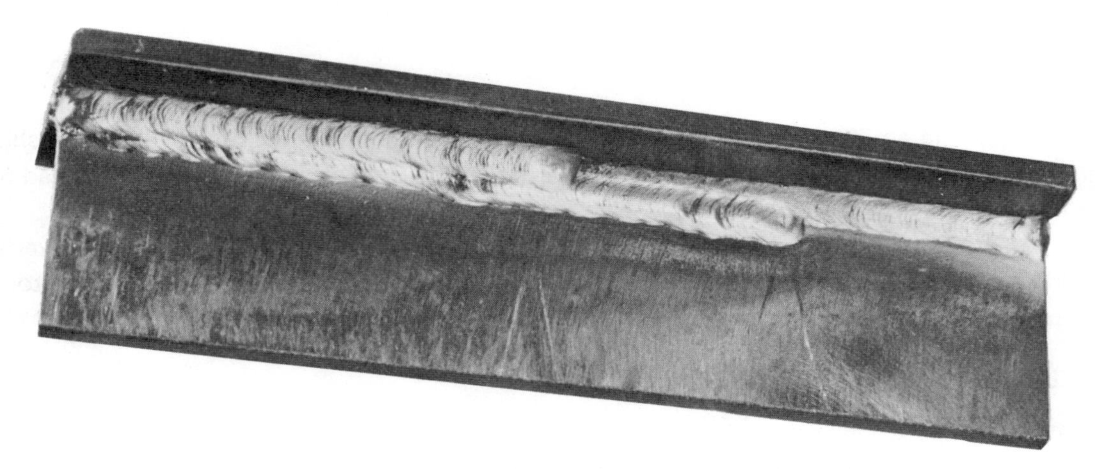

Fig. 25-2 Three-pass overhead fillet weld using E-7018 electrode.

6. The electrode should be at an angle of about 45 degrees with the surface of the metal. It should point in the direction of travel at an angle of about 10 degrees to 20 degrees.

   Note: The electrode angle can be changed to force the weld metal to go where it should.

7. The second pass of the E-6010 electrode is also applied with a whipping motion as shown in figure 25-1.

   Note: There must be a pause at the top and back of the crater on each weave. If there is a buildup of weld metal on the vertical plate, the amount of pause and length of the arc must be adjusted.

8. Set up another pair of plates and make an overhead fillet weld using the stringer pass method with 1/8-inch E-7018. No whipping motion is required. Apply the beads in the order shown in figure 25-2.

REVIEW QUESTIONS

1. Why is the overhead welding position more difficult than the horizontal position?

2. What should be done if there is trouble in starting the arc?

3. What is the difference in the way the E-7018 and E-6010 electrodes are used on this joint?

4. Why is welding in the overhead position more dangerous than welding in other positions?

5. What should be the difference between a weld made in the overhead position and a weld made in the flat position?

UNIT 26 VERTICAL WELDING

Probably the most difficult welding position for the beginner is the vertical position. The welding student should plan to spend considerable time practicing the various applications of this position. As in any out-of-position welding, safety equipment and clothing are very important.

Vertical welding can be done by welding uphill or by welding downhill. Welding vertical down is performed on sheet metal through 3/16-inch thickness and on some piping applications. Welding vertical up is performed on plate and structural parts 1/4 inch and heavier.

In the construction field, vertical welding ability is necessary because most weldments are so large they can't be positioned.

As in the preceding units, 1/8-inch E-6010 and 1/8-inch E-7018 electrodes are used here. Pictures are used to show the student what the different beads should look like. The electrode motion, where applicable, is shown on the pictures.

VERTICAL-UP BEADS

Materials

DC welding machine

1/4-inch or heavier plate

1/8-inch E-6010 and 1/8-inch E-7018 electrodes

Procedure Using E-6010 Electrodes

1. Position the plate in the positioner with the surface vertical at eye level.
2. At approximately 110 amps, start the arc at the bottom of the plate. The electrode angle should be straight out from the plate and pointing upward from 5 to 10 degrees.

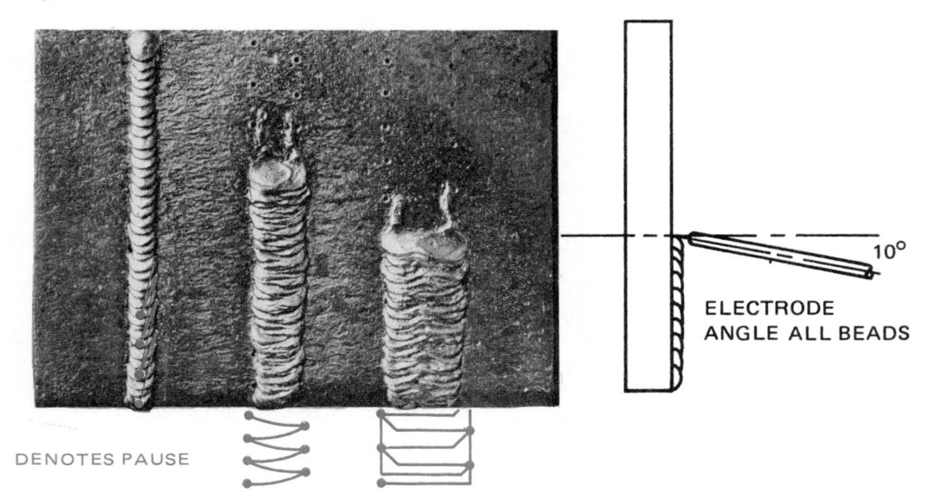

Fig. 26-1 Vertical up stringer and weave beads using E-6010 electrodes.

First pass stringer, no motion, straight up. Second pass inverted slightly U-shaped weave with hesitation at sides. Third pass straight side-to-side weave with pause at each side.

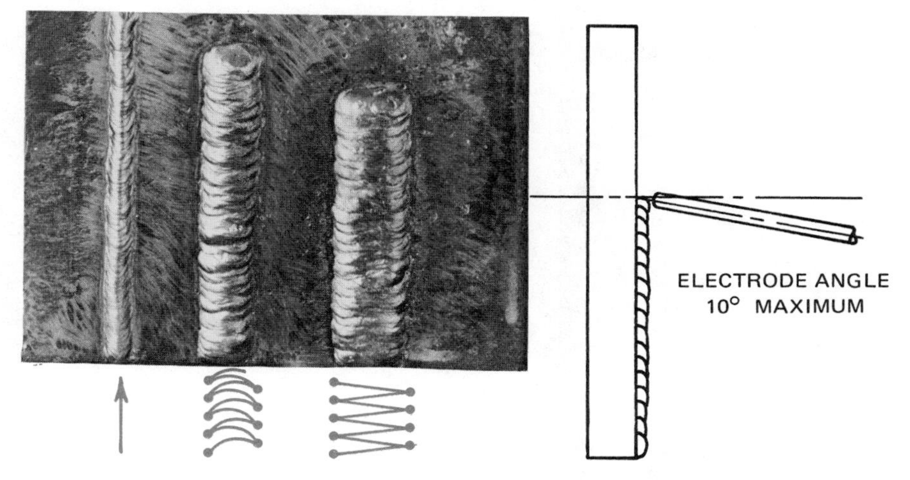

Fig. 26-2 Vertical-up stringer and weave beads using E-7018 electrode.

Note: The stringer bead is applied with a straight up long arc movement and a straight down short arc. The weaves are applied with a side-to-side motion.

3. Always chip and wire brush the slag from each bead.
4. Practice each type of bead separately, by padding a plate.

Procedure Using E-7018, Low-hydrogen Electrodes

1. Position the plate in the positioner with the surface vertical at eye level.
2. At approximately 125 amps, start the arc at the bottom of the plate. Keep a short arc for the E-7018 electrode.
3. Stringer beads are applied with no motion other than the smooth forward progress of the bead.
4. The weave beads are applied with a slight inverted U-shaped side-to-side motion.
5. Practice each type of bead by padding a plate.

VERTICAL UP MULTI-PASS FILLET WELD

Materials

DC welding machine

1/4-inch or heavier plate

1/8-inch E-6010 and 1/8-inch E-7018 electrodes

Procedure

1. Tack weld pieces of plate to form a T joint.
2. Apply the stringer bead and two weave beads, as shown in figures 26-3 and 26-4.

Weaves have a slight pause at the sides to prevent undercut.

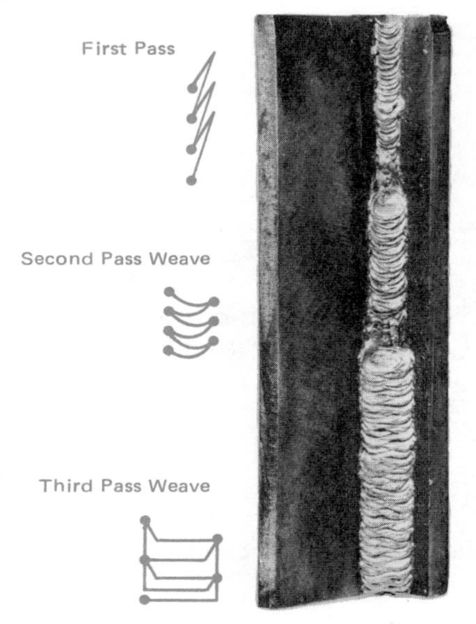

Fig. 26-3 Three-pass vertical up fillet weld using E-6010 electrode

Hold a very close arc.

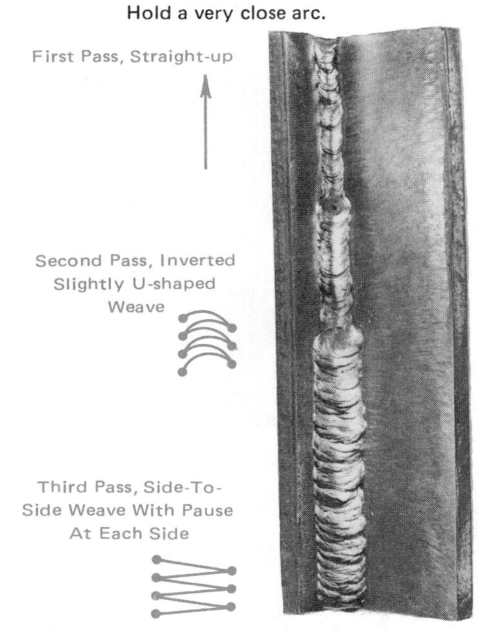

Fig. 26-4 Three-pass vertical up fillet weld using E-7018 electrode.

3. The welding current may have to be set higher because of the greater amount of plate being welded.
4. The beads should be positioned so that the weld is centered in the joint.

Note: All beads should be flat and uniform in appearance and have no undercut or pinholes. This is accomplished by using a proper length arc and good timing when motion is used.

VERTICAL DOWN WELDING

1. E-6010, E-6011, E-6013 and E-7014 electrodes are all satisfactory for welding vertical down.
2. Stringer beads with no side motion are generally used at higher amperages. See fig. 26-5, 1/8-inch E-6013 vertical down T joint.
3. Weaves are only used as cover passes with a slight inverted U-type motion.
4. The penetration is not as great when welding vertical down.

No motion required. Hold a close arc.

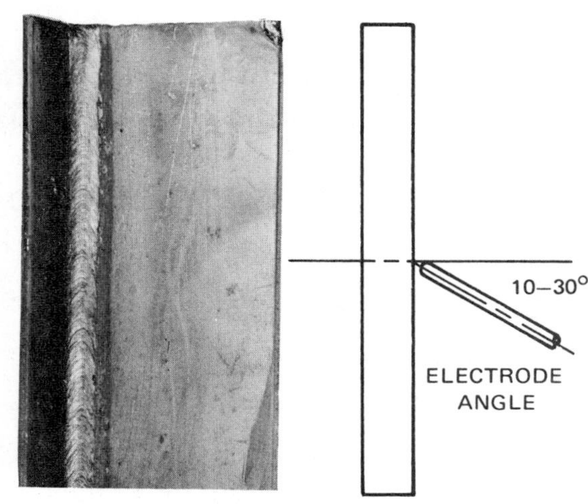

Fig. 26-5 Vertical down fillet weld using E-6013 electrode and DC straight polarity.

REVIEW QUESTIONS

1. When is it an advantage to weld vertical down instead of vertical up?

2. What is the difference in the way the E-6010 and E-7018 electrodes are handled to apply vertical-up stringer beads?

3. When welding vertical up with an E-6010 electrode, does undercut become a problem at the edges of the weld?

4. What electrode is similar to E-6010 in operation and can be used on an AC welding machine?

5. Is the degree of penetration greater with vertical-down welding or vertical-up welding?

ACKNOWLEDGMENTS

The authors wish to express their appreciation and acknowledge the following organizations for their assistance in the development of this text:

- The American Welding Society, for permission to use and adapt the Chart of Standard Welding Symbols.
- Hobart Brothers Company, Troy, Ohio, for comparison charts.
- Lincoln Electric Company, Cleveland, Ohio.

The following members of the staff at Delmar Publishers assisted in the preparation of this edition:

Publications Director — Alan N. Knofla

Source Editor — Mark W. Huth

Director of Manufacturing and Production — Frederick Sharer

Production Specialists — Alice Schielke, Jean LeMorta, Sharon Lynch, Patti Manuli, Betty Michelfelder, Debbie Monty, Lee St. Onge

Illustrators — Tony Canabush, George Dowse, Michael Kokernak

This material has been used in the classroom by the Oswego County Board of Cooperative Educational Services, Mexico, New York.

BASIC OXYACETYLENE WELDING

PREFACE

Oxyacetylene processes are used by personnel employed in a wide range of trades. These processes include cutting, welding, brazing and silver soldering through the use of oxygen and acetylene gases. *Basic Oxyacetylene Welding* emphasizes performance in all of these areas.

The instructional pattern employed is one of learning by doing. The exercises included in this book are considered to be the most useful ones for helping students to acquire the essential skills for welding. Extensive industrial experience and teaching experience has guided the selection of practical applications of the theory learned. These activities may be used to teach welding as a separate occupation, or they may be combined with the content from another area of instruction, such as automotive mechanics or the building trades. This book provides for easy readability, a very logical sequence of presentation and high illustration. This text should provide the foundation upon which advanced welding techniques may be built.

CONTENTS

CHARTS

APPLICATION CHART – BASIC OXYACETYLENE WELDING

UNIT NUMBER	AUTO MECHANIC	BOILERMAKER	BRICKLAYER	CARPENTER	ELECTRICIAN	FARM EQUIPMENT REPAIR	GENERAL WELDING	IRON WORKER (ORNAMENTAL)	IRON WORKER (STRUCTURAL)	MACHINIST	PLUMBER	SHEET METAL WORKER	STEAMFITTER
1	■	■	■	■	■	■	■	■	■	■	■	■	■
2	■	■	■	■	■	■	■	■	■	■	■	■	■
3	■	■	■	■	■	■	■	■	■	■	■	■	■
4	■	■	■	■	■	■	■	■	■	■	■	■	■
5	■	■	■	■	■	■	■	■	■	■	■	■	■
6	■	■	■	■	■	■	■	■	■	■	■	■	■
7	■	■	■	■	■	■	■	■	■	■	■	■	■
8	■	■	■	■	■	■	■	■	■	■	■	■	■
9	■	■	■	■	■	■	■	■	■	■	■	■	■
10		■	●			■	■	●	●	●	■		■
11		■	●			■	■	●	●	●	■		■
12		■	●			■	■	●	●	●	■		■
13	■	■	■	■	■	■	■	■	■	■	■	■	■
14	■	■	■	■	■	■	■	■	■	■	■	■	■
15	■	■	■	■	■	■	■	■	■	■	■	■	■
16	■	●	●	●		■	■	■	●	■	■	■	■
17	■	●	●	●		■	■	■	●	■	■	■	■
18	■	●	●	●		■	■	■	●	■	■	■	■
19	■	●	●	●		■	■	■	●	■	■	■	■
20	●	■	●			■	■	●	●	●	■	●	■
21	●	■	●			■	■	●	●	●	■	●	■
22	●	■	●			■	■	●	●	●	■	●	■
23	●	■	●			■	■	●	●	●	■	●	■
24	●	■	●			■	■	●	●	●	■	●	■
25	■		●			■	■	●	●	■	■	■	●
26	■		●			■	■	●	●	■	■	■	●
27	■		●			■	■	●	●	■	■	■	●
28	■		●			■	■	●	●	■	■	■	●
29	■		●			■	■	●	●	■	■	■	●
30	●		●			■	■	●	●	■	■		●
31	●		●			■	■	●	●	■	■		
32	●		●			■	■	●	●	■	■		
33	●		●		■	■	■	●	●	■	■	■	■
34	■		●		■	■	■	●	●	■	■	■	■
35	■	■	■	■	■	■	■	■	■	■	■	■	■
36	■	■	■	■	■	■	■	■	■	■	■	■	■
37	■	■	■	■	■	■	■	■	■	■	■	■	■

LEGEND

- ■ – **Required**
- ● – **Optional**
- (blank) – **Not Necessary**

Note: The above chart is in terms of suggested minimums only. The final choice of course content is a function of the individual instructor, often with the advice of an industry advisory committee.

UNIT 1 THE OXYACETYLENE WELDING PROCESS

Oxyacetylene welding is one of the three basic nonpressure processes of joining metals by *fusion* alone. The process of joining two pieces by partially melting their surfaces and allowing them to flow together is called fusion. The other two fusion processes are electric arc welding and thermit welding. Each of the three types has advantages and disadvantages.

In the oxyacetylene process, the metal is heated by the hot flame of a gas-fed torch. The metal melts and fuses together to produce the weld. In many cases, additional metal from a welding rod is melted into the joint which becomes as strong as the base metal.

EQUIPMENT AND SAFETY

The basic equipment and materials for welding by this process are:

1. Oxygen and acetylene gas supplied from cylinders to provide the flame.
2. Regulators and valves to control the flow of the gases.
3. Gages to measure the pressure of the gases.
4. Hoses to carry the gases to the torch.
5. A torch to mix the gases and to provide a handle for directing the flame.
6. A tip for the torch to control the flame.

The above equipment is described in some detail in the next several units. A thorough understanding of the equipment is highly important so that welding may be done safely as well as efficiently. The hazards arising from lack of understanding and improper use of the equipment are:

1. Burns to the operator or nearby persons.
2. Fires in buildings or materials.
3. Explosions resulting in personal injury and property damage.
4. Damage to expensive welding equipment.

ADVANTAGES AND DISADVANTAGES

Oxyacetylene welding, brazing and soldering operations, which are carried out with similar equipment, have certain advantages and disadvantages.

1. Oxyacetylene welding is a process which can be applied to a wide variety of manufacturing and maintenance situations.
2. The equipment is portable.
3. The cost and maintenance of the welding equipment is low when compared to that of some other welding processes.
4. The cost of welding gases, supplies and operator's time, depends on the material being joined and the size, shape, and position in which the weld must be made.

5. The rate of heating and cooling is relatively slow. In some cases, this is an advantage. In other cases where a rapid heating and cooling cycle is desirable, the oxyacetylene welding process is not suitable.
6. A skilled operator can control the amount of heat supplied to the joint being welded. This is always a distinct advantage.
7. The oxygen and nitrogen in the air are kept from combining with the metal to form harmful oxides and nitrides.

In general, the oxyacetylene process can be used to advantage in the following situations:

- When the materials being joined are thin;
- When excessively high temperatures, or rapid heating and cooling of the work would produce unwanted or harmful changes in the metal.
- When extremely high temperatures would cause certain elements in the metal to escape into the atmosphere.

HAZARDS

Many of the hazards in oxyacetylene welding can be minimized by careful consideration of the following points:

1. The welding flame and the sparks coming from the molten puddle can cause any flammable material to ignite on contact. Therefore,
   - Flame-resistant clothing must be worn by the operator and his hair must be protected.
   - Welding and cutting should not be done near flammable materials such as wood, oil, waste or cleaning rags.
2. In addition to the risk of eye injury from flying molten metal, there is also the danger of radiation burns due to the infrared rays given off by red hot metal. The eyes may be burned if these rays are not filtered out by proper lenses. Therefore,
   - The eyes should be protected at all times by approved safety glasses and the proper shield.
   - Sunglasses are not adequate for this purpose.
3. Fluxes used in certain welding and brazing operations produce fumes which are irritating to the eyes, nose, throat, and lungs. Likewise, the fumes produced by overheating lead, zinc, and cadmium are a definite health hazard when inhaled even in small quantities. The oxides produced by these elements are poisonous. Therefore,
   - Welding should be done in a well-ventilated area.
   - The operator should not expose others to fumes produced by welding.

REVIEW QUESTIONS

1. What is the difference between fusion welding and other welding processes?

2. What features of the oxyacetylene welding process make it useful in a wide variety of jobs?

3. What are four distinct hazards that must be guarded against when oxyacetylene welding?

4. What type of rays come from the oxyacetylene flame and red hot metal?

5. What is the function of a regulator?

UNIT 2 OXYGEN AND ACETYLENE CYLINDERS

Hazards are always present when gases are compressed, stored, transported, and used under very high pressures. Oxygen and acetylene, are delivered to the user under high pressure in steel cylinders. These cylinder are made to rigid specifications.

A simple demonstration of the effects of compressing gas can be shown with an ordinary toy balloon. When the balloon is blown up and held tightly at the neck so the air cannot escape, it resembles the compressed gas cylinder. What happens if the balloon is punctured, heated, or suddenly released? The explosive burst when punctured or heated, or the sudden flight of the balloon when released shows that compressed gas, even the small amount in a toy balloon, has considerable force.

OXYGEN CYLINDERS

The most common size of oxygen cylinder, when fully charged with gas, contains 244 cubic feet of oxygen. This oxygen is at a pressure of 2,200 pounds per square inch when the temperature is 70 degrees F.

The steel walls of these cylinders are only slightly more than one-quarter inch thick, .260 inch. Dropping such a cylinder, hitting it with heavy or sharp tools, or striking an electric arc on it can cause the cylinder to explode with enough force to cause serious injury and death.

The general size and shape of an oxygen cylinder is indicated in figure 2-1. As a safety precaution, the cylinder valve is protected by a removable steel cap. This cap must be on the cylinder at all times when it is being stored or transported. The cylinder valve should always be closed when not in use, even when the cylinder is empty.

The oxygen cylinder valve is designed to handle the highly compressed oxygen gas safely. The essential parts of the valve are shown in figure 2-2. The threads on the nozzle must be protected at all times.

The *bursting disc* and *safety cap* are designed to allow the gas in the cylinder to escape if the cylinder is subjected to undue heat and the pressure in the tank begins to rise.

The double-seating valve is designed to seal off any oxygen that might leak around the valve stem. When the valve is fully open there is no leakage.

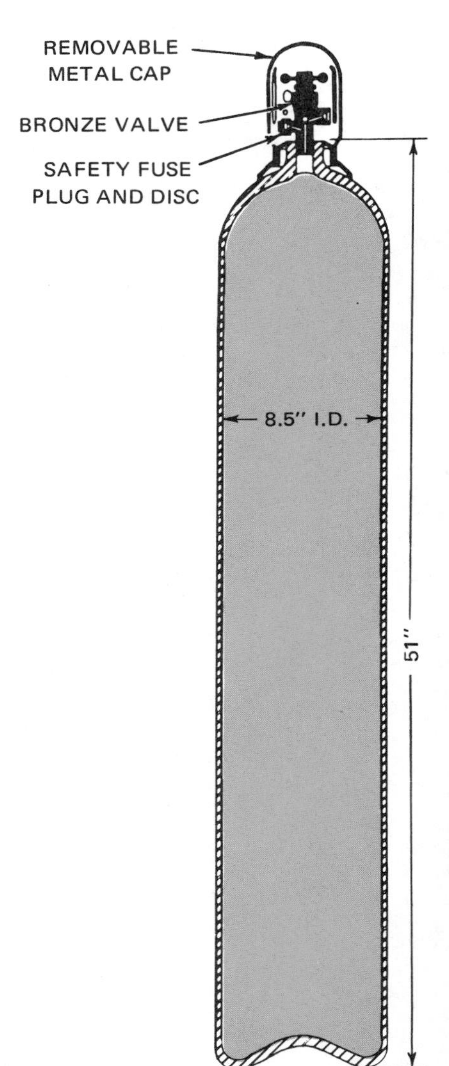

Fig. 2-1 Oxygen Cylinder

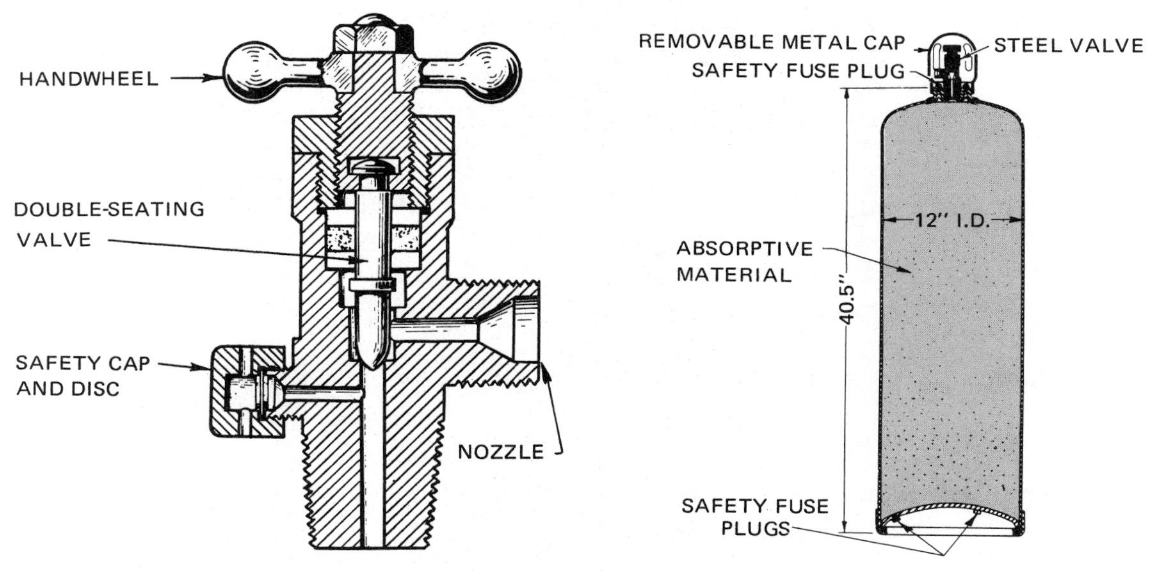

Fig. 2-2 Oxygen Cylinder Valve

Fig. 2-3 Acetylene Cylinder

ACETYLENE CYLINDERS

The acetylene cylinder is a welded steel tube. It is filled with a spongy material such as balsa wood or some other absorptive material which is saturated with a chemical solvent called acetone. Acetone absorbs acetylene gas in much the same manner as water absorbs ammonia gas to produce common household ammonia.

The cylinder is equipped with a valve which can only be opened with a special wrench. Safety regulations require this type of valve on all containers carrying flammable, explosive, or toxic gases. The wrench must be in place whenever the cylinder is in use. Acetylene cylinders are also equipped with a number of *fusible plugs* designed to melt at 220 degrees F. These melt and release the pressure in the event the cylinder is exposed to excessive heat.

Figure 2-3 is a cross section of a common acetylene cylinder. The construction details may vary from one manufacturer to another, but all acetylene cylinders are made to very rigid specifications.

Acetylene cylinders are usually charged to a pressure of 250 pounds per square inch; the large size contains about 275 cubic feet. The steel walls of these cylinders are only .175 inch thick. The precautions set forth for oxygen cylinders should be observed with acetylene cylinders. Escaping acetylene mixed with air forms a highly explosive mixture.

REVIEW QUESTIONS

1. What prevents unauthorized persons from opening acetylene valves?

2. Why must an oxygen cylinder valve be fully opened?

3. What happens to the pressure in a cylinder as the temperature is raised?

   What happens when the temperature is lowered?

4. Suppose the protective cap is left off a fully-charged oxygen cylinder and an accident causes the valve to be broken off:

   a. What is the force of the gas per square inch?

   b. Is this enough force to cause the cylinder to move?

5. What is the function of the cylinder valve?

UNIT 3 WELDING GASES

OXYGEN

Flame is produced by combining oxygen with other materials. When the air we breathe, which is only one-fifth oxygen, combines with other elements to produce a flame, this flame is low in temperature and the rate of burning is rather slow.

However, if pure oxygen is substituted for air, the burning is much more rapid and the temperature is much higher. Oil in the regulators, hoses, torches, or even in open air burns with explosive rapidity when exposed to pure oxygen.

CAUTION: Oxygen must never be allowed to come in contact with any flammable material without proper controls and equipment. The use of oxygen to blow dust and dirt from working surfaces or from a worker's hair or clothing is extremely dangerous.

Most of the oxygen produced commercially in the United States is made by liquefying air and then recovering the pure oxygen. The oxygen thus produced is of such high purity that it can be used not only to produce the most efficient flame for welding and flame cutting operations, but also for medical purposes.

Oxygen is a colorless, odorless, tasteless gas which is slightly heavier than air. The weight of 12.07 cubic feet of oxygen at atmospheric pressure and 70 degrees F. is one pound.

ELECTRIC ARC	19,832 F	11,000 C
SURFACE OF SUN	10,832 F	6,000 C
OXYACETYLENE FLAME	5,900 F	3,260 C
OXYHYDROGEN FLAME	4,752 F	2,900 C
INTERIOR OF INTERNAL COMBUSTION ENGINE	3,272 F	1,800 C
COPPER MELTS	1,976 F	1,080 C
MAGNESIUM MELTS	1,204 F	651 C
WATER BOILS	212 F	100 C
ICE MELTS	32 F	0 C
LIQUID AIR BOILS	-292 F	-180 C
LIQUID HELIUM BOILS	-452 F	-269 C
ABSOLUTE ZERO	-459.4 F	-273 C

Fig. 3-1 Some comparative temperatures

ACETYLENE

Acetylene gas is a chemical compound composed of carbon and hydrogen. It combines with oxygen to produce the hottest gas flame known. Unfortunately, acetylene is an unstable compound and must be handled properly to avoid explosions.

Unstable acetylene gas tends to break down chemically when under a pressure greater than 15 pounds per square inch. This chemical break down produces great amounts of heat; the resulting high pressure develops so rapidly that a violent explosion may result.

Acetylene gas which is dissolved in acetone does not tend to break down chemically and can be used with complete safety. However, any attempt to compress acetylene in a free state in hoses, pipes, or cylinders at a pressure greater than 15 pounds per square inch can be very dangerous.

Acetylene is produced by dissolving calcium carbide in water. This process should be carried out only in approved generators. One pound of calcium carbide produces 4.5 cubic feet of acetylene gas. Acetylene is made up of two atoms of carbon and two atoms of hydrogen. It has a distinctive odor. The weight of 14.5 cubic feet is one pound. The amount dissolved in an acetylene cylinder is determined by weighing the cylinder and contents, subtracting the weight of the empty cylinder, and multiplying the remainder, which is the weight of the gas, by 14.5. The empty cylinder weight is always stamped into the cylinder.

REVIEW QUESTIONS

1. Why is it dangerous to place calcium carbide and water in a closed container and generate acetylene gas?

2. What is the probable effect if oil or grease is allowed to come in contact with oxygen in the regulators or cylinders?

3. What element must always be present if a flame is to be produced and maintained?

4. At what pressure will acetylene gas become unstable in a free state?

5. Can oxygen be referred to as air? Why?

UNIT 4 OXYGEN AND ACETYLENE REGULATORS

REGULATORS

Oxygen and acetylene *regulators* reduce the high cylinder pressures, safely and efficiently, to usable working pressures. Regulators also maintain these pressures within very close limits under varying conditions of demand.

Figure 4-1 shows the relatively simple operation of a regulator. The pressure in the hoses is controlled by applying pressure to the spring through an adjusting screw. The spring applies pressure to a flexible rubber diaphragm which is connected to the high pressure valve. The gas from the cylinder flowing through this valve builds up pressure behind the diaphragm. When this pressure equals the pressure of the spring, the valve closes and shuts off the flow of gas to the diaphragm area. When the pressure in this area is reduced by drawing gas from the regulator to the torch the spring opens the valve again.

Regulators are made to rigid specifications from the finest of materials, and are equipped with safety devices to prevent injury to the operator or the equipment. All regulators are equipped with ball-check safety valves or bursting discs to prevent pressure buildup within the regulator, hoses or torch.

GAGES

Most regulators are equipped with gages which indicate the amount of pressure in the cylinder and the working pressure in the hoses and torch.

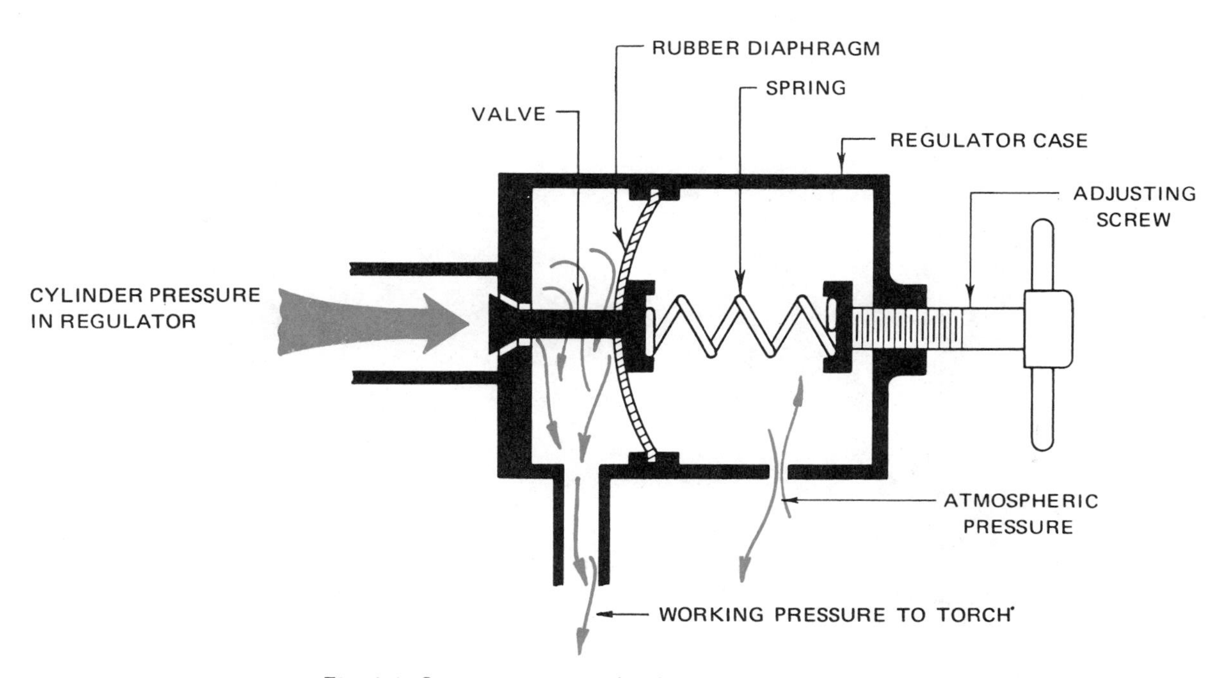

Fig. 4-1 Construction details of a single-stage regulator

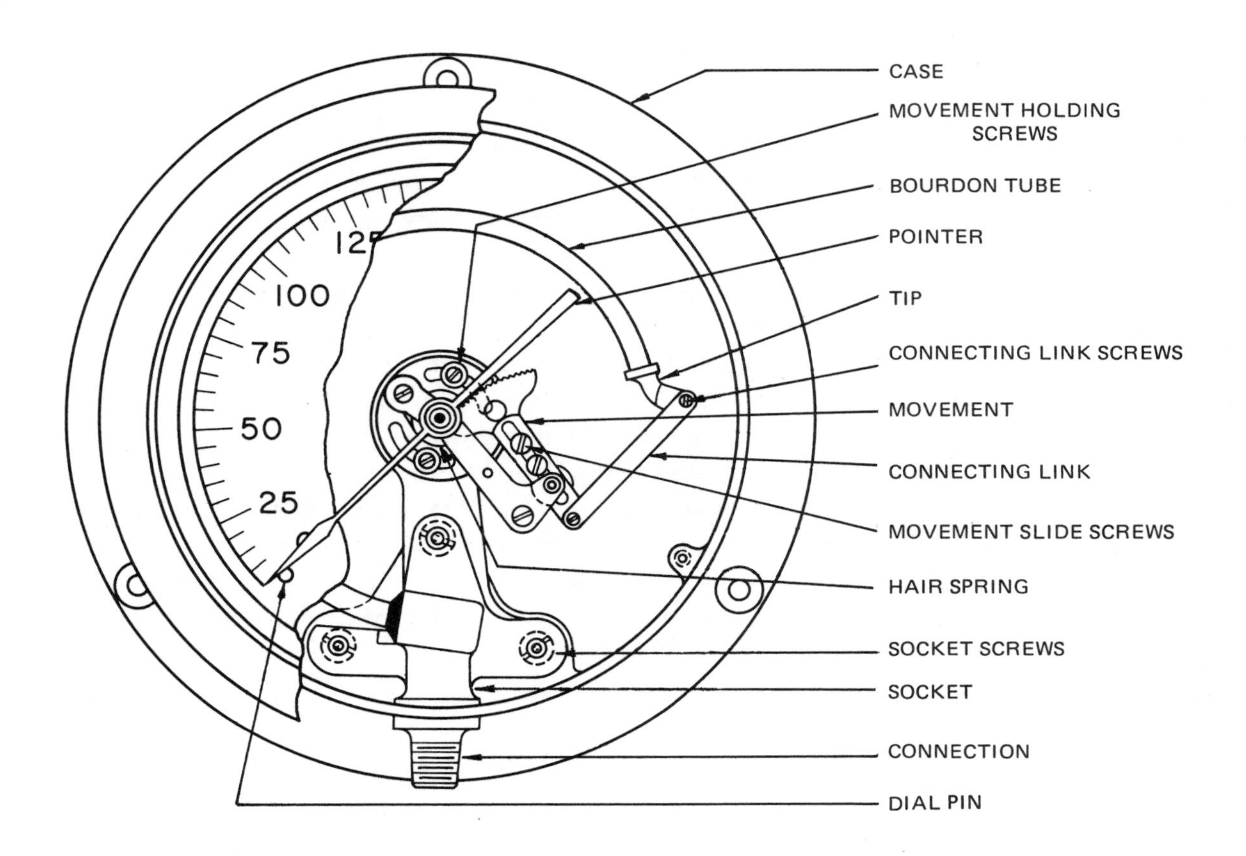

Fig. 4-2 Construction details of a gage

Fig. 4-3 Oxygen Regulator

Fig. 4-4 Acetylene Regulator

These gages have very thin backs which open to release the pressure if the Bourdon tube ruptures. This tube is essential to the operation of each gage. If this precaution were not taken, excessive pressure in the gage case could cause the glass front of the gage to explode and injure the operator.

Since gages frequently get out of calibration, they are only indicators of cylinder and working pressures. Regulators work regardless of the accuracy of the gages.

Unit 8 describes the procedure that should be followed to insure safety and efficiency when adjusting the regulators, regardless of the pressures indicated on the gages.

REVIEW QUESTIONS

1. What two purposes do oxygen and acetylene regulators serve?

2. What purpose do gages on regulators serve?

3. Should a regulator-adjusting screw be turned all the way in when the regulator is to be turned off?

4. What safety devices guard against excessive pressure in regulator cases and in hoses?

5. Can a regulator be accurate if the gage is damaged?

UNIT 5 TYPES AND USES OF WELDING TORCHES

The body of the welding torch serves as a handle so the operator can hold and direct the flame. Beyond the handle, the torch is equipped with a means of attaching the mixing head and welding tip, figure 5-1.

The accurately sized holes in welding and cutting tips are called *orifices.* The purpose of the *mixing head* is to combine the two welding gases into a usable form. The only mixed oxygen and acetylene is that amount contained in the space from the mixing head to the tip orifice. In most cases, it represents a very small portion of a cubic inch. This keeps the amount of this highly explosive mixture within safe limits. Any attempt to mix greater amounts may result in violent explosions.

Two types of torches are in common use. In the *injector-type torch,* the acetylene at low pressure is carried through the torch and tip by the force of the higher oxygen pressure through a venturi-type device, shown in figure 5-2. The mixing head and injector are usually a part of the tip which the operator changes according to the size needed.

In the *medium-pressure torch,* figure 5-3, both gases are delivered through the torch to the tip at equal pressures. In this type of torch, the mixer or mixing head is usually a separate piece into which a variety of tips may be fitted.

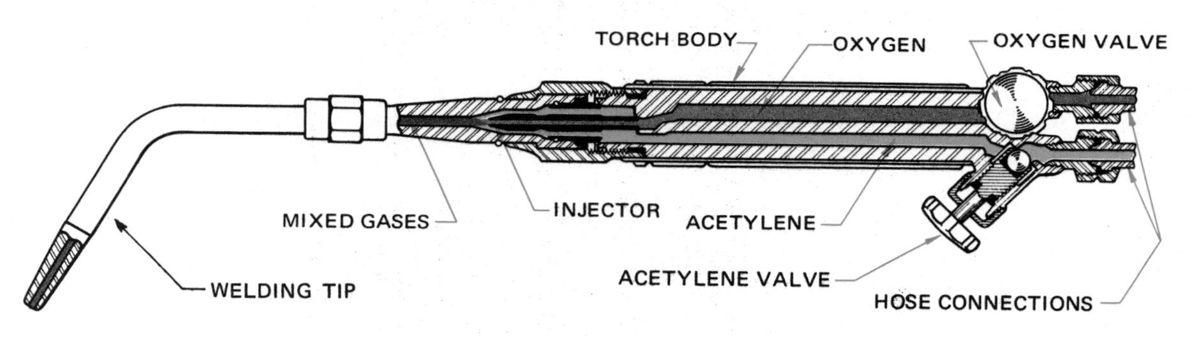

Fig. 5-1 Welding torch

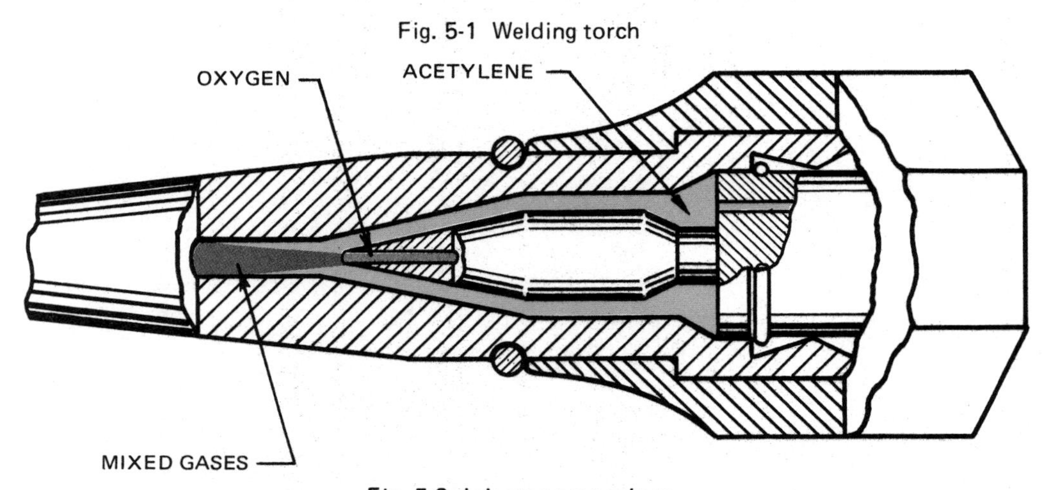

Fig. 5-2 Injector-type mixer

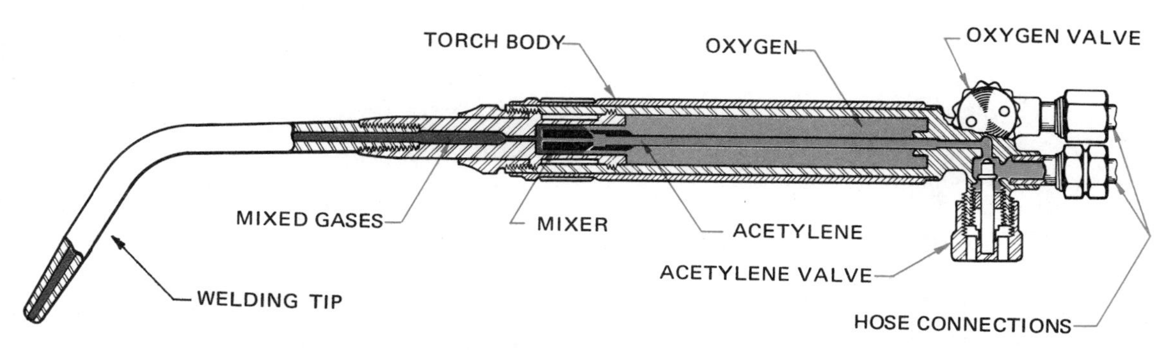

Fig. 5-3 Medium-pressure torch

Fig. 5-4 Welding torch

All types of torches are equipped with a pair of needle valves to turn the welding gases on and off, and to make small pressure adjustments.

REVIEW QUESTIONS

1. What is the chief difference between the two types of torches?

2. a. What is the basic purpose of torch needle valves?

   b. What secondary purpose do they serve?

3. From the construction details indicated in the text, which type of torch is probably more costly?

4. What is the proper name for the holes in welding and cutting tips?

5. How many valves do most torches have?

UNIT 6 WELDING TIPS

The purpose of all welding tips is to provide a safe, convenient method of varying the amount of heat supplied to the weld. They also provide a convenient method of directing the flame and heat to the exact place the operator chooses.

SELECTION OF SIZES

To provide for different amounts of heat, welding tips are made in various sizes. The size is determined by the drill size of the orifice in the tip. As the orifice size increases, greater amounts of the welding gases pass through and are burned to supply a greater amount of heat. However, the temperature of the neutral oxyacetylene flame always remains at 5,900 degrees F., regardless of the quantity of heat provided.

The choice of the proper tip size is very important to good welding. All manufacturers of welding torches supply a chart of recommended sizes for various thicknesses of metal. They also recommend oxygen and acetylene pressures for various types and sizes of tips. These tables provide orifice sizes and the proper drill size for cleaning each orifice, figure 6-1.

CARE OF TIPS

All welding tips are made of copper and may be damaged by careless handling. Dropping, prying, or hammering, the tips on the work may make them unfit for further use. It is im-

PLATE THICKNESSES		ORIFICE SIZE	GAS PRESSURES IN P.S.I.			
			INJECTOR-TYPE TORCH		EQUAL-PRESSURE TORCH	
GAGE	INCHES	NO. DRILL	ACETYLENE	OXYGEN	ACETYLENE	OXYGEN
32	.010	74	5	5 - 7	1	1
28	.016	70	5	7 - 8	1	1
26	.019	70	5	7 - 10	1	1
22	$\frac{1}{32}$	65	5	7 - 18	2	2
16	$\frac{1}{16}$	56	5	8 - 20	3	3
13	$\frac{3}{32}$	56 - 54	5	15 - 20	4	4
11	$\frac{1}{8}$	54 - 53	5	12 - 24	4	4
8	$\frac{3}{16}$	53 - 50	5	16 - 25	5	5
	$\frac{1}{4}$	50 - 46	5	20 - 29	6	6
	$\frac{3}{8}$	46 - 44	5	24 - 33	7	7
	$\frac{1}{2}$	40	5	29 - 34	8	8
	$\frac{5}{8}$	30	5	30 - 40	9	9
	$\frac{3}{4}$	30 - 29	5	30 - 40	10	10
	1	23	5	30 - 42	12	12

DRILL SIZE NO.	AIRCO - ALL	CRAFTSMAN - ALL	DOCKSON - 4EC, 4SC	7EC	GASWELD - G25, G35	G55	AVG	HARRIS - 13, 14, 16, 17, 50	OTHERS	K-G - EUS, KUS, KS	MARQUETTE - A, AL	B, BL	MECO - ALL	MILBURN - W-200	W-11	W-600	NATIONAL - G	P	R	OXWELD - W-29	W-17	PUROX - 33	34	35	REGO - ALL	SMITH - LIFETIME	NO. 5	NO. 2	VICTOR - ALL
80													00	00		0000													
79						00	00										00					0							
78			1	1				00				00B		0														18	
77						0	0									000													
76								0					0	1			0	0										19	
75	00									75	00	0B							000			1							000
74					00	1	1		00							00				1								A20	
73			2	2																			1						000½
72	0	0			0					72		1B					1	1							72				
71									0						000											B60	50	A21	
70		1									0		1						00	2									00
69					1	2	2							2		0													
68	1							1		68		2B			00							2			68			A22	
67																													00½
66			3	3					1		1						2	2					2	2					
65					2		3	2					2			1			0	4	4							A23	0
64																													
63	2								2						0											B61	51		0½
62		2								62	2	3B		3		2									62				
61																													
60							4												1			3							1
59																													
58			4	4	3	3		3	3		3					3									58	B62	52	A25	
57							5						3	4	1									3					1½
56	3								4	56		4B							2	6	6	4						A26	2
55			5	5	4	4		4	5				4	5		4	3	3					4		55				2½
54		3			5	5	6				4	5B			2				3	9	9					B63	53	A27	
53	4							5	6	53						5				12	12	5		5	53				3
52			6	6	6	6	7						5													B64	54	A28	
51	5							6				6B		6	3		4	4											3½
50							8		7	50	5					6				15	15	6	6		50			A29	
49																													4
48	6	4	7	7	7			7				7B		7	4						20					B65	55	A210	
47									8		6		6						4			7							
46																	5	5		20					46				
45								8		45																			
44	7		8	8	8	8									5	7					30	8				B66	56		
43								9	9		7			8					5				8						5
42		5			9								7							30					42				
41			9												6	8	6	6											
40	8				10			10	10	40	8			9							40					B67	57		
39																		7						9					
38					11										7														
37		6																											
36			10		12						9		8	10	8			8					10		36	B68			6
35	9				13					35						9													
34																													
33		7			14										9				6										
32			11						12		10							9								B69			
31					15									11		10								11	31				
30	10		12						15	30	11			12	10			10	7		55			13					7
29		8			16								9	13					8		70								8
28			13										10		12				9							B610			9
27									19		12								10										10
26			14																11							B611			11
25	11									25				14	13				12						25				12
24									22		14															B612			
23			15																		90			15					
22													11		14														
21														15															
20	12									20					15										20				
19														16															
18													12																

NOTE: USE A DRILL ONE SIZE SMALLER FOR CLEANING ORIFICES

Fig. 6-2 Comparison guide for welding tip sizes

portant to clean the tip orifice with the proper tip drill. The use of an incorrect drill or procedure can ruin a tip.

REVIEW QUESTIONS

Note: Determine the proper orifice size from figure 6-1. Then find the tip number closest to this size in figure 6-2.

1. What size Craftsman® tip should be used for welding 22-gage steel?

2. What size Rego® tip should be used for welding 1/8-inch plate?

3. What size Victor® tip should be used for welding 16-gage steel?

4. What size Airco® tip should be used for welding 1/4-inch steel?

5. What size Oxweld® W-17 tip should be used for welding 1/2-inch steel?

6. What size Smith-Lifetime® tip should be used for welding 8-gage steel?

UNIT 7 THE OXYACETYLENE WELDING FLAME

The flame is the actual tool of oxyacetylene welding. All of the welding equipment merely serves to maintain and control the flame.

The flame must be of the proper size, shape, and condition in order to operate with maximum efficiency. The oxyacetylene flame differs from most other types of tools in that it is not ready-made. The operator must produce the proper flame each time he lights the torch.

Once the operator masters the adjustment of the flame, his ability as a welder increases in direct proportion to the amount of practice he has.

TYPES OF FLAMES

The oxyacetylene flame can be adjusted to produce three distinctly different types of flame. Each of these types has a very marked effect on the metal being fused or welded. In the order of their general use, the flames are *neutral, carburizing,* and *oxidizing.* Figure 7-1 illustrates their shapes and characteristics.

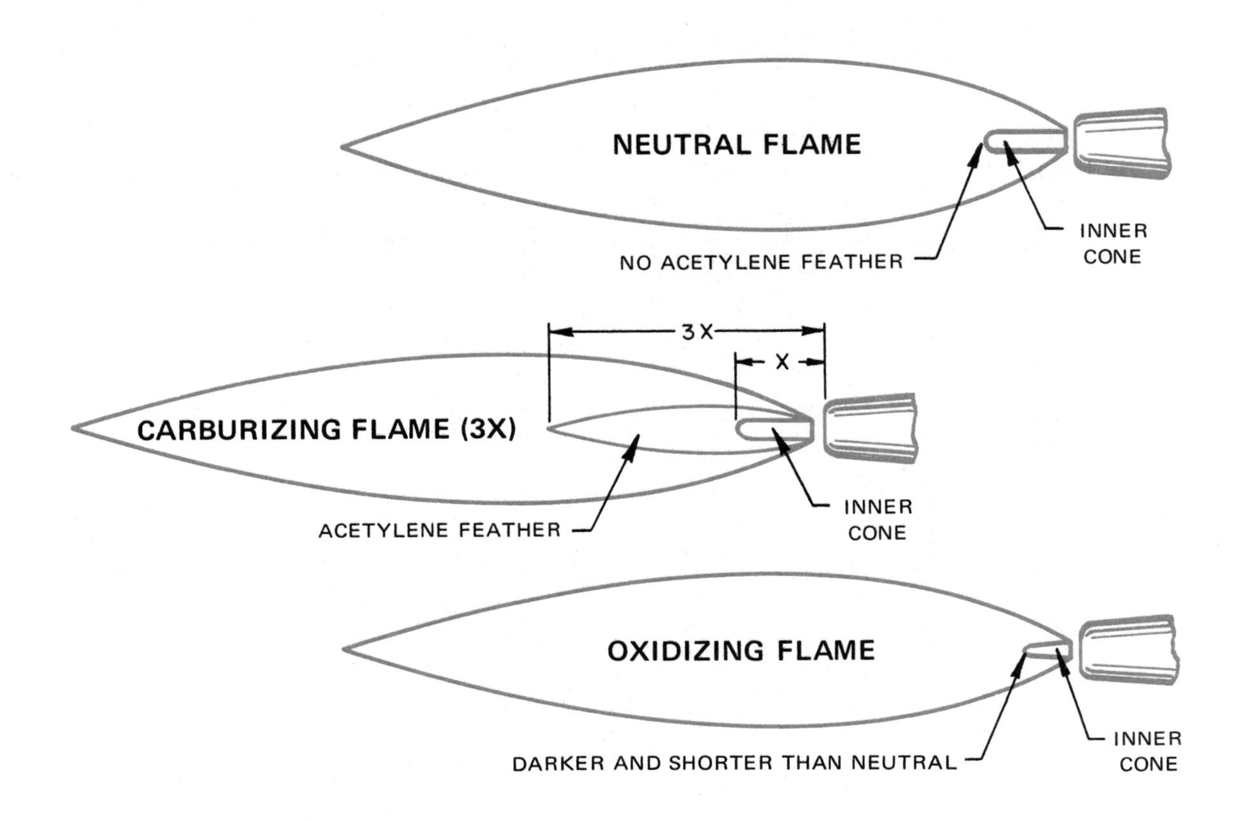

Fig. 7-1 Types of flames

The *neutral* flame is one in which equal amounts of oxygen and acetylene combine in the inner cone to produce a flame with a temperature of 5,900 degrees F. The inner cone is light blue in color. It is surrounded by an outer flame envelope, produced by the combination of oxygen in the air and superheated carbon monoxide and hydrogen gases from the inner cone. This envelope is usually a much darker blue than the inner cone. The advantage of the neutral flame is that it adds nothing to the metal and takes nothing away. Once the metal has been fused, it is chemically the same as before welding.

The *carburizing* flame is indicated by streamers of excess acetylene from the inner cone. These streamers are usually called *feathers* of acetylene, or simply the *acetylene feather.* The feather length depends on the amount of excess acetylene. The outer flame envelope is longer than that of the neutral flame and is usually much brighter in color. This excess acetylene is very rich in carbon. When carbon is applied to red-hot or molten metal, it tends to combine with steel and iron to produce the very hard, brittle substance known as iron carbide. This chemical change leaves the metal in the weld unfit for many applications in which the weld may need to be bent or stretched. While this type of flame does have its uses, it should be avoided when fusion welding those metals which tend to absorb carbon.

The carburizing flame in figure 7-1 shows the relation of the acetylene streamers to the inner cone. Job conditions sometime require an excess of acetylene in terms of the length of the inner cone. Figure 7-1.

The *oxidizing* flame, which has an excess of oxygen, is probably the least used of any of the three flames. In appearance, the inner cone is shorter, much bluer in color, and usually more pointed than a neutral flame. The outer flame envelope is much shorter and tends to fan out at the end. The neutral and carburizing envelopes tend to come to a sharp point. The excess oxygen in the flame causes the temperature to rise as high as 6,300 degrees F. This temperature would be an advantage if it were not for the fact that the excess oxygen, expecially at high temperatures, tends to combine with many metals to form hard, brittle, low-strength oxides. For this reason, even slightly oxidizing flames should be avoided in welding.

REVIEW QUESTIONS

1. What chemical change takes place when a carburizing flame is used to weld steel?

2. What substance is produced when an oxidizing flame is used to weld steel?

3. Make a labeled and dimensioned sketch of a 2X flame. What are the significant parts?

4. Make a labeled and dimensioned sketch of a 3 1/2X flame. What are the significant parts?

5. What is the temperature of a neutral flame?

UNIT 8 SETTING UP EQUIPMENT AND LIGHTING THE TORCH

Oxyacetylene welding equipment must be set up frequently and it must be done efficiently. Since hazards are present, each step must be performed correctly. The proper sequence must be followed to insure maximum safety to personnel and equipment.

The cylinder caps are removed and put in their proper place. The cylinders should be fastened to a wall or other structure with chains, straps or bars, to prevent them from being tipped over. To use oxygen and acetylene cylinders and equipment without this safety precaution is to invite damage to the equipment and injury to the operator.

PROCEDURE

1. Aim the cylinder nozzle so it does not blow toward anyone. Crack the valve on each cylinder by opening the valve and closing it quickly. This blows any dust or other foreign material from the nozzle.
2. Attach the regulators to the cylinder nozzles.

   Note: All oxygen regulators in commercial use have a standard *right-hand* thread and fit all standard oxygen cylinders. Acetylene regulators may have *right- or left-hand* threads and may have either a male or female connection, depending on the company supplying the gas. Adapters of various types may be needed to fit the existing regulators to different acetylene cylinders.
3. Attach the hoses to the regulators.

   Note: All oxygen hose connections have *right-hand* threads. All acetylene hose connections have *left-hand* threads. The acetylene hose connection nuts are distinguished from the oxygen nuts by a *groove* machined around the center of the nut figure 8-1.

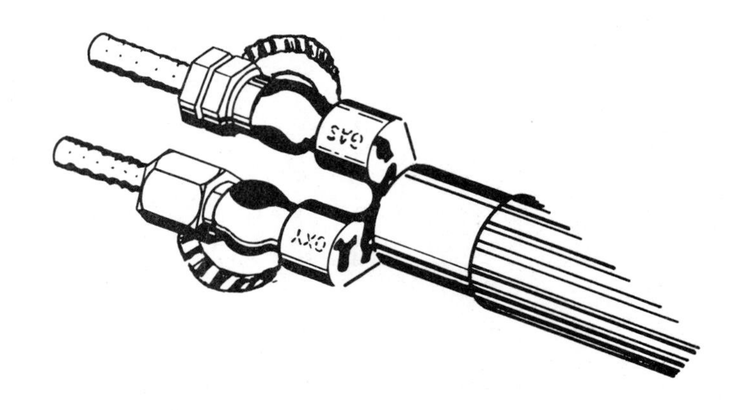

Fig. 8-1 Oxygen and Acetylene Hose Connections

4. Attach the torch to the other end of the hoses noting that while the hose connections may be a different size at the torch than at the regulators, they still have right- and left-hand threads.

   Note: Use only the wrenches provided for attaching hoses and regulators. These wrenches are designed to give the proper leverage to tighten the joints without putting undue strain on the equipment. If the joints cannot be properly tightened, something is wrong.

5. Select the proper tip and mixing head and attach it to the torch. Position the tip so that the needle valves are on the side or bottom of the torch when the tip is in the proper welding position.
6. Back off the regulator screws on both units until the screws turn freely. This is necessary to eliminate a sudden surge of excessive pressure on the working side of the regulator when the cylinder is turned on.
7. Be sure both torch needle valves are turned off (clockwise). This is an added safety precaution to make sure excessive pressure cannot be backed through the mixing head and into the opposite hose.
8. Open the acetylene cylinder valve 1/4 to 1/2 turn. Open the oxygen cylinder valve all the way.
9. Open the acetylene needle valve one full turn. Turn the adjusting screw on the acetylene regulator clockwise until gas comes from the tip. Light this gas with a spark lighter.
10. Adjust the regulator screw until there is a gap of about 1/4 inch between the tip and the flame. This is the proper pressure for the size of tip being used regardless of the gage pressure shown on the working pressure gage.
11. Open the oxygen needle valve on the torch one full turn. Turn the oxygen regulator adjusting screw clockwise until the flame changes appearance as oxygen is mixed with the acetylene.
12. Continue to turn the adjusting screw until the feather of acetylene just disappears into the end of the inner cone. This produces a neutral flame which is used in most welding.

This procedure for adjusting the oxyacetylene flame is the safest method of insuring the proper working pressures in both hoses and tip. Working pressure gages are delicate and easily get out of calibration. If this happens, excessive pressure can be built up in the hoses before it is discovered. However, if the pressures are adapted to the flame as indicated, there are equal pressures in both hoses which eliminates the possibility of backing gas from one hose to the other to form an explosive mixture. With the regulators properly adjusted, minor flame adjustments are made with the torch needle valves.

When the welding or cutting operation is finished, close the torch acetylene valve first, then the torch oxygen valve.

To shut down the equipment for an extended period of time, such as overnight, it should be purged. Use the following procedure:

1. Close the oxygen cylinder valve.
2. Open the torch oxygen valve to release all pressure from the hose and regulator.

3. Turn out the pressure adjusting screw of the oxygen regulator.
4. Close the torch oxygen valve.
5. Follow the same sequence for purging acetylene.

REVIEW QUESTIONS

1. What are the steps necessary to turn on the welding gases properly and safely and adjust them to a suitable flame?

2. What steps are necessary to assemble an oxyacetylene outfit for welding?

3. Why should the needle valves on the torch be turned off at a particular step in the sequence rather than at some other time?

4. Could excessive oxygen pressure backed through the torch cause an explosion in the acetylene hose without outside ignition? Explain. (Refer to Unit 3 – Acetylene Gas.)

5. How is a left-hand nut different from a right-hand nut in appearance?

UNIT 9 FLAME CUTTING

One of the fastest ways of cutting ferrous metals is by the use of the oxyacetylene torch. Other advantages of this cutting method are:

1. A relatively smooth cut is produced.
2. Very thick steel (over 4 feet) can be cut.
3. The equipment is portable.
4. Underwater cutting is possible with some adaptations.
5. The equipment lends itself to automatic processes in manufacturing.

The terms "cutting" and "burning" are used interchangeably to describe this process.

THE BURNING PROCESS

Oxyacetylene flame cutting is actually a burning process in which the metal to be cut is heated on the surface to the kindling temperature of steel, (1,600 - 1,800 degrees F.). A small stream of pure oxygen is then directed at the work. The oxygen causes the metal to ignite and burn to produce more heat. This additional heat causes the nearby metal to burn so that the process is continuous once it has started.

Only those ferrous metals which oxidize rapidly can be flame-cut. These metals include all the straight carbon steels and many of the alloys. Stainless steels and most of the so-called high-speed steels cannot be flame-cut.

EQUIPMENT

Cutting is done with a special torch fitted with interchangeable tips so that it can be adapted to cut a wide variety of metal thicknesses. The torch and tips are constructed so that they can preheat the work to the kindling temperature. The torch also includes a lever for turning on and stopping the stream of high-pressure cutting oxygen as required. The torch is usually made of forged brass and brass tubing, figure 9-1.

For hand-manipulated flame-cutting operations, the tips are made of copper. If the tip of the torch is used as a hammer, lever, or crowbar, permanent damage is done.

COMPARISON CHARTS

Because cutting torch tips are interchangeable, chart 9-1 may be used for the torch tips of all major manufacturers.

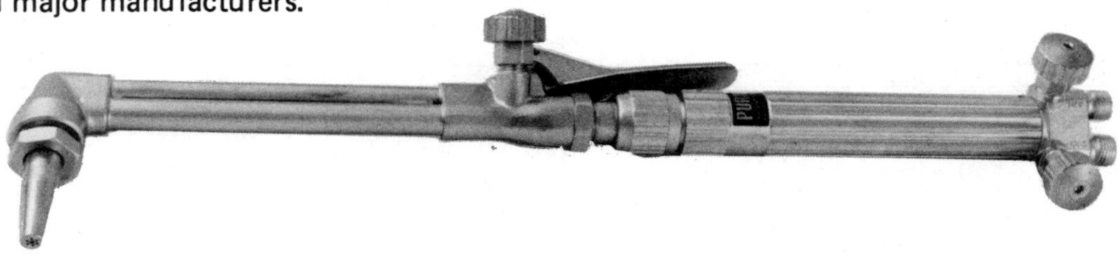

Fig. 9-1 Cutting torch

DRILL SIZE NO. FOR OXYGEN CUTTING ORIFICE	AIRCO-ALL EXCEPT #45	CRAFTSMAN-B	DOCKSON-ALL	GASWELD-HC-31,32	HC-39	WC-20, 35	WC-10, 55	HARRIS 2890-F	6290	7490-A	K-G, M4, M5	MARQUETTE E	C,D,DI, 4 PREHEAT	C,D,DI, 6 PREHEAT	MECO-ALL	MILBURN X100	X2000	X2300	NATIONAL-ALL	OXWELD CW-29	CW-23, C-31, 32	PUROX-33	34, 35	REGO-AW	SMITH LIFETIME 4 PH	LIFETIME 6 PREHEAT	VICTOR-ALL
76																				2	2						
75																											
74																											
73																											000
72																											
71								00																			
70																											
69					00		0		00																		
68	00										68										3			68			
67															0												
66					0																	0					
65									0			1B															00
64																										0	
63																00	00										
62	0						1			00	62											1		62			
61			1					0																			
60		1										2B	0A	0A		0	0		0	4	4		1		0		
59							2																				
58																											0
57															1		1	1	1								
56	1				1			1	1	1	56		1A	1A										56	1	1	
55		2	2													1											
54	2									2					2		2	2	2				2		2		1
53		3									53										6			53			
52	3				2			2						2A		2									3	2	2
51									2															51			
50											50						3	3	3				3				
49	4	4	3							3			2A														3
48																3	4								4	3	
47																											
46															3				4	8				46			
45	5	5			3			3	3		45					4	5	4									4
44																							4				
43			4											3A													
42										4						5			5					42			
41	6																	5									
40		6									40					6	6									4	5
39																				10							
38													3A														
37														4A		7	7		6								
36			5		4										4			6									
35									4	5	35					8	8							35			6
34	7	7			5			4																			
33																											
32			6						5							9			7								
31																				12							
30	8							5	6		30		4A	5A	5	10		7	8					30		5	
29																11											
28			7												6	12											7
27			8																								
26	9													6A				8									
25											25													25			
24																										6	
23																											
22																		9									
21																											
20																											
19																											
18	10																	10									
17																		11									

NOTE: USE A DRILL ONE SIZE SMALLER FOR CLEANING ORIFICES.

Chart 9-1 Comparison guide of cutting tip sizes

THICKNESS OF STEEL	1/4"	3/8"	1/2"	3/4"	1"	1/4"	1 1/2"	2"	2 1/2"	3"	4"	5"	6"
AIRCO TIP SIZE	0	1	1	2	2	2	3	3	4	5	5	6	6
GAGE PRESSURE OXYGEN P.S.I.	30	30	40	40	50	60	45	50	50	45	60	50	55
GAGE PRESSURE ACETYLENE P.S.I.	3	3	3	3	3	3	3	3	3	4	4	5	5
SPEED IN INCHES PER MIN.	20	19	17	15	14	13	12	10	9	8	7	6	5
OXYGEN CON-SUMPTION CU. FT. PER HOUR	50	75	90	120	140	160	185	200	250	310	385	460	495
ACETYLENE CON-SUMPTION CU. FT. PER HOUR	9	12	12	14	14	14	16	16	17	22	22	28	28
APPROXIMATE WIDTH OF KERF IN INCHES	.075	.095	.095	.110	.110	.110	.130	.130	.145	.165	.165	.190	.190
CUTTING ORIFICE CLEANING DRILL SIZE	64	57	57	55	55	55	53	53	50	47	47	42	42
PREHEAT ORIFICE CLEANING DRILL	71	69	69	68	68	68	66	66	65	63	63	61	61

NOTE: This chart pertains to Airco Style 124 tips only. If equipment from other manufacturers is used, refer to the chart, "Comparison Guide of Cutting Tip Sizes" and choose a tip with a cutting or orifice size comparable to that indicated above.

Chart 9-2 Relation of cutting tip size to plate thickness

Chart 9-2, refers to Airco® style 124 tips only. This chart gives proper tip sizes, cutting oxygen pressures, and rate of travel for the various thicknesses of metal. If equipment from other manufacturers is used, refer first to the Comparison Chart and choose a tip with a cutting orifice size close to the Airco® size.

HAZARDS

The operator must protect his eyes at all times with goggles fitted with proper lenses, usually shade 5 or 6. Gauntlet-type gloves and any other equipment necessary to give protection from the molten iron oxide, should be worn.

Since the high-pressure stream of cutting oxygen can throw small bits of molten oxide, at a temperature of 3,000 degrees F. and up, for distances of 50-60 feet, the operator should check before starting the burning operation to be sure that all flammable and explosive materials have been removed to a safe place.

The operator should insure that all personnel in the area are warned of the shower of molten metal that will occur so that they may take the necessary precautions.

The International Acetylene Association and the Underwriters' Laboratories recommend that an additional workman with fire-fighting equipment be assigned to each unit during cutting and for 2 hours after completion of cutting to guard against fires.

REVIEW QUESTIONS

1. What are the limitations of flame cutting?

2. What chemical change takes place during the burning process?

3. What special safety precautions must be taken?

4. How wide must a piece of one-inch steel plate be so that 10 strips each 3 inches wide can be cut from it? Make the proper allowance for the width of the kerf (cut width) from chart 9-2.

5. Using chart 9-2 determine how much time is required to cut the plates in question 4 if they are each 10 feet long? Figure actual cutting time only.

6. How much oxygen and acetylene are used in problems 4 and 5? Figure only the actual time required to make the cuts.

UNIT 10 STRAIGHT LINE CUTTING

Several things affect the speed, smoothness, and general quality of a cut made by an oxyacetylene flame. This unit provides practice in changing these variables to determine the best methods for flame cutting.

The actual cutting process demonstrates the danger of personal burns and fires which might cause property damage.

MATERIALS

1/4-inch or 3/8-inch thick steel plate, approximately 4 in. x 10 in.

Cutting torch fitted with an Airco® #0 or #1 cutting tip or comparable equipment. See chart 1, Comparison of Cutting Tip Sizes, page 24.

PROCEDURE

1. Draw a series of straight parallel lines about 2 inches apart on the plate. Use soapstone for marking so that the lines show up when the cutting goggles are being worn.
2. Light and adjust the preheating flame to neutral using the data supplied in chart 9-2, "Relation of Cutting Tip Size to Plate Thickness."
3. Start the cut by holding the tip over the edge of the metal so that the vertical centerline of the tip is square with the work and in line with the edge of the plate. The tip is positioned in the torch as indicated in figure 10-1.

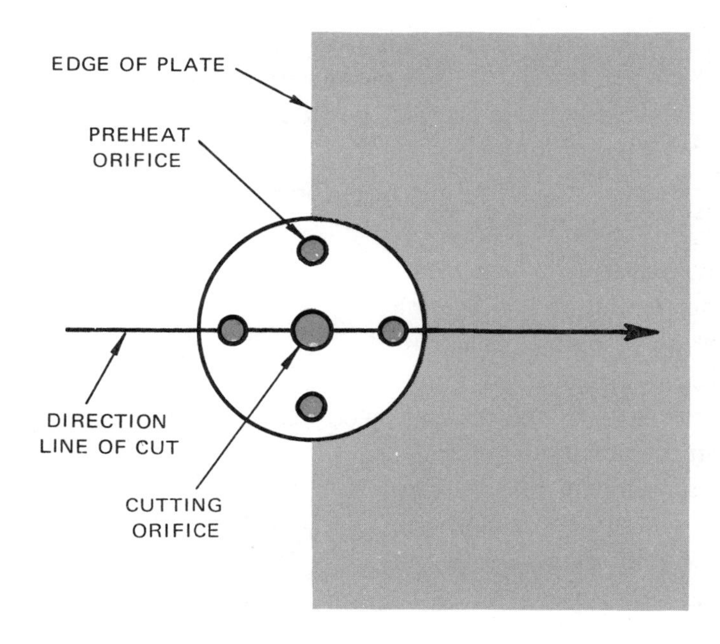

Fig. 10-1 Tip alignment for square cuts

1. This is a correctly made cut in 1-in. plate. The edge is square and the draglines are vertical and not too pronounced.
2. Preheat flames were too small for this cut with the result that the cutting speed was too slow, causing gouging at the bottom.
3. Preheat flames were too long with the result that the top surface has melted over, the cut edge is rough, and there is an excessive amount of adhering slag.
4. Oxygen pressure was too low with the result that the top edge has melted over because of the too slow cutting speed.
5. Oxygen pressure was too high and the nozzle size too small with the result that the entire control of the cut has been lost.
6. Cutting speed was too slow with the result that the irregularities of the draglines are emphasized.
7. Cutting speed was too high with the result that there is a pronounced break to the dragline and the cut edge is rough.
8. Torch travel was unsteady with the result that the cut edge is wavy and rough.
9. Cut was lost and not carefully restarted with the result that bad gouges were caused at the restarting point.
10. Correct procedure was used in making this cut.
11. Too much preheat was used and the nozzle was held too close to the plate with the result that a bad melting over of the top edge occured.
12. Too little preheat was used and the flames were held too far from the plate with the result that the heat spread opened up the kerf at the top. The kerf is too wide at the top and tapers in.

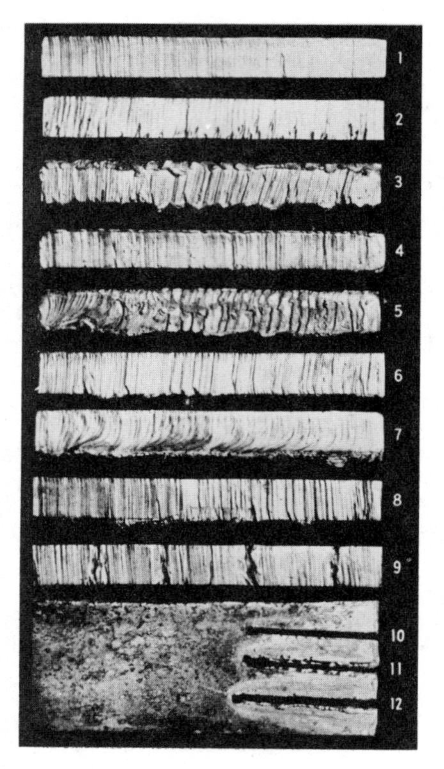

Fig. 10-2 Common faults that occur in hand cutting

4. When the edge of the work becomes bright red, turn the cutting oxygen on with the lever. Note that the oxygen makes a cut through the plate at the same angle that the centerline of the tip makes with the work.
5. Continue the cut, making sure that the tip is square with the work. Observe that when the rate of travel is right, the slag or iron oxide coming from the cut makes a sound like cloth being torn. The tearing sound serves as a guide to the correct rate of travel in most manual flame-cutting operations.
6. Finish the cut and check the flame-cut edge for smoothness, straightness, and amount of slag on the bottom edge of the cut surfaces.
7. Make more cuts but vary the amount of preheating by decreasing and increasing the acetylene pressure before each cut. Observe the finished cut for smoothness, melting of the top edge of the plate, amount of slag on the bottom edge of the plate, and ease of removal of this slag. Compare the plates cut and determine which amount of preheat produces the best results.
8. Make more cuts but vary the rate of travel from very slow to normal to very fast. Ob-

Fig. 10-3 Straight cut

serve these finished cuts and check the appearance of the top and bottom of each plate, and also the ease of slag removal. Determine which rate of travel produces the best results.

9. Make more cuts but vary the amount of cutting oxygen pressure from low to normal to high and check the results as in step 8.
10. Make one or two cuts with the tip perpendicular to the work but move the torch so that the tip zigzags along the straight line drawn on the plate. Notice that the surface of the cut edge follows the amount and direction the tip moves from the straight line.

REVIEW QUESTIONS

1. A number of variables have been tried out in this unit. What conclusions can be drawn about the importance of each of these variables with regard to the ability to make straight line cuts?

   a. Tip angle

   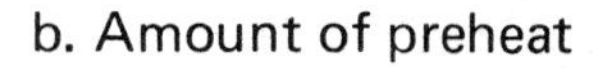

   b. Amount of preheat

   c. Amount of cutting oxygen pressure

   d. Rate of travel

   e. Direction of travel

UNIT 11 BEVEL CUTTING

Making bevel cuts on steel plate is a common cutting operation. The technique is similar to that used for making straight cuts.

Skill in cutting operations is gained only through practice and with a definite goal in mind. Unguided wanderings over a plate add very little to an operator's skill and waste material and gas.

MATERIALS

3/8-inch thick steel plate

Airco® cutting torch fitted with a #1 tip or comparable equipment

PROCEDURE

1. Draw a series of parallel straight lines spaced on 2-inch centers on the work with soapstone.

   Note: These lines are sometimes centerpunched at close intervals to improve visibility. This improves accuracy, but usually results in a slight loss of quality in the cut. The center punch marks cause the high-speed, high-pressure cutting oxygen to stray somewhat.

2. Hold the cutting tip at an angle of 45 degrees with the work and keep this angle when bevel cutting.

3. Proceed with the cut in the same manner as in unit 10. The cut progresses better if the preheating orifices are aligned as in figure 11-1.

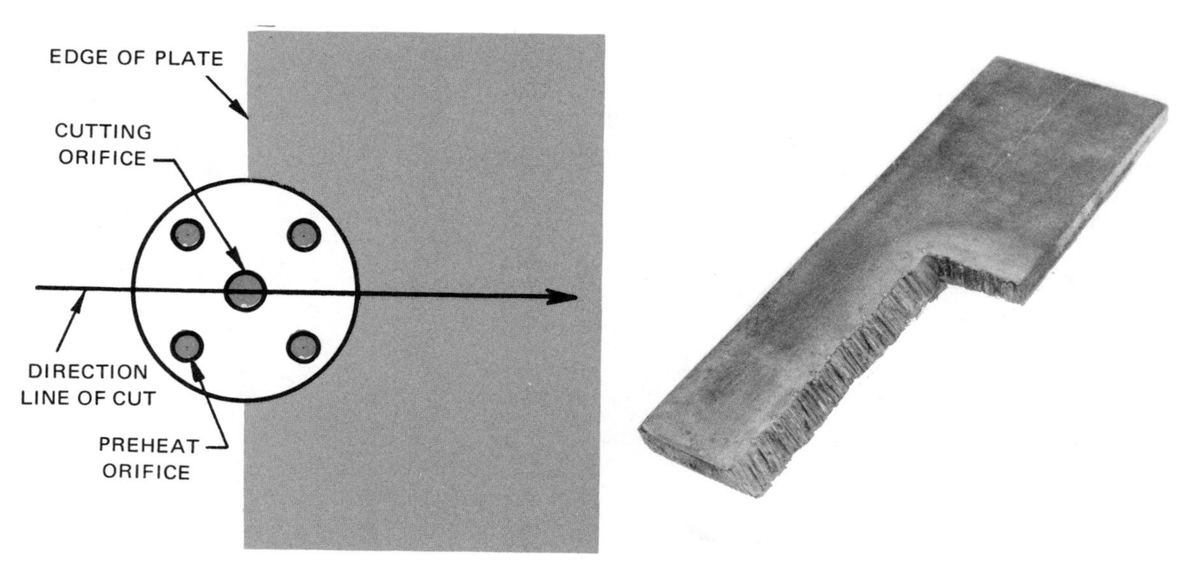

Fig. 11-1 Tip alignment for bevel cuts

Fig. 11-2 Bevel cut

4. Inspect the finished cut for smoothness and uniformity of angle. Check the amount and ease of removal of slag.
5. Make more bevel cuts, but correct the variables until good results are obtained each time.
6. Make another cut, but as the tip moves forward, bring it alternately closer and farther from the work and observe the results.

REVIEW QUESTIONS

1. What variable is responsible for the slag being hard to remove from the bottom of the plate?

2. How does the distance of the tip from the work during the burning process affect the appearance of the finished cut?

3. In making the cut on the plate in this unit, how is the proper tip size, gas pressure, and rate of travel determined?

4. Why is a line to be cut center punched?

5. Why is bevel cutting done?

UNIT 12 PIERCING AND HOLE CUTTING

Holes are easily cut in steel plate by the oxyacetylene flame-cutting procedure. This is a fast operation, adaptable to plates of varying thicknesses and holes of varying sizes. It is useful in cutting irregular shapes.

It is recommended that a diamond point chisel be used to turn up a burr at the point at which the cut is to be started. This burr reaches the kindling temperature much faster than the surface of a flat plate. If a large number of holes are to be pierced, this procedure saves large amounts of preheating gas and operator time.

CAUTION: If particular care is not used in this operation, molten metal may be blown in the face of the operator or into the tip of the torch.

MATERIALS

3/8-inch thick steel plate
Airco® #1 cutting tip or comparable equipment

PROCEDURE

1. Pierce the plate, using the sequence of operations shown in figure 12-1.
   a. Hold the tip about 1/4 inch from the work until the surface reaches the kindling temperature, figure 12-1A.
   b. Open the cutting oxygen valve slowly and, as the burning starts, back the tip away from the work to a distance of about 5/8 inch. The tip must be tilted slightly so that the oxide blows away from the operator and does not blow directly back at the tip, figure 12-1B and C.
   c. Hold the tip in this position until a small hole is pierced through the plate, figure 12-1D.
   d. Lower the tip to the normal burning distance and be sure that it is exactly square with the plate. Then move the torch to enlarge the hole, figure 12-1E.

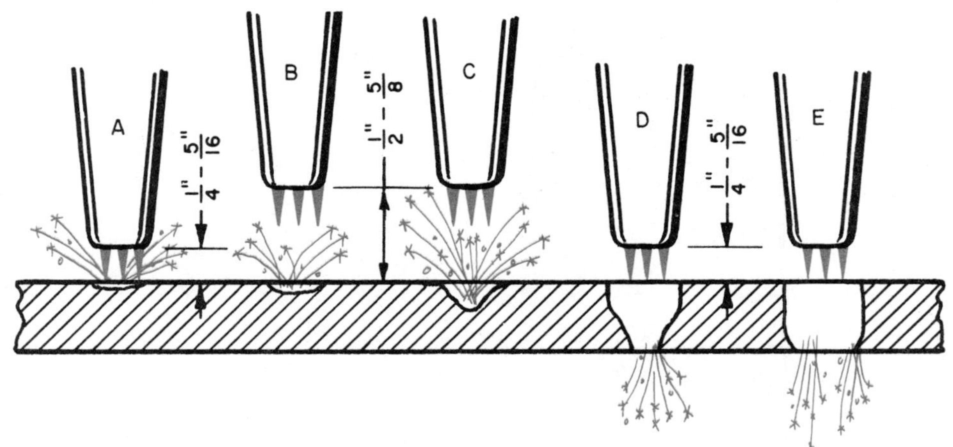

Fig. 12-1 Sequence for piercing plate

2. Continue to move the tip until the hole is the desired size.

   Note: This piercing procedure is recommended for two reasons:

   1. The tip is not ruined by burning the end. This occurs when the tip is held too close to the work for a long period.
   2. The possibility of the molten slag blowing back and clogging the cutting and preheating orifices is reduced.

3. After piercing the plate, move the tip in a circular path to cut a hole of the desired diameter. Considerable practice is necessary to become skillful in making holes which have straight sides, are reasonably round, and close enough to the given diameter to be acceptable.
4. With a pair of dividers, lay out some holes from 1/2 inch to 1 inch in diameter. Center punch the layout line at close intervals to serve as a guide for the cutting operation.
5. Cut the holes, but remember that when cutting to a line the plate should be pierced some distance inside the line. The hole can then be enlarged to the line and the operation completed.
6. Lay out and cut some holes 2 inches and 3 inches in diameter.

   Note: When making large diameter holes, pierce the plate and then move the torch in a straight line until the cut reaches the layout line. Then proceed with the circular cut.

7. Make some round discs. In this case, the piercing operation is performed at a distance outside the layout line. If these discs are to be turned to a specified diameter after cutting, enough material for this operation must be provided.

REVIEW QUESTIONS

1. If a 3-inch round shaft is to be cut off, what preparation is necessary to insure a quick start of the cutting action?
2. How does the procedure vary from the above if the shaft to be cut is square instead of round?
3. If it is desirable to save both the disc and hole when cutting large diameter holes, what procedure should be followed?
4. Why is the tip tilted slightly when starting the cut?
5. Can hole piercing be dangerous for the operator?

UNIT 13 WELDING SYMBOLS

DESCRIBING WELDS ON DRAWINGS

Welding symbols form a shorthand for the draftsman, fabricators, and welding operators. A few good symbols give more information than several paragraphs.

The American Welding Society has prepared a pamphlet, "Symbols for Welding and Nondestructive Testing" (AWS 2.4-76), which indicates to the draftsman the exact procedures and standards to be followed so the fabricators and welding operators may read and understand all the information necessary to produce the correct weld.

The standard AWS symbols for arc and gas welding are shown in figure 13-1.

EXAMPLES OF THE USE OF SYMBOLS

Each of the symbols in this unit should be studied and compared with the drawing which shows its significance. They should also be compared with the symbols shown in figure 13-1.

Throughout this book a symbol related to the particular job is shown together with its meaning. A study of each of these examples will clarify the meaning of the welding symbols.

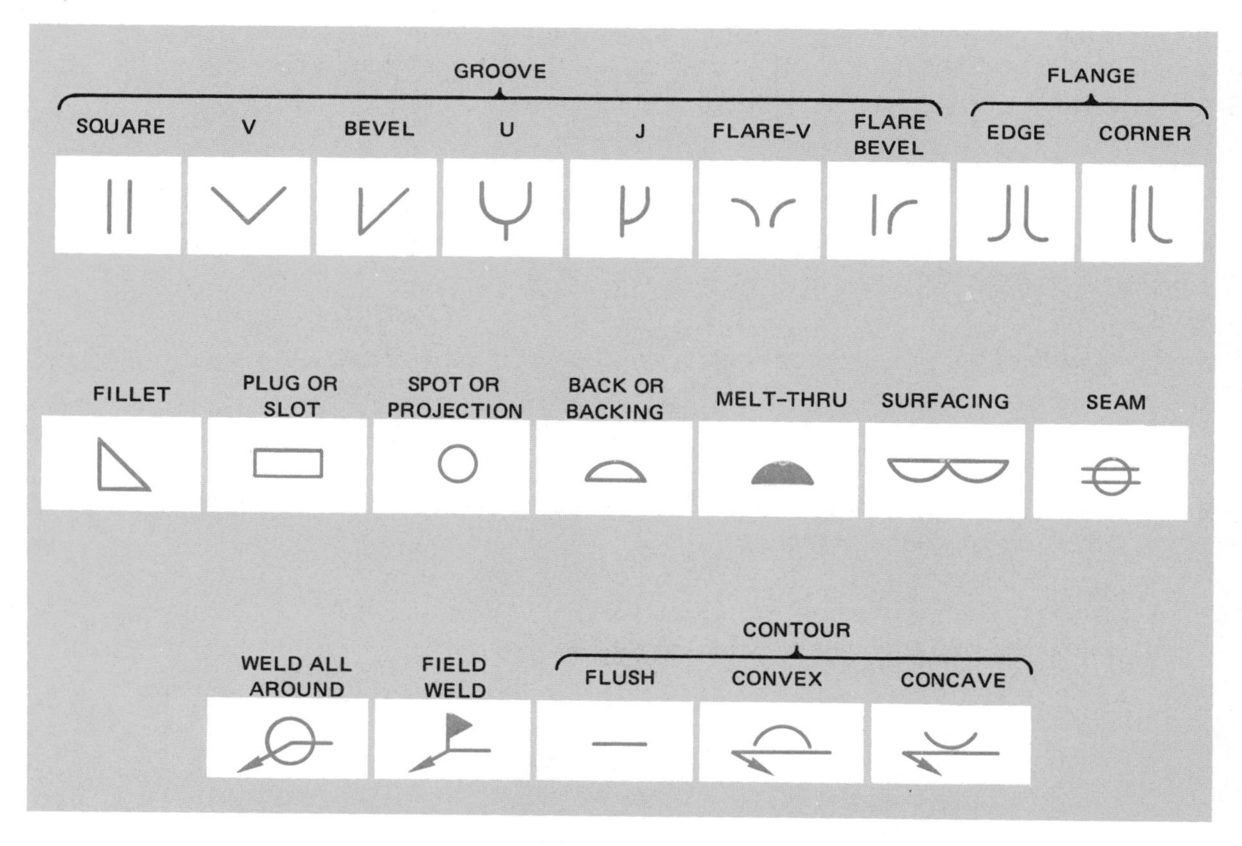

Fig. 13-1 Standard welding symbols

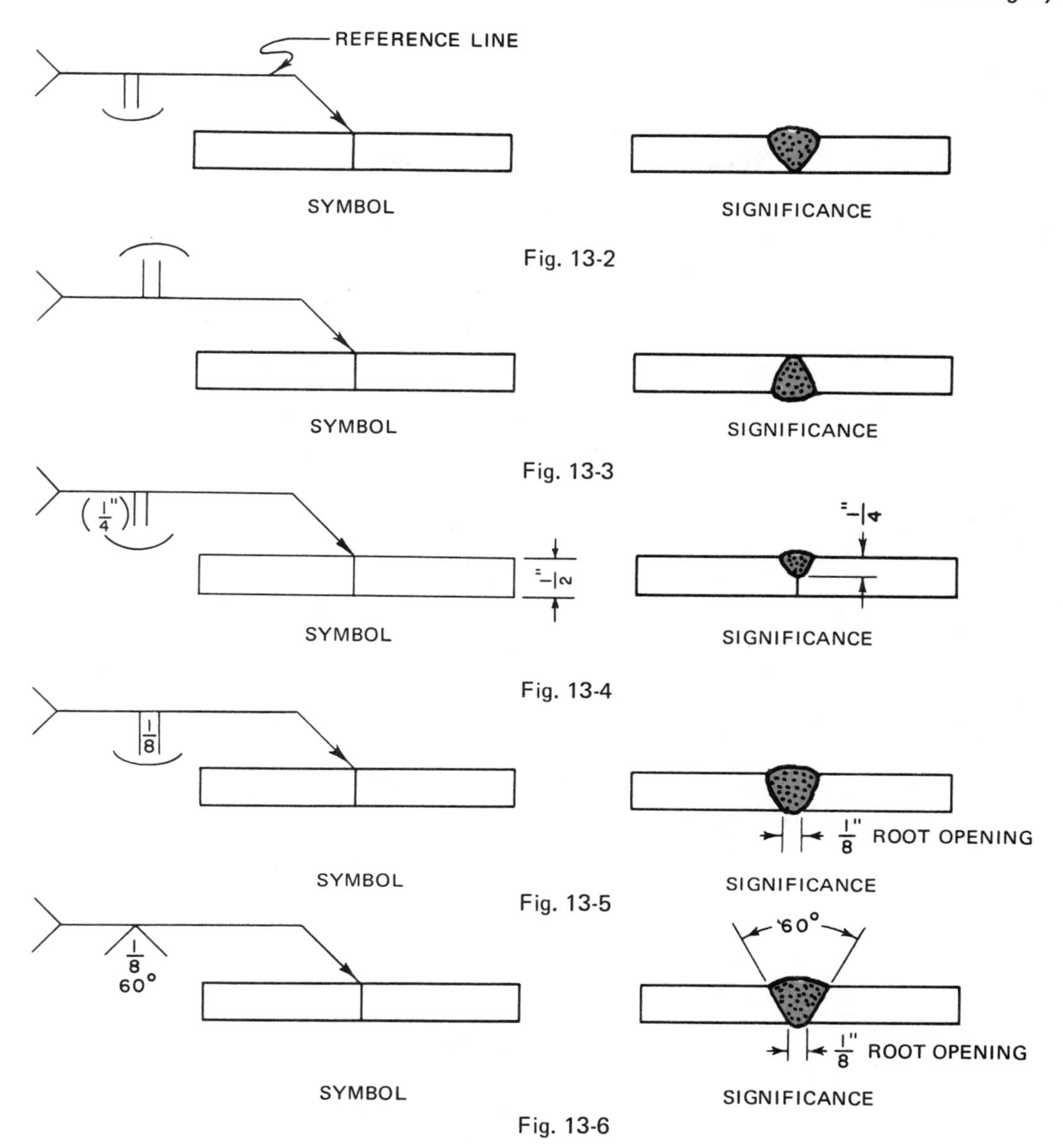

Fig. 13-2

Fig. 13-3

Fig. 13-4

Fig. 13-5

Fig. 13-6

The symbols from the chart are placed at the mid-point of a reference line. When the symbol is on the near side of the reference line the weld should be made on the arrow side of the joint as in figure 13-2.

If the symbol is on the other side of the reference line, as in figure 13-3, the weld should be made on the far side of the joint or the side opposite the arrowhead.

All penetration and fusion is to be complete unless otherwise indicated by a dimension positioned as shown by the (1/4) in figure 13-4.

To distinguish between root opening and depth of penetration, the amount of root opening for an open square butt joint is indicated by placing the dimension within the symbol, figure 13-5, instead of within parentheses as in the preceding drawing.

The included angle of beveled joints and the root opening is indicated in figure 13-6. If no root opening is indicated on the symbol, it is assumed that the plates are butted tight, unless the manufacturer has set up a standard for all butt joints.

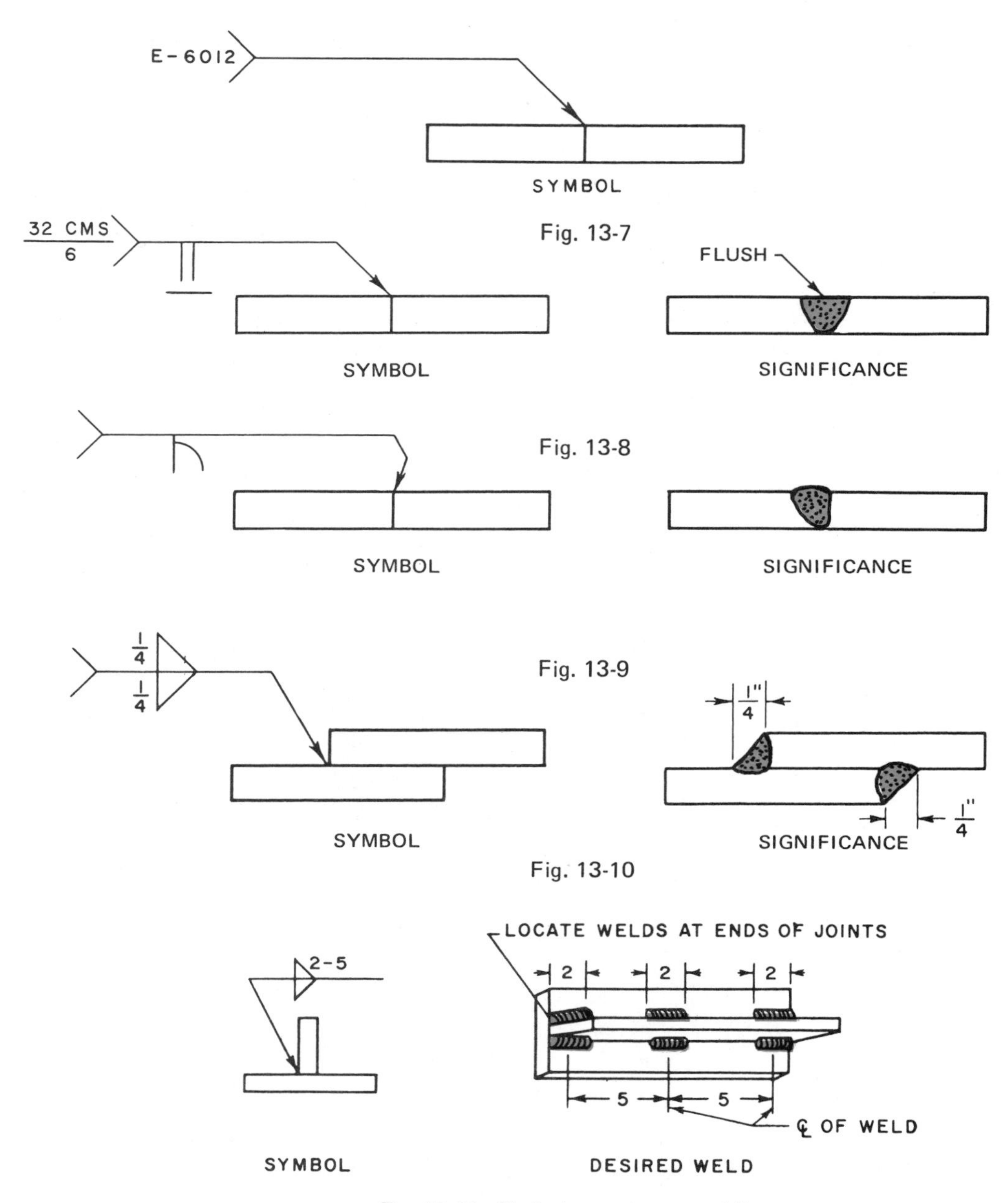

Fig. 13-7

Fig. 13-8

Fig. 13-9

Fig. 13-10

Fig. 13-11 Chain intermittent welding

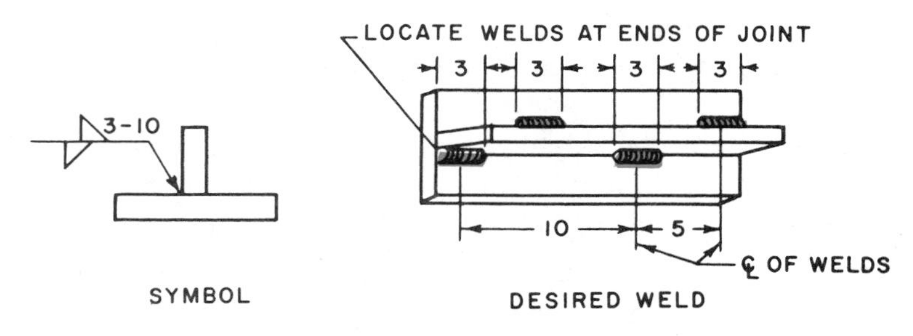

Fig. 13-12 Staggered intermittent welding

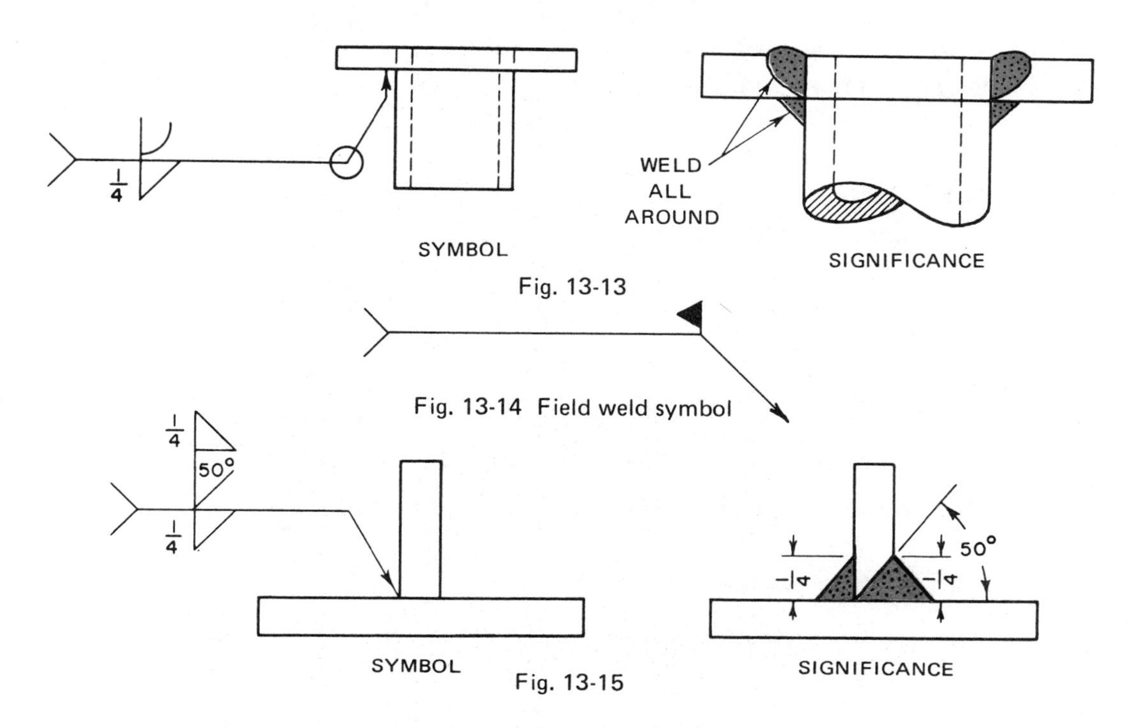

Fig. 13-13

Fig. 13-14 Field weld symbol

Fig. 13-15

The tail of the arrow on reference lines is often provided so that a draftsman may indicate a particular specification not otherwise shown by the symbol. Such specifications are usually prepared by individual manufacturers in booklet or loose-leaf form for their engineering and fabricating departments. These specifications cover such items as the welding process to be used (i.e. arc or gas), the size and type of rod or electrode, and the preparation for welding, such as preheating.

Many manufacturers are using the AWS publication, "Symbols for Welding and Nondestructive Testing" which gives very complete rules and examples for welding symbols as well as a complete set of specifications with letters and numbers to indicate the process.

One method of indicating the type of rod to be used is shown in figure 13-7. This shows that the butt weld is to be made with an AWS classification E-6012 electrode.

In figure 13-8, the rod to be used is indicated as a number 32 CMS (carbon mild steel) type; the 6 indicates the size of the rod in 32nds of an inch. In this case it is a 3/16-inch diameter rod. In addition, the symbol indicates that the finished weld is to be flat or flush with the surface of the base metal. This may be accomplished by: G = Grinding, C = Chipping, or M = Milling.

When only one member of a joint is to be beveled, the arrow makes a definite break back toward the member to be beveled, figure 13-9.

The size of fillet and lap beads is indicated in figure 13-10. In all lap and fillet welds, the two legs of the weld are equal unless otherwise specified.

If the welds are to be chain intermittent, the length of the welds and the center-to-center spacing is indicated, as in figure 13-11.

When the weld is to be staggered, the symbol and desired weld is made as in figure 13-12.

An indication that the joint is to be welded all around is shown by using the weld all around symbol at the break in the reference line, as in figure 13-13.

Field welds (any weld not made in the shop) are indicated by placing the field weld symbol at the break in the reference line, as in figure 13-14.

Several symbols may be used together when necessary, figure 13-15.

REVIEW QUESTIONS

1. What is the symbol for a 60-degree closed butt weld on pipe?

2. What is the symbol for a U-groove weld with a 3/32-inch root opening?

3. What is the symbol for a double V, closed butt joint in plate?

4. What is the symbol for a 1/2-inch fillet weld in which a column base is welded to an H-beam all around?

5. What is the symbol for a J-groove weld on the opposite side of a plate joint?

UNIT 14 RUNNING BEADS AND OBSERVING EFFECTS

The quality of the finished weld depends to a large extent on the correct adjustment and use of the flame. This unit provides an opportunity to weld with different kinds of flames and to compare the results. At the same time some acutal welding skill is acquired.

MATERIALS

16- or 18-gage mild steel, 2 to 4 in. wide X 6 to 9 in. long
Airco® #2 welding tip or equivalent

PROCEDURE

1. Light the torch and adjust the flame to neutral.
2. Hold the tip of the inner cone of the flame about 1/8 inch above the work and pointed in the exact direction in which the weld is to proceed. The centerline of the flame should make an angle of 45 to 60 degrees with the work, figure 14-2.
3. Hold the flame in one spot until a puddle of metal 1/4 inch to 3/8 inch in diameter is formed.
4. Proceed with the weld, advancing the flame at a uniform speed in order to keep the molten puddle the same diameter at all times. This keeps the weld or *bead* the same width throughout its length. Start this bead 1/2 inch from the near edge of the plate being welded and proceed in a straight line parallel to this edge.

   Note: The width of the bead is directly related to the thickness of the plate being welded. The accepted standard for welds in aircraft tubing and light sheet metal requires the weld to be six times as wide as the thickness of the metal.
5. After the weld has been completed, examine it for uniformity of width and smoothness of appearance. Turn the plate over and examine the bottom for uniformity of *penetration.*

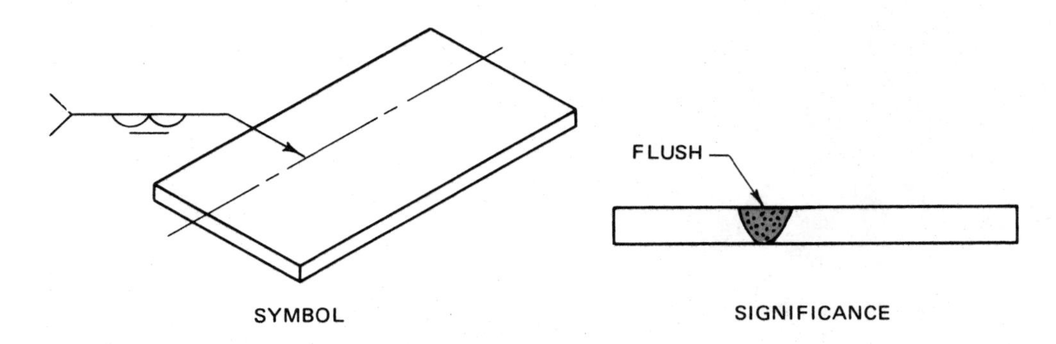

Fig. 14-1 Bead weld

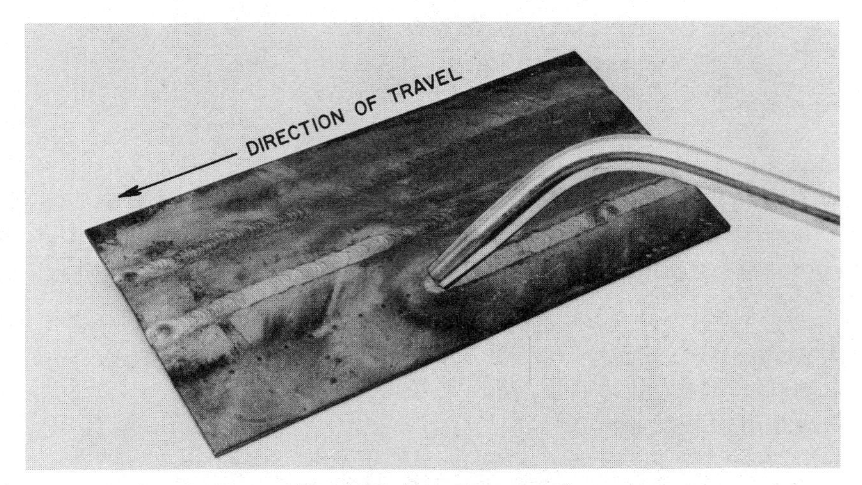

Fig. 14-2 Running a bead

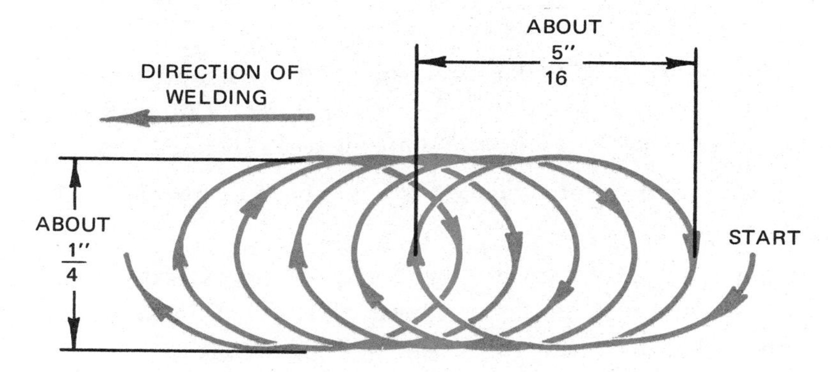

Fig. 14-3 Torch manipulation. Note: right hand operator

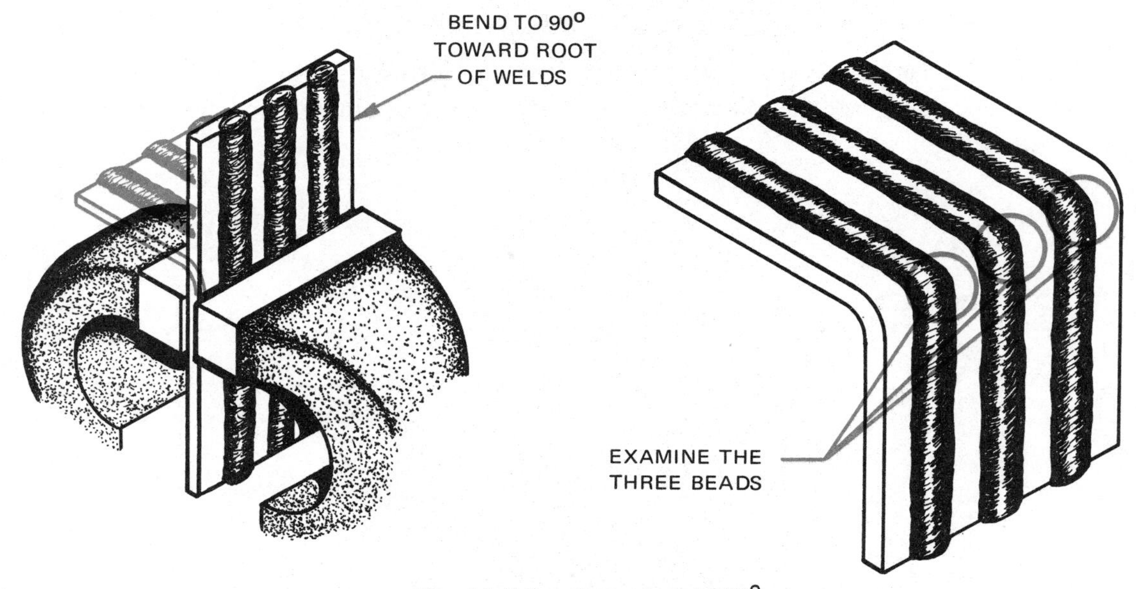

Fig. 14-4 Bending test weld 90°

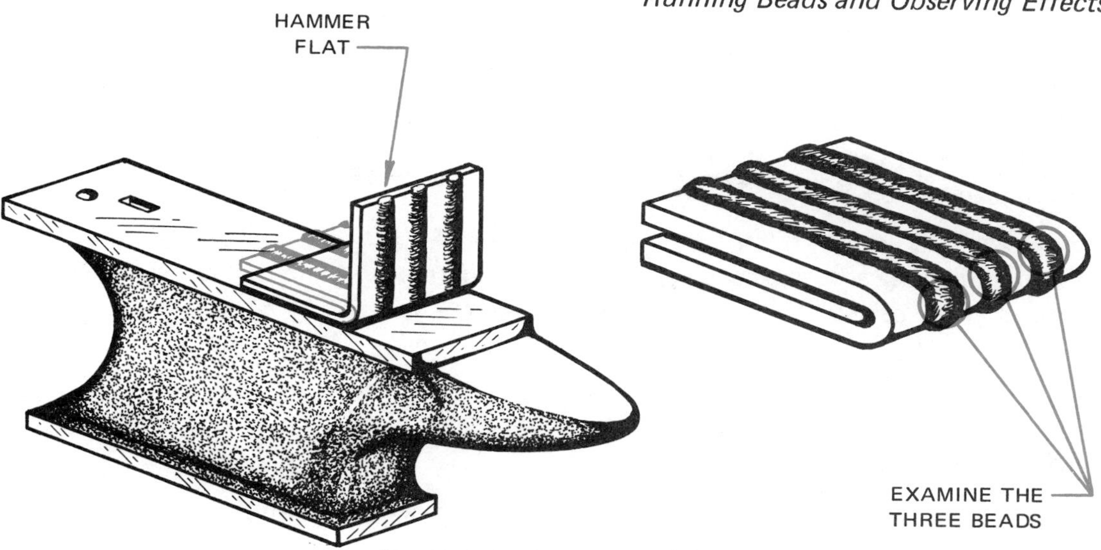

Fig. 14-5 Bending test weld flat

Note: The term penetration refers to the depth to which the parent metal is melted and fused. Fusion is the important factor. It is possible to obtain complete melting of the base metal and less than complete fusion. The welder cannot expect to produce a welded joint equal in strength to the base metal if the fusion is less than 100 percent.

6. Make more beads on the same plate parallel to the first bead and 1/2 inch apart. Vary the angle the flame makes with the work for each new bead. Observe the finished welds for appearance and penetration.
7. Continue to make more beads on additional plates until good appearance and penetration are attained. Manipulate the flame to obtain better results. The simplest manipulation is to rotate the flame in a small circle with a clockwise motion so that the flame is alternately closer and farther away from the work, figure 14-3. The frequency and length of these cycles give the welder added control of the amount of heat applied to the work. In most cases, a better appearing weld results when the flame is manipulated in this way.
8. Run a bead on another plate, using a neutral flame. Then adjust the flame to carburizing and run a bead parallel to the first and about 1/2 inch from it. Note that the sparks coming from the molten puddle tend to break into bushy stars. Notice the cloudy appearance of the molten puddle.

   Note: When steel is heated in a carbon rich atmosphere, such as that produced by a carburizing flame, it tends to absorb the carbon in direct proportion to the temperature and the amount of carbon present. This carbon combines with the steel to form the hard, brittle substance known as iron carbide.
9. Adjust the flame to highly oxidizing and run a third bead parallel to the second and 1/2 inch from it. Note that the molten puddle is violently agitated and that the molten iron oxide has an incandescent frothy appearance. Iron oxide is a hard, brittle, low-strength material of no structural value. Note, also, that the oxidized bead is much narrower than either of the others.

10. Cool the finished test plate and grasp it in a vise across the center of the three welds. Bend this plate 90 degrees toward the root of the welds, figure 10-4. Notice that the carburized bead has cracked and most of the oxidized material has fallen from the oxidized bead.
11. Hammer the test plate on an anvil until it is bent flat upon itself, figure 10-5. Notice that while the carburized and oxidized beads have cracked, the neutral bead has bent as much as the original material with no indication of failure.

REVIEW QUESTIONS

1. How does uniformity of procedure affect appearance of the finished weld?
2. What effect does flame angle have on penetration?
3. What effect does bead width have on fusion?
4. Does the force of the flame blow the molten puddle along the plate or does the molten puddle have to follow the direction the flame takes?
5. What type of flame is best for oxyacetylene welding operations? Why?
6. Why is the bead produced by the oxidizing flame narrower than that produced by the other two flames?

UNIT 15 MAKING BEADS WITH WELDING ROD

In most oxyacetylene welding, additional metal is added to the weld by melting a filler rod into the puddle to produce a stronger weld. These rods are available in various diameters and materials.

The use of the filler rod requires the operator to manipulate not only the torch but also the rod. The proper coordination of the torch and the rod is necessary for the production of good welds. This unit provides an opportunity for practicing this manipulation and observing the results.

MATERIALS

16- or-18-gage steel plate, approximately 4 in. X 9 in.
3/32-inch diameter steel welding rod
Airco® #1 or #2 welding tip or equivalent

PROCEDURE

1. Light the torch and adjust the flame to neutral.
2. Melt the base metal near one end of a plate until a puddle of the proper size is obtained as in unit 14.
3. Place the welding rod in the puddle, making sure the rod is aimed in the direction of travel of the weld, figure 15-2.
4. Proceed with the weld, making sure the welding rod and the tip of the torch make the correct angles with the work. Attempt to make a straight, uniform bead parallel to the edges of the base metal.
5. Make more welds in this manner but vary the angle that the rod makes with the work. Note the effect on the height of the bead.

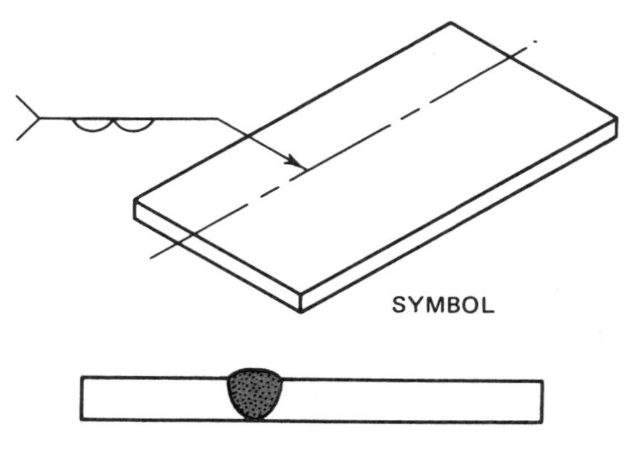

Fig. 15-1 Bead weld with welding rod

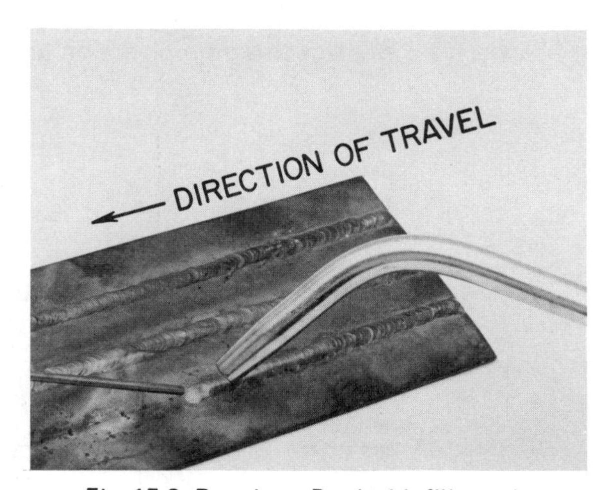

Fig. 15-2 Running a Bead with filler rod

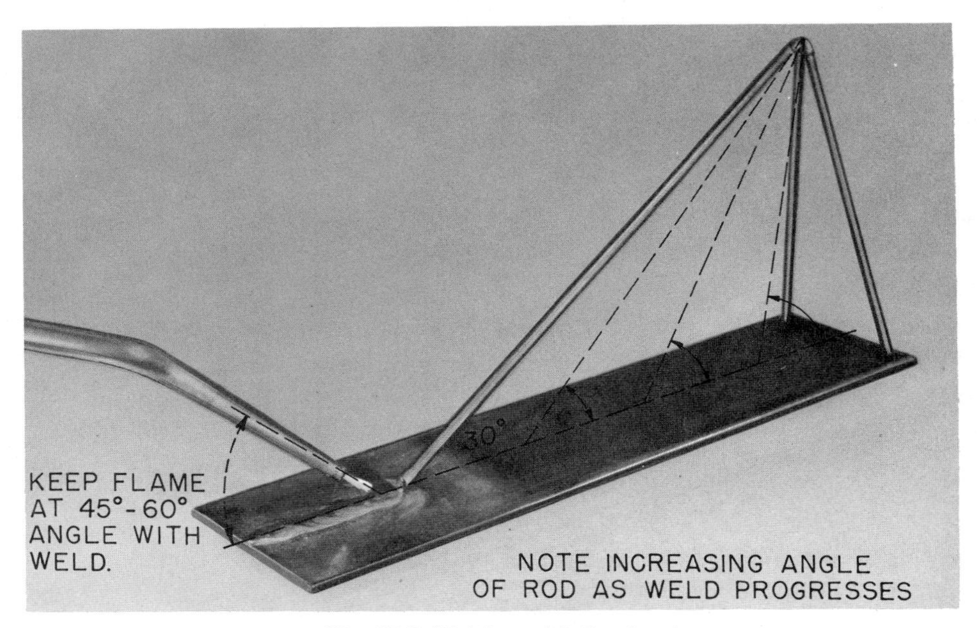

Fig. 15-3 Welding with fixed rod

6. Obtain another plate and set up the plate and rod according to figure 15-3.

   Note: The rod should be twice as long as the legs and welded to them. This produces a starting angle of 30 degrees between the rod and the welding line. As the weld progresses and the rod melts, this angle gradually becomes greater. Observation of the finished weld shows the effect the rod and angle have on the height of the finished bead.

7. Make the weld and observe that in this case, the rod manipulation is not necessary to make a weld with good appearance.

8. Obtain more plates and rods and make more welds; but, as the welding progresses, dip the rod end into the molten puddle with a regular rhythm, figure 15-4. Try one second in the puddle and one second out, and then increase the rhythm until the dipping action is rather rapid.

9. Observe the effect this dipping action has on bead height and uniformity.

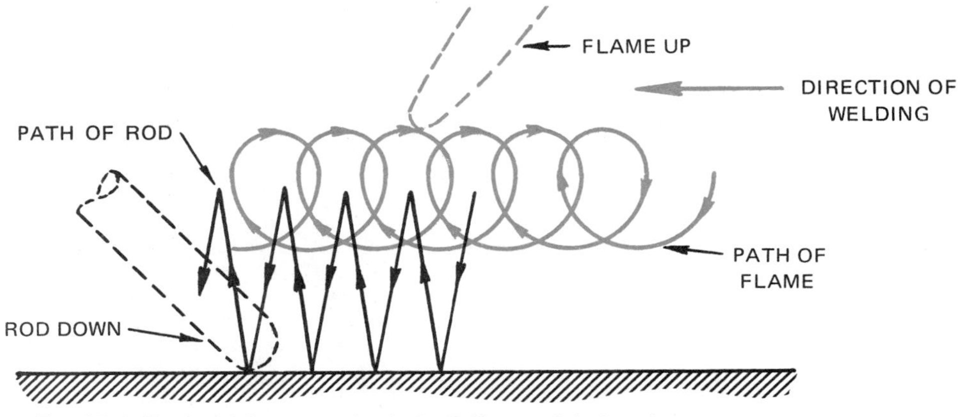

Fig. 15-4 Rod and flame manipulation. Note: right hand operator

REVIEW QUESTIONS

1. What effect does rod angle have on the finished weld?

2. What effect does rod manipulation have on the weld?

3. What effect does uniformity of flame manipulation have on the weld?

4. What is wrong if the weld surface is flat in appearance?

5. What causes washout at the start and finish of a weld?

UNIT 16 TACKING LIGHT STEEL PLATE AND MAKING BUTT WELDS

When making a butt weld, the metal expands as heat is applied and contracts as it cools. This may distort the metal and cause an unsatisfactory job.

To avoid such distortions, several precautions may be taken. One of the most common is to *tack* the two pieces in position. Tack welds are small temporary welds to hold the work in place and control the distortion. A skilled operator must know how to place tack welds, and what effect they have on the finished job.

MATERIALS

Two pieces of 16- or 18-gage mild steel plate, 1 1/2 to 2 in. X 9 in. each
3/32-inch diameter steel welding rod
Airco® #1 or #2 welding tip or equivalent

PROCEDURE

1. Place the plates in position on the welding bench as indicated in figure 16-2.
2. Make tack welds approximately every two inches from right to left. The distance between tacks may be greater for thicker plates.

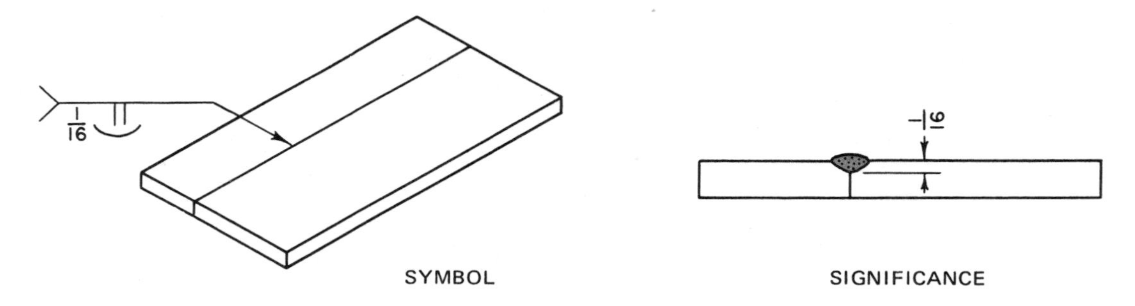

Fig. 16-1 Butt weld

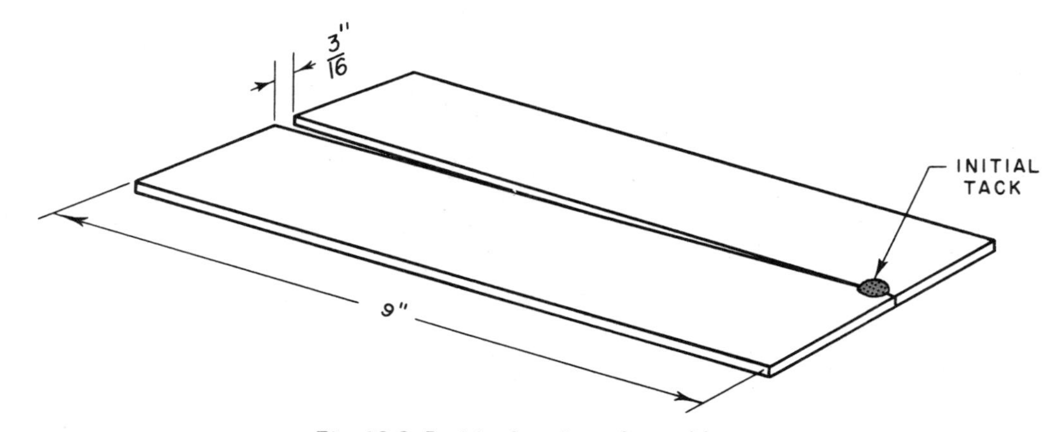

Fig. 16-2 Positioning plates for tacking

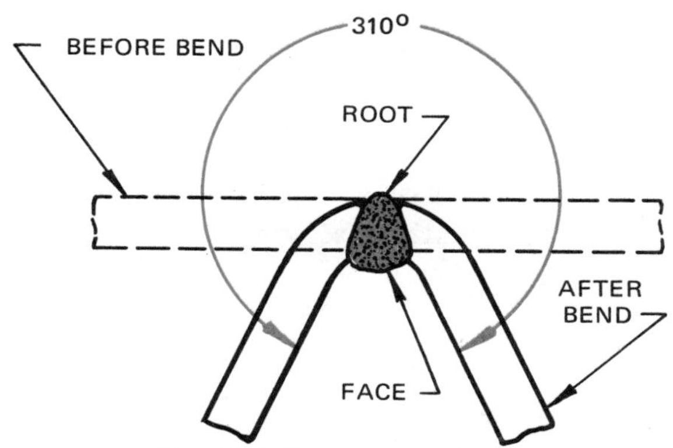

Fig. 16-3 Bend test for butt weld

3. Make the butt weld in much the same manner as in unit 15. However, the weld must be straight and the center of the bead must be on the exact center of the joint.
4. Examine the finished weld for uniformity; inspect the reverse side for penetration.

   Note: On light metal, penetration should be complete from one side of the plate. If this penetration is not obtained, secure more plates and make additional butt welds with wider beads. Practice this until penetration is complete. This happens when the width of the bead is about six times the thickness of the plate.
5. Test the butt welds, as shown in figure 16-3, by holding the finished weld in a vise with the centerline of the weld 1/8 inch above the jaws. Hammer the plate toward the face of the weld. A good weld shows no evidence of root cracks.
6. Obtain two more plates and tack both ends. Then try a third tack weld midway between the first two. Observe the effect of this procedure on plate alignment and ease of tacking.

REVIEW QUESTIONS

1. What happens if the plates are placed in contact for their entire length and then tacked?

2. What effect does plate thickness have on plate spacing?

3. What effect does weld width have on penetration?

4. What effect does tacking only the ends of the joint have on plate alignment during welding?

5. What is penetration?

UNIT 17 OUTSIDE CORNER WELDS

The welded shape in this unit is easily tested by a simple method to determine the quality of the fusion. Welds may be tested to discover poor welds and errors in the procedure used.

MATERIALS

Two pieces of 16- or 18-gage steel plate, 1 1/2 to 2 in. X 6 in. each
Airco® #1 or #2 welding tips or equivalent

PROCEDURE

1. Set up and tack the plates every two inches starting at the end.
2. Weld the plates by placing the flame on the work so that it is split by the sharp corner of the assembly, figure 17-4.
3. Make the weld, trying at all times to make a smooth uniform bead, figure 17-4.
4. Examine the finished bead for uniformity and complete penetration.
5. Check the finished bead by placing the assembly on an anvil and hammering the bead until the plates lie perfectly flat, figure 17-5. Examine the underside for cracks and lack of fusion.
6. Weld more joints of this type, varying the size of the puddle until complete penetration is obtained.

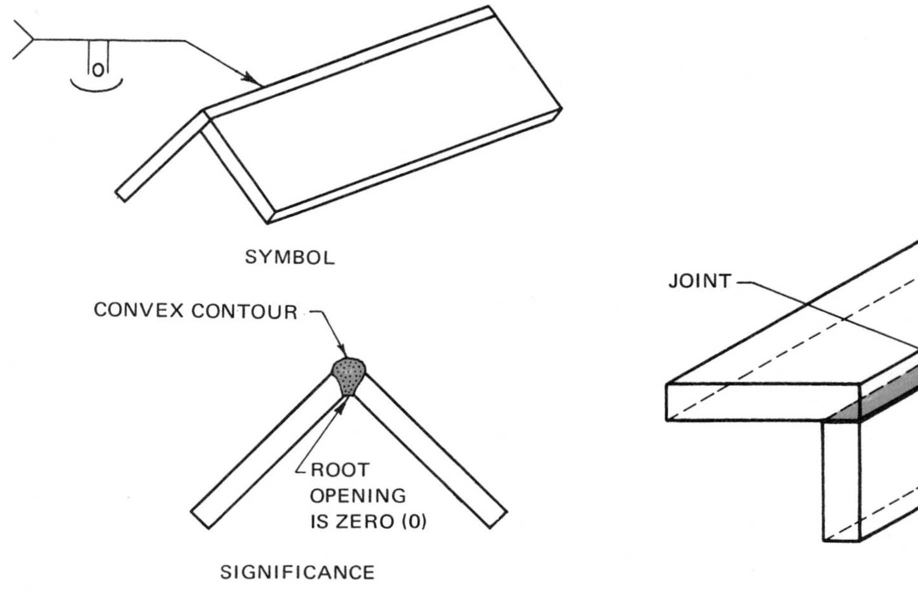

Fig. 17-1 Corner joint

Fig. 17-2 Outside corner weld

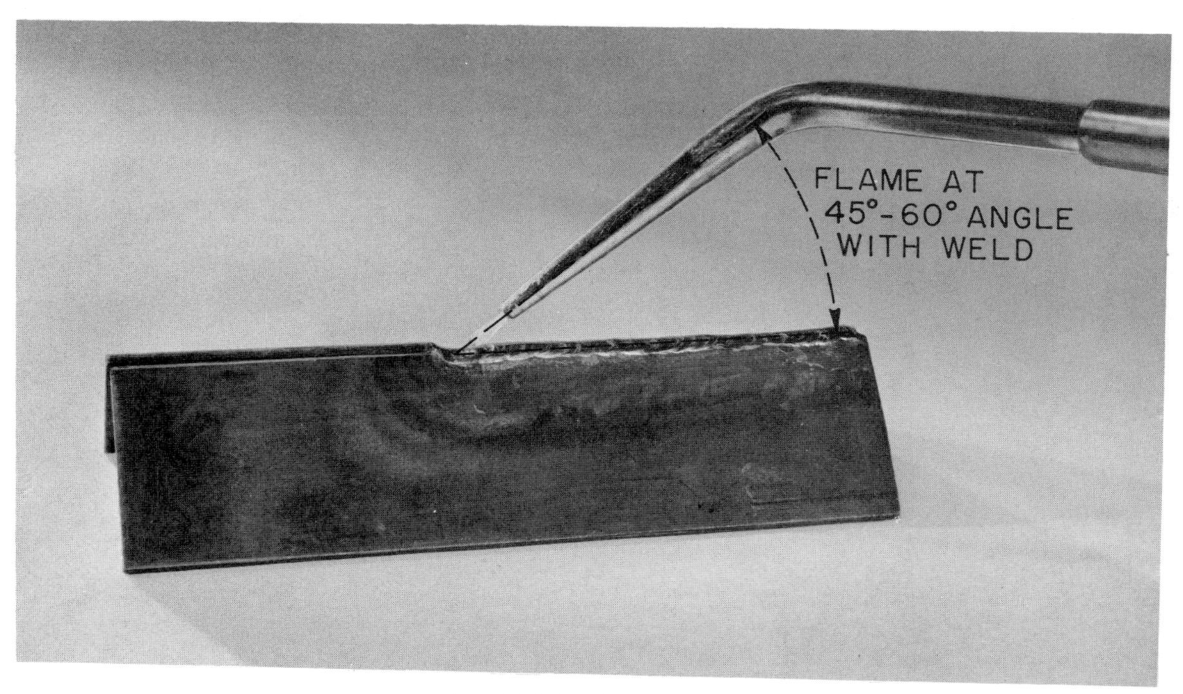

Fig. 17-3 Welder's eye view

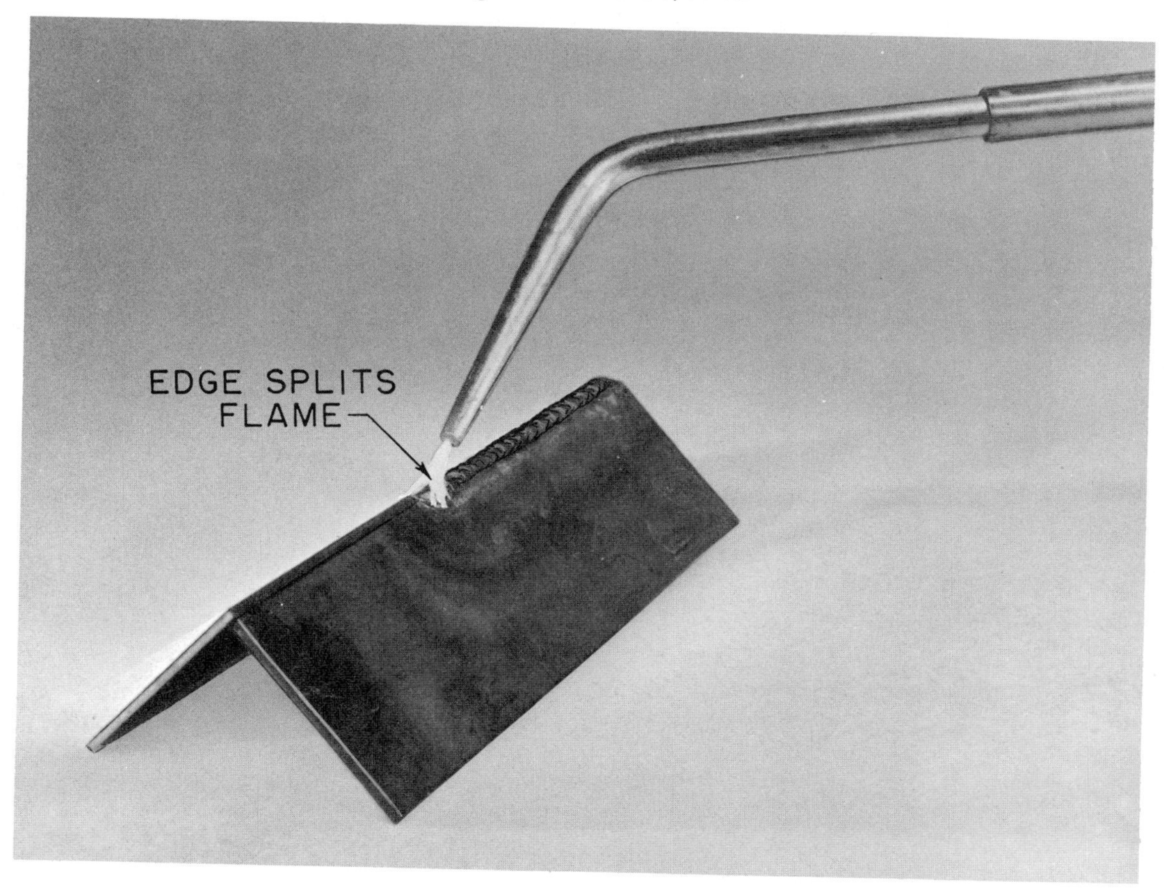

Fig. 17-4 Corner Weld

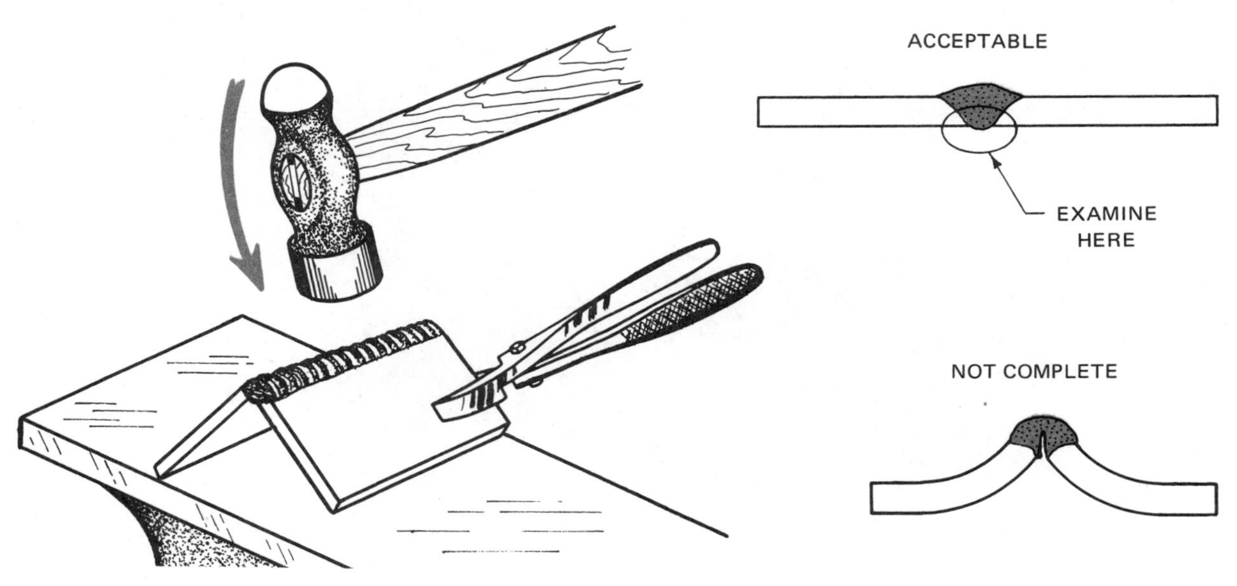

Fig. 17-5 Testing the weld

Note: When welding mild steel with a neutral flame, one way to check for complete penetration is the number of sparks coming from the molten puddle. Very few sparks are produced when the penetration is not complete, and large numbers of sparks are noted when the penetration is excessive. Good welders learn to use this indication to insure proper welds.

REVIEW QUESTIONS

1. In hammer-testing this type of joint, what color is the break if the penetration is not complete?

2. How does the strength of a properly made outside corner weld compare with the base metal?

3. What effect does too much penetration have on the appearance of the finished bead?

4. Is the angle of the torch tip critical, when making this weld? Why?

5. What is the function of a tack weld?

UNIT 18 LAP WELDS IN LIGHT STEEL

The lap weld introduces some new difficulties from possible uneven melting of the two lapped plates. This type of welded joint illustrates the importance of the proper distribution of heat on the surface to be welded. One part of the joint may be melting while the other part may be far below melting temperature. This uneven heating prevents good fusion.

A simple test of the welded lap joint shows the quality of the weld. This shows the student where more practice is needed, and which welding procedures need to be changed to produce a better weld.

MATERIALS

Two pieces of 16-gage mild steel plate,
2 in. X 6 or 9 in. each
3/32-inch diameter steel welding rod
Airco® #3 welding tip or equivalent

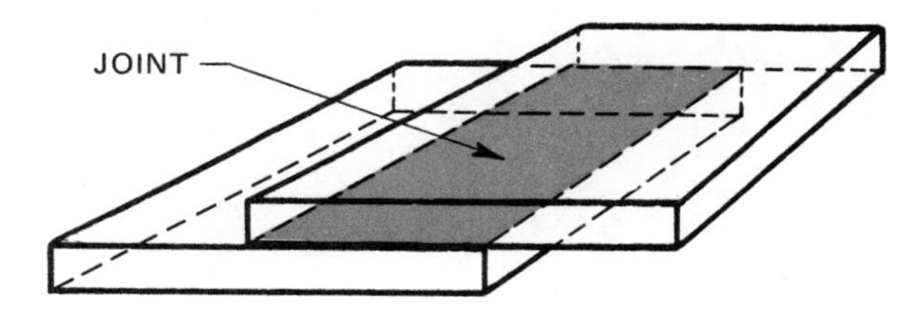

Fig. 18-1 Lap joint

PROCEDURE

1. Place the plates on the welding bench in the position shown in figure 18-3.

   Note: Tack weld the two pieces at the ends so the tacks will not interfere with the weld.

2. Weld the plates, holding the rod and flame as shown in figure 18-3.

   Note: The edge of a plate melts more readily than the center of the plate. Therefore, in this weld there is a tendency for the top plate to melt back too far. This is overcome by placing the rod in the puddle as shown in figure 18-4. The rod is tilted slightly toward the top plate. The rod then absorbs some of the heat and eliminates excessive melting of the top plate.

3. Examine the finished weld for uniformity and for excessive melting of the top plate.
4. Place the weld in a vise, figure 18-5, and test by hammering the lapping plate until it forms a T with the bottom plate.

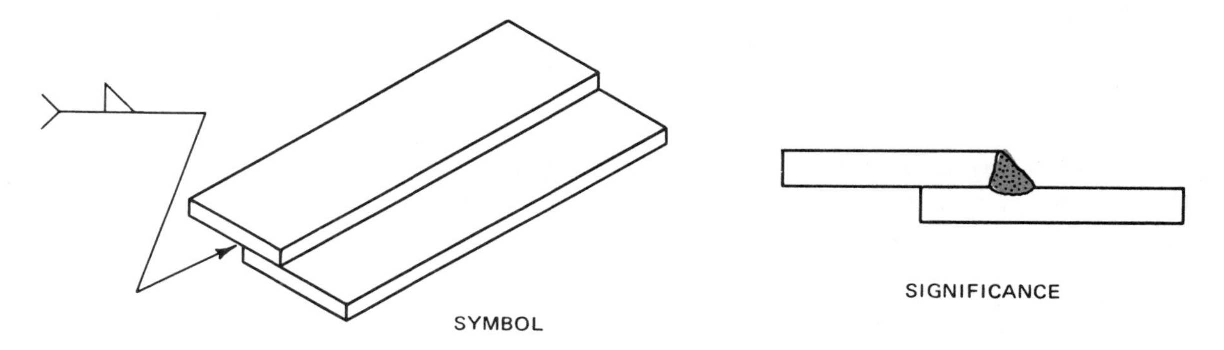

Fig. 18-2 Lap weld

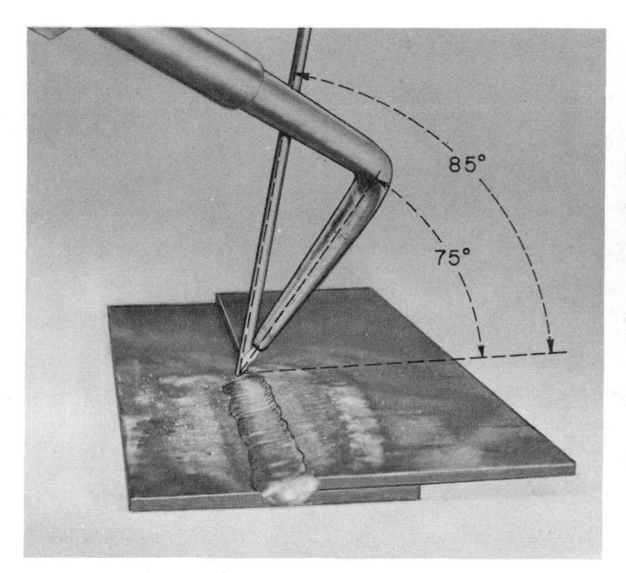

Fig. 18-3 Making the lap weld

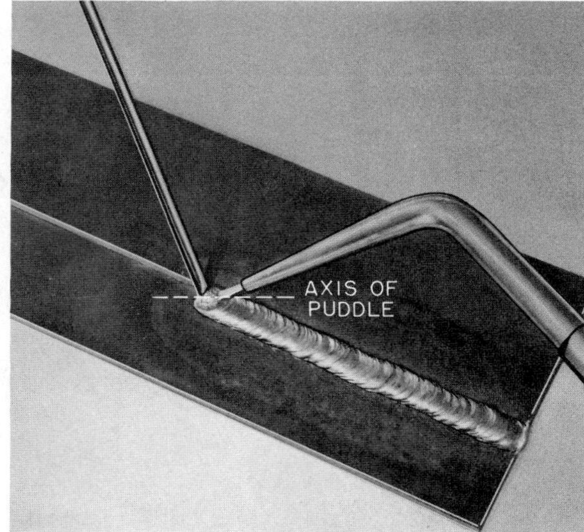

Fig. 18-4 Welder's eye view

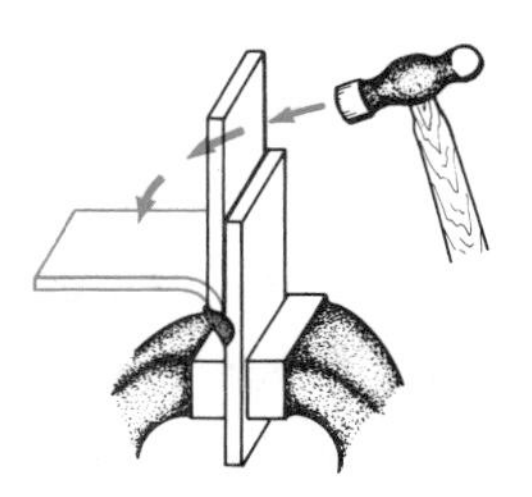

Fig. 18-5 Testing the weld

5. Examine the root of the weld for complete penetration.
6. Practice this type of joint until good surface appearance is obtained and root penetration is complete.

REVIEW QUESTIONS

1. Where should the greatest amount of heat be directed in the lap weld?
2. What effect does the position of the rod in the puddle have on the melting of the lapping plate?
3. What is the relative position of the top and bottom of the molten puddle while the weld is being made?
4. Are the rod and torch angles more or less critical in this job than in the previous jobs?

UNIT 19 TEE OR FILLET WELDS ON LIGHT STEEL PLATE

A tee or fillet weld in light steel plate provides experience in welding two steel plates set at right angles to each other. The angle of the flame to each of the two plates is important. The positions of the puddle and the rod also have an important effect on the quality of the weld. This job presents a new problem for the beginner — the possibility of *undercutting.*

MATERIALS

Two pieces of 16- or 18-gage steel plate,
1 1/2 or 2 in. X 9 in. each
3/32-inch diameter steel welding rod
Airco® #2 or #3 welding tip or equivalent

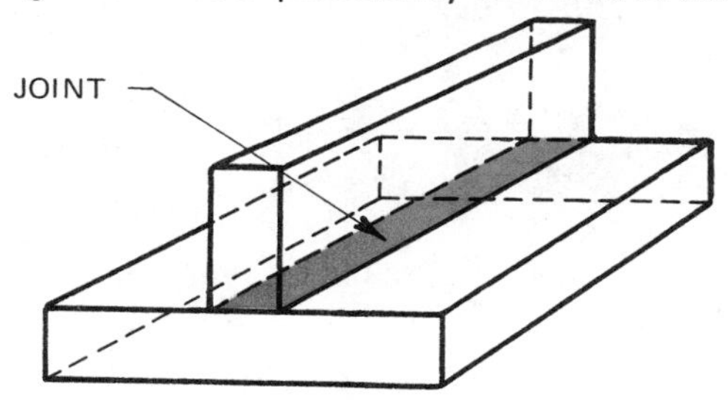

Fig. 19-1 Tee joint

PROCEDURE

1. Set up and tack the plates as shown in figure 19-4.
2. Establish the size and shape of the weld.
3. Proceed with the welding, and pay particular attention to the following points:
   a. The centerline of the flame should make an angle of 45 degrees or less across the bottom plate.
   b. The angle the flame should make with the weld centerline varies from 60 degrees to 80 degrees. For thicker plate the flame should be pointed more directly into the weld.
   c. The puddle of molten metal should be positioned so that the bottom of the puddle is slightly ahead of the top. This is done by rotating the flame in a clockwise direction so the flame follows an oval path.
   d. The rod is usually placed near the top of the puddle so that it comes between the flame and the upstanding plate. In this position, the rod absorbs some of the heat and prevents excessive melting (burning through), or undercutting the vertical leg.

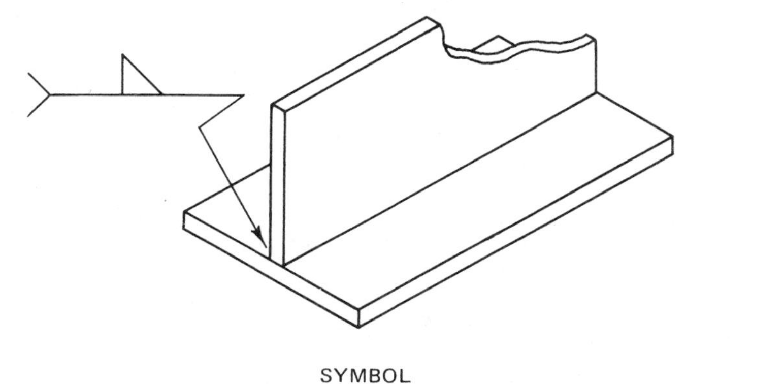

SYMBOL

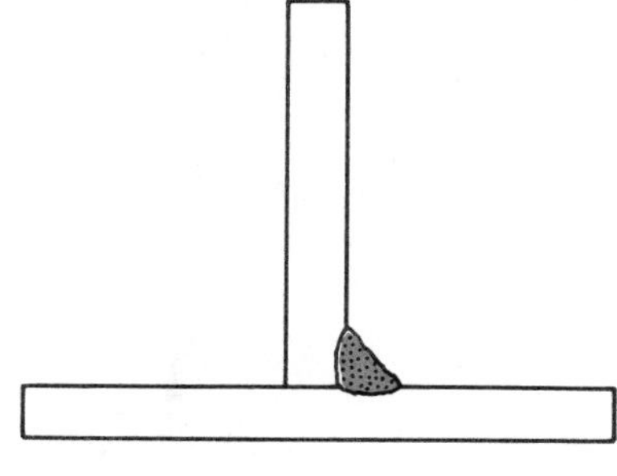

SIGNIFICANCE

Fig. 19-2 Tee weld

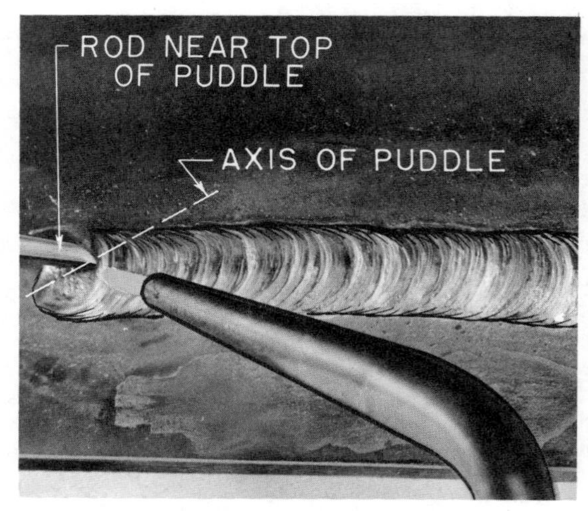

Fig. 19-3 Welder's eye view

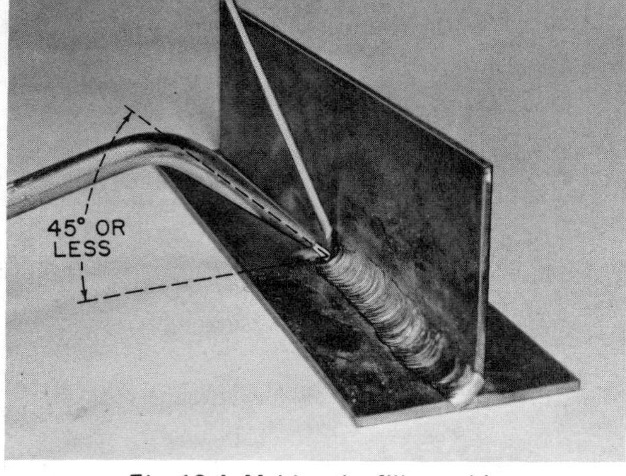

Fig. 19-4 Making the fillet weld

Note: Undercutting may be encountered, figure 19-5. This is an absence of metal along the top edge of the weld. It is caused by too much heat or poor rod movement. It should be avoided at all times.

4. Check the finished fillet weld by placing the assembly on an anvil and hammering the upstanding leg flat toward the face of the weld, figure 19-6. Examine for cracks in the root of the weld.
5. Make more joints of this type until smooth uniform welds are made. It should be possible to bend these welds 90 degrees in either direction without cracking.
6. After a fillet weld has been made on one side of the assembly, make a weld on the opposite side.

   Note: The first weld has produced oxide or scale on the reverse side. This is removed easily by playing the flame rapidly back and forth along the back surface of the joint. The flame causes the oxide to expand and pop from the surface.

CAUTION: When using the flame descaling procedure, extreme care must be used to protect the eyes and skin from burns caused by the hot, flying scale.

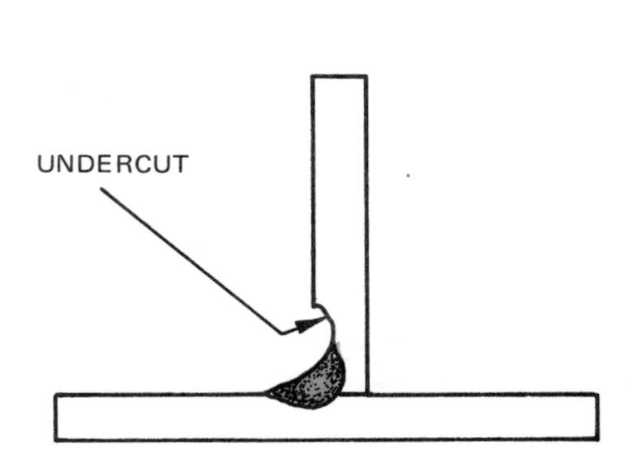

Fig. 19-5 Undercutting

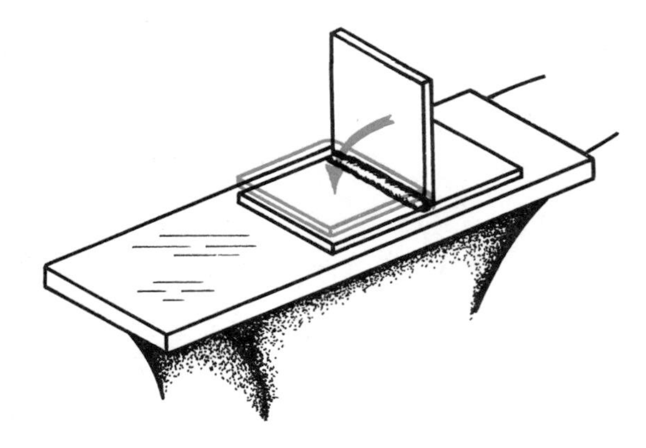

Fig. 19-6 Testing the fillet weld

REVIEW QUESTIONS

1. Is it possible to develop the full strength of the joint when undercutting is present? Explain.

2. What effect does flame and rod movement have on root penetration and appearance of the weld?

3. When making fillet welds on both sides of the joint, how does the amount of heat required for the second weld compare to the first? Why?

4. What effect does melting the excess oxide on the plates and fusing it with the second weld (in step 6) have on the finished bead?

5. Define undercutting.

UNIT 20 BEADS OR WELDS ON HEAVY STEEL PLATE

Although the principles of welding on heavy steel plate are the same as with lighter plate, the problems are greater because more heat is required. To distribute this heat properly, attention must be paid to torch motion and flame angle.

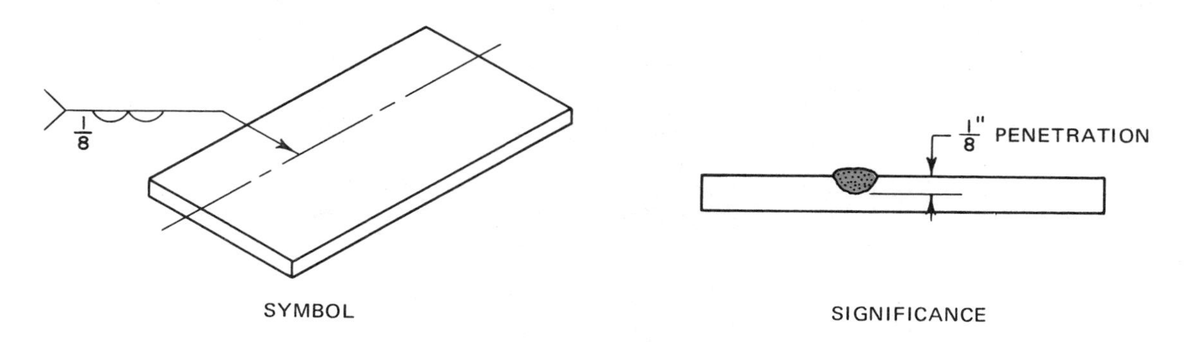

Fig. 20-1 Bead on heavy steel plate

MATERIALS

3/16-inch thick mild steel plate, 4 in. X 9 in.
1/8-inch diameter steel welding rod
Airco® #5 or #6 welding tip or equivalent

Fig. 20-2 Torch motion

PROCEDURE

1. Adjust the flame to neutral.
2. Apply the flame to the work with the tip at an angle of 75 to 80 degrees along the line of the weld.
3. Weld a bead the length of the plate as in unit 15. The flame should be moved in a half-moon weave to produce a weld of adequate width and depth of penetration, figure 20-2. Make the bead 1/2 inch to 5/8 inch wide.
4. Observe the finished bead for appearance, particularly the spacing of ripples, edge of fusion, and penetration.
5. Run more beads on the same plate by varying the angle that the tip makes with the work with each bead.
6. Observe these beads for uniformity. Determine the effect of too little or too great a tip angle on the penetration of the base metal and the appearance of the finished weld.

   Note: This job requires a much larger tip size than any used in previous jobs. As a result, the gas consumption rises very rapidly with a corresponding rise in the hourly cost of operation. The gas should be shut off as soon as the welding is completed so that the cost of operation can be kept down.

REVIEW QUESTIONS

1. What effect does flame angle have on the size and shape of the puddle and the bead ripples?

2. How does the puddle look when the flame angle is too small? Sketch below.

3. What effect does flame angle have on penetration in heavy plate?

4. Is the tip manipulation the same for heavy plate as it is for light plate?

5. Does penetration become more of a problem on heavy plate? How is it helped?

UNIT 21 MANIPULATION OF WELDING ROD ON HEAVY STEEL PLATE

Considerable practice is required to develop skill in manipulating the welding rod and flame in the molten puddle when welding heavy steel plate. The relationships of the rod, flame, and puddle are particularly important. This unit provides practice in manipulating all three.

MATERIALS

3/16- or 1/4-inch thick mild steel plate, 4 in. X 9 in.
1/8-inch diameter steel welding rod
Airco® #5 or #6 welding tip or equivalent

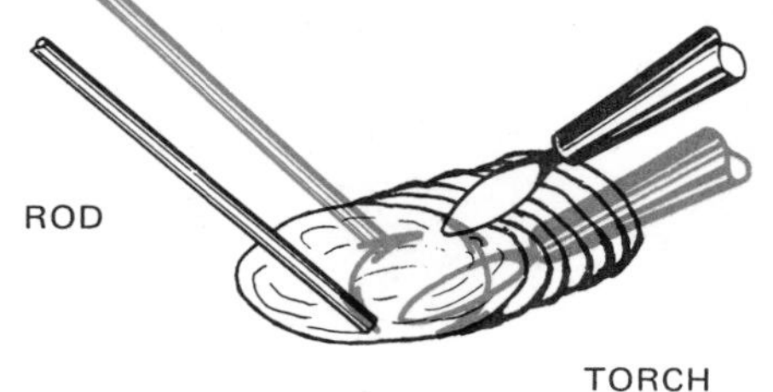

Fig. 21-1 Torch and rod motion

PROCEDURE

1. Apply a neutral flame to the work as in unit 20.
2. The molten puddle should be 1/2 inch to 5/8 inch wide.
3. Weld as in unit 20 except that both the rod and the flame should be moved alternately. In other words, the rod and flame should be moved so that they are on opposite sides of the molten puddle at all times, figure 21-1.
4. Inspect the finished weld for appearance.
5. Make more parallel welds on the same plate, but vary the angle of the rod and the flame. Observe the effect that this variation has on the height of the weld, the depth of penetration, and the face of the weld.
6. Compare these beads or welds with those made in unit 20.

REVIEW QUESTIONS

1. What is the advantage of moving the flame and rod, rather than the flame alone?
2. Which weld has the more uniform appearance?
3. Can a weld bead on heavy plate be too large?
4. Will the surface appearance of the bead be more uneven on heavy plate than light plate welding?
5. On heavy plate is the penetration the same as, less than or more than that on light plate if the correct tip is used?

UNIT 22 BUTT WELDS ON HEAVY STEEL PLATE

Butt welding heavy steel plate is a basic welding operation. The quality of the weld is determined by: the positioning of the plates, tacking the plates, preparation of the edges, and the fit of the plates to each other. All of these factors also affect the ease with which the weld may be made.

Besides providing another opportunity to acquire more skill in butt welding, this unit points out the importance of plate edge spacing.

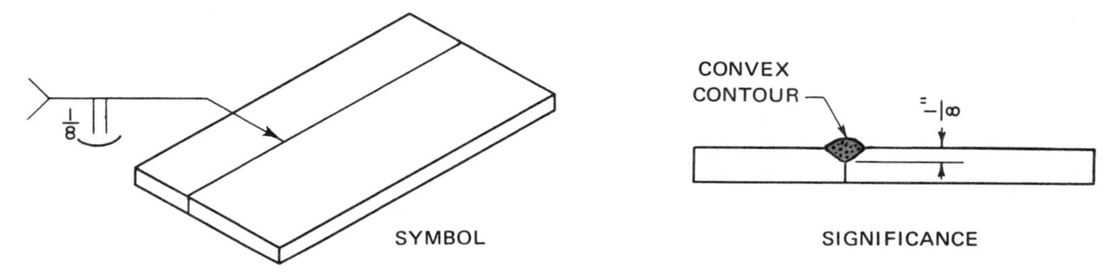

Fig. 22-1 Butt weld on heavy steel plate

MATERIALS

Two pieces of 3/16-inch or 1/4-inch thick mild steel plate, 1 1/2 to 2 in. X 9 in. each
1/8-inch diameter steel welding rod
Airco® #5 or #6 welding tip or equivalent

PROCEDURE

1. Align the plates and tack as in making butt welds in thin plate, unit 16. The distance between tacks may be greater here than on the thinner material.
2. Proceed with welding as in unit 21.
3. Obtain more plates and tack them so that they are spaced 1/16 inch apart for one pair, 3/32 inch apart for another pair, and 1/8 inch apart for a third pair, figure 22-2.

   Note: When the plate edges are touching as in step 1, the joint is called a *closed square butt joint.* When they are spaced as in step 3, the joint is an *open square butt joint.*

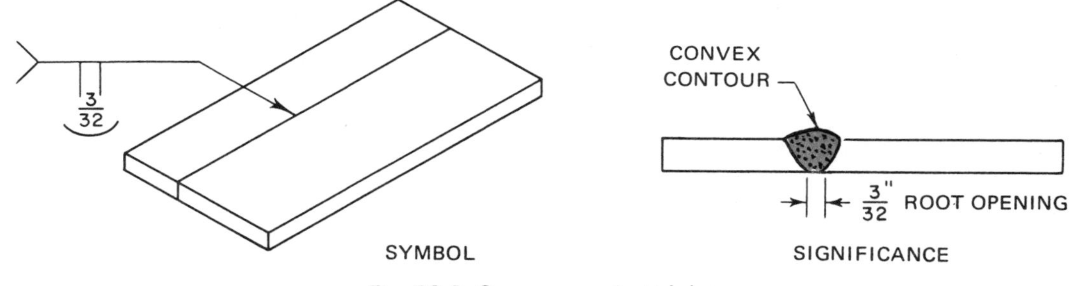

Fig. 22-2 Open square butt joint

4. Weld each pair of plates.
5. Examine and compare the finished butt joints.

REVIEW QUESTIONS

1. What is the relationship between plate thickness and distance between tacks?

2. What effect does plate edge spacing have on:

   a. Penetration?

   b. General appearance of the welds?

3. Draw a sketch of a cross section of a closed square butt joint.

4. Draw a sketch of a cross section of a open square butt joint.

5. Why is there a tendency for the open square butt joint to melt away at the plate edges?

UNIT 23 LAP WELDS ON HEAVY STEEL PLATE

The procedure for making a lap or fillet weld on heavy steel plate is much the same as that required for making this weld on light plate.

In the job performed in this unit, more metal must be deposited because of the thicker plate. Since the heat must cover a larger area, more torch movement is involved. In addition, the larger molten puddle requires more rod movement.

MATERIALS

Two pieces of 3/16 or 1/4-inch thick steel plate, 2 in. X 9 in. each
1/8-inch diameter steel welding rod
Airco® #5 or #6 welding tip or equivalent

PROCEDURE

1. Align the plates so that they lap approximately halfway in the long direction, figure 23-1.

   Note: Tack weld the two pieces at the ends as in unit 18.
2. Weld the plates, but point the flame toward the joint more than when lap-welding light steel plate, figures 23-2 and 23-3. Keep the flame pointed more toward the top plate. The tendency to overmelt the top plate is much less than in welding light plate.
3. As the welding proceeds, rotate the flame clockwise so that the bottom of the molten puddle is slightly ahead of the top. Try varying the position of the rod in the molten puddle.
4. Obtain more plates and repeat this type of joint; but alternate the flame and rod in the puddle in much the same manner as in the butt joint. Manipulate the flame and rod to keep the bottom of the puddle slightly in advance of the top.
5. Make more joints of this type. In each weld, vary the amount that the bottom of the puddle leads the top.
6. Inspect the finished welds for appearance.

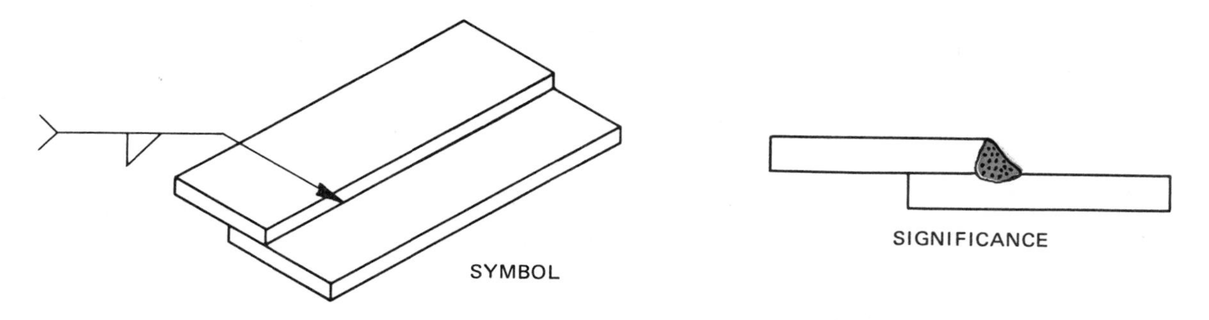

Fig. 23-1 Lap weld

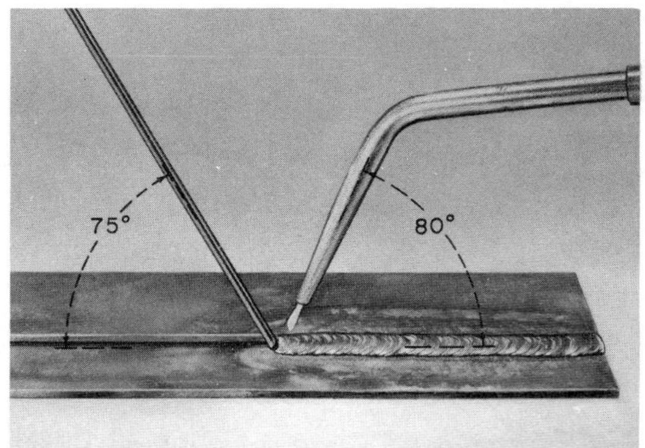

Fig. 23-2 Welder's eye view

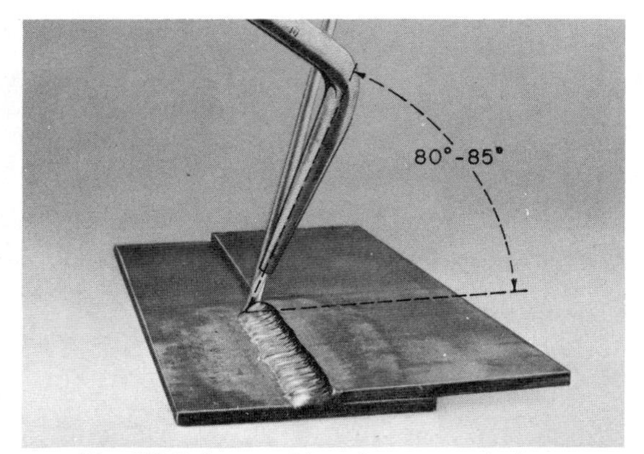

Fig. 23-3 Lap weld on heavy steel plate

7. Break these joints in a vise and examine the welds. Inspect for complete fusion of both plates. Note the size of the crystals or grain of the weld metal in the break.
8. In a piece of steel about the same size, make a saw cut part way through, so it breaks when bent. Compare the grain structure or crystal size of this break with the grain structure of the broken weld metal.

REVIEW QUESTIONS

1. What effect does the tip angle have on the appearance of the finished bead?

2. What effect does the tip angle have on the fusion of the two plates?

3. What effect does the heat of fusion have on the size of the grain structure in the weld and nearby metal?

4. Does undercut become more of a problem with heavy plate? Why?

5. Is it possible for a lap weld to look good and not be strong? Explain.

UNIT 24 FILLET OR TEE JOINTS IN HEAVY STEEL PLATE

Fillet or tee joints are welded in much the same way as the lighter plates in unit 19. However, changes must be made in the flame angle, rod position, and the molten puddle to produce a good weld.

MATERIALS

Two pieces of 3/16- or 1/4-inch thick mild steel plate, 2 in. X 9 in. each
1/8-inch diameter steel welding rod
Airco® #5 or #6 welding tip or equivalent

PROCEDURE

1. Position the plates so that the upstanding plate is about in the center of the flat plate. Tack the plates so that the upstanding leg makes an angle of exactly 90 degrees with the bottom plate.
2. Make the weld, using much the same technique as when making the weld in unit 23. The flame angle and rod position are very important and must be correct to avoid undercutting the upstanding leg, figure 24-2.
3. Check the angle between the vertical and horizontal plates after the weld has cooled.
4. Make more joints, tacking the plates at slightly varying angles. This will allow for shrinkage as the weld cools. Check the angle between the vertical and horizontal plates after the weld has cooled.
5. Make more joints of this type, varying the flame angle, rod position, and amount of lead of the bottom of the puddle.

   Note: The size of the legs of the weld must equal the thickness of the plate being welded if the joint is to be full strength.

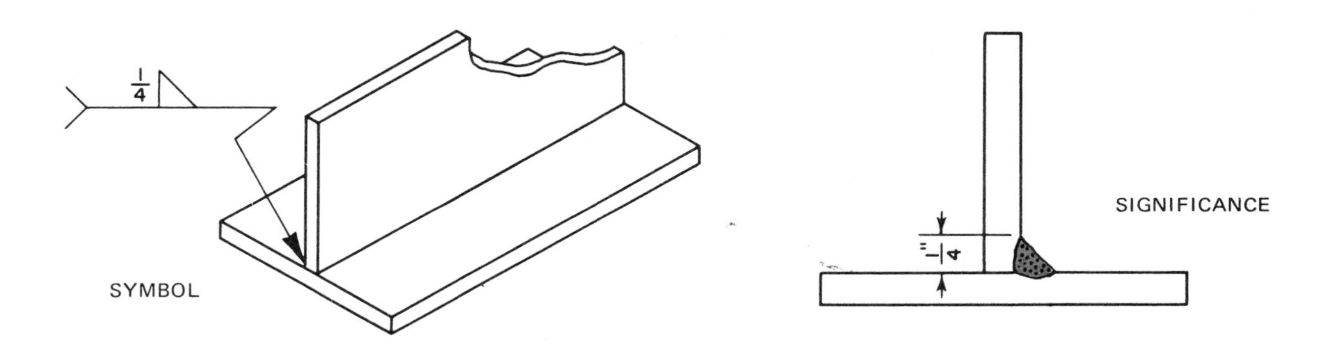

Fig. 24-1 Fillet weld

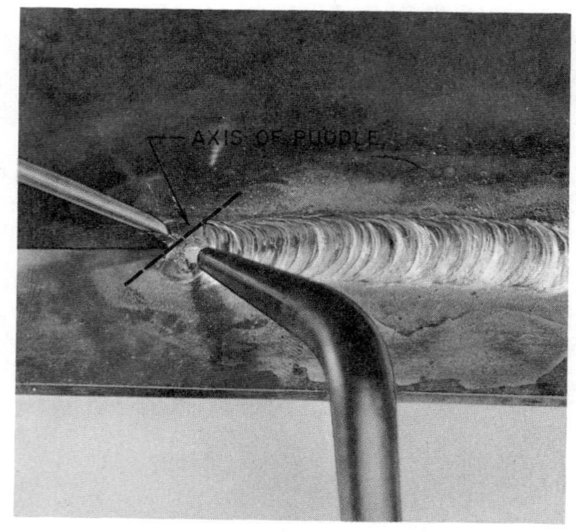

Fig. 24-2 Fillet weld on heavy steel plate

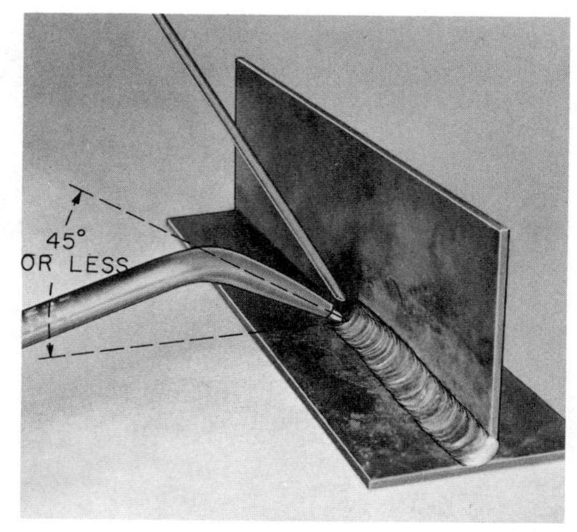

Fig. 24-3 Welder's eye view

6. Break the joints by placing the bottom plate in a vise and bending the upstanding leg toward the weld face.
7. Examine the welds for fusion and penetration, especially at the root of the weld.

REVIEW QUESTIONS

1. What effect does flame angle have on the appearance of the finished bead?

2. What effect does flame angle have on the tendency to undercut?

3. What effect does rod position in the molten puddle have on appearance and the tendency to undercut?

4. How are the plates set up so the final angle is exactly 90 degrees, after shrinkage? Make a cross section sketch of this joint as set up ready for welding.

5. Why is correct hand protection so important when making this weld?

UNIT 25 BEVELED BUTT WELD IN HEAVY STEEL PLATE

Complete fusion and penetration are of great importance to the welding operator. Joints which have been V'd or beveled make penetration in heavy plate much easier. This unit provides practice in flame cutting, flame and rod movement, and multilayer welding.

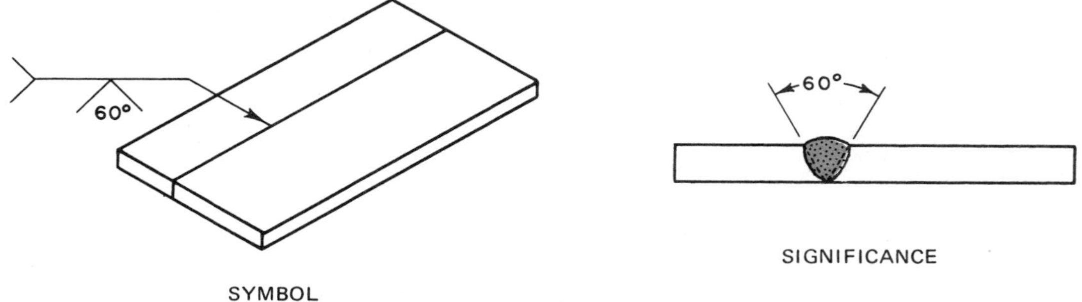

Fig. 25-1 Beveled butt weld

MATERIALS

Two pieces of 5/16- or 3/8-inch thick mild steel plate, 3 in. X 9 in. each
1/8-inch diameter steel welding rod
Airco® #7 welding tip or equivalent

PROCEDURE

1. Flame-cut one edge of each of the plates so that the included angle is 60 degrees.
2. Align and tack the plates with 3/32-inch root opening, figure 25-1.
3. Make the first weld in the bottom of the groove. Fusion between the plates should be complete. The flame should make an angle of 60 to 75 degrees with the work, figure 25-2.

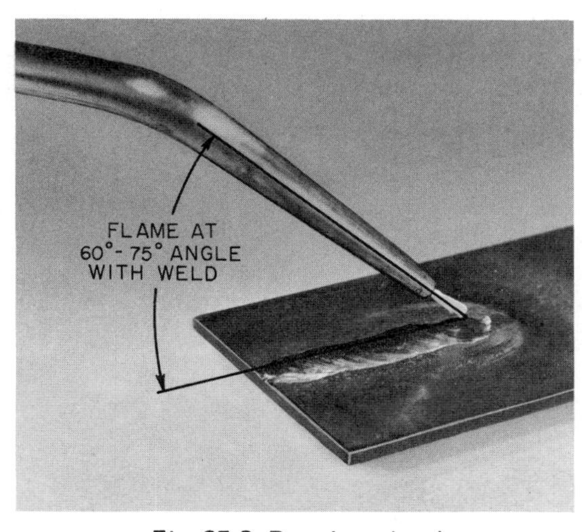

Fig. 25-2 Running a bead

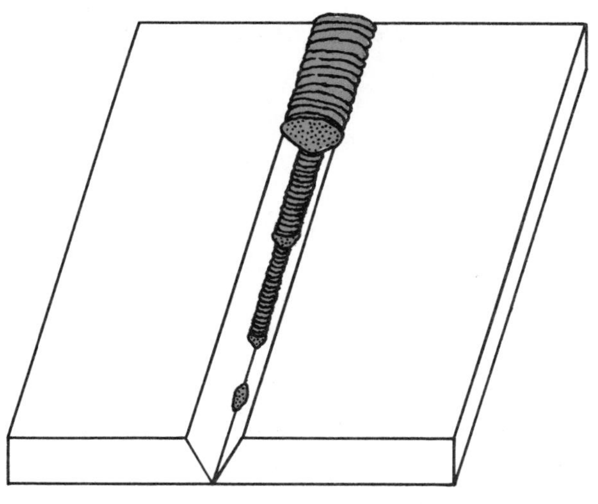

Fig. 25-3 Multilayer welding

If there is a tendency to burn through at the bottom of the V, try a smaller flame angle. If the bead is irregular and the fusion is poor, try a greater flame angle.

4. Apply the second bead as shown in figure 25-3. Increase the flame angle and weave the torch slightly. Apply only enough rod to make a weld that is flat or somewhat concave.
5. Make the third bead, weaving the flame and rod alternately, as in step 3, unit 21. Control of the flame is important at this point so the fusion is complete along the top edges of the V. However, the melting must not be so great as to make the weld too wide. The weld should not be over 1/8 inch wider than the top of the opening.
6. Make two saw cuts across the finished weld to produce a sample 1 1/2 inch wide. Examine these sawed surfaces for complete fusion and absence of gas pockets or holes in the weld.
7. Break this sample by hammering it toward the face of the weld. Examine the break for lack of fusion, oxide in the weld, and gas pockets.

REVIEW QUESTIONS

1. What effect does too small a flame angle have on the first bead?

2. How can you overcome the tendency of the first and second beads to pile up or become convex?

3. How do the flame and rod motion, and the flame angle in this unit compare with those in unit 21?

4. How can poor fusion at the root of the weld be corrected?

5. Is the root spacing of the joint critical for good penetration? Why?

UNIT 26 BACKHAND WELDING ON HEAVY PLATE

All of the welds made so far have been made by the *forehand* method with the flame pointing in the direction of welding. In the *backhand* method, the torch and rod are held in the same position but the flame points opposite the direction of welding. The welding flame is directed at the completed portion of the weld. The welding rod is placed between the completed weld and the flame.

MATERIALS

Two pieces of 1/4-inch thick mild steel plate, 3 in. X 9 in. each
1/8-inch diameter steel welding rod
Airco® #7 welding tip or equivalent

PROCEDURE

1. Align and tack the plates as shown in figure 26-1.
2. Start a puddle at the end of the joint and proceed with the weld by the backhand method. Use a weaving motion similar to that in units 21 and 25.
3. As the weld progresses, observe the bead and alter the flame angle until the weld has good appearance.
4. Make joints of this type until welds are produced with uniform ripples and complete fusion throughout the entire length.

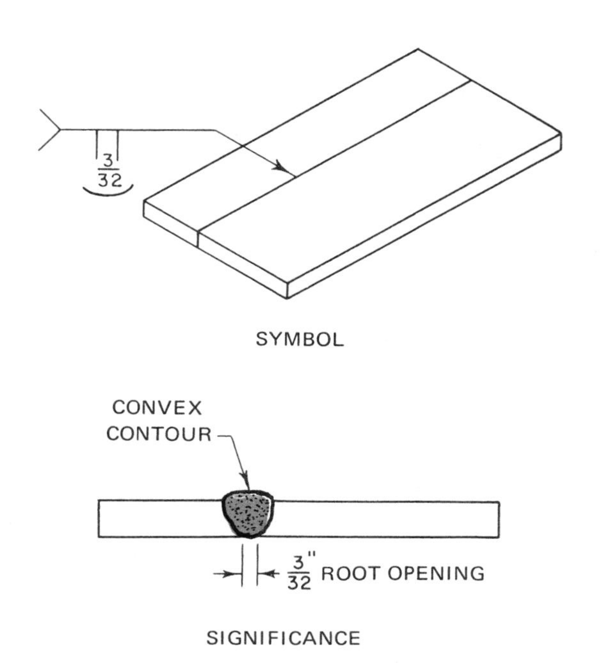

Fig. 26-1 Butt joint in heavy plate

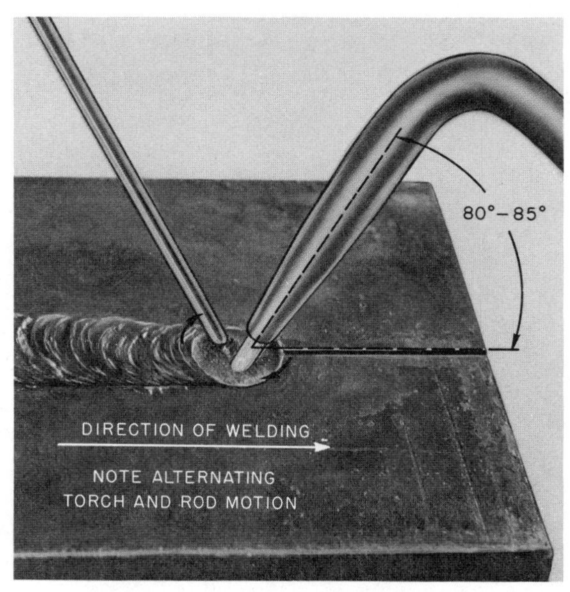

Fig. 26-2 Welder's eye view: backhand welding

5. As these welds proceed, observe the amount of penetration and the width of the beads. Compare these welds with those made in unit 21.
6. Make two saw cuts across the finished weld to produce a sample 1 1/2 inches wide. Examine these sawed surfaces for complete fusion and absence of gas pockets in the weld.
7. Break this sample by hammering it toward the face of the weld. Examine the break for lack of fusion, oxides in the weld, and gas pockets.

REVIEW QUESTIONS

1. How does the width of a backhand weld compare with the width of a forehand weld on a plate of similar thickness?

2. How does the flame angle compare with forehand welding?

3. Sketch a lengthwise cross section of this type of weld showing the finished bead, the rod and flame penetration.

4. What is the advantage of backhand welding over forehand?

5. Which type of welding is faster, forehand or backhand?

UNIT 27 BACKHAND WELDING OF BEVELED BUTT JOINTS

Backhand welding of beveled butt joints is used to great advantage in oxyacetylene welding of steel pipe. This unit provides practice in making this type of joint. Although this joint is not easy to make, it is much less difficult than a similar one in pipe. This joint should be mastered before pipe joints are welded.

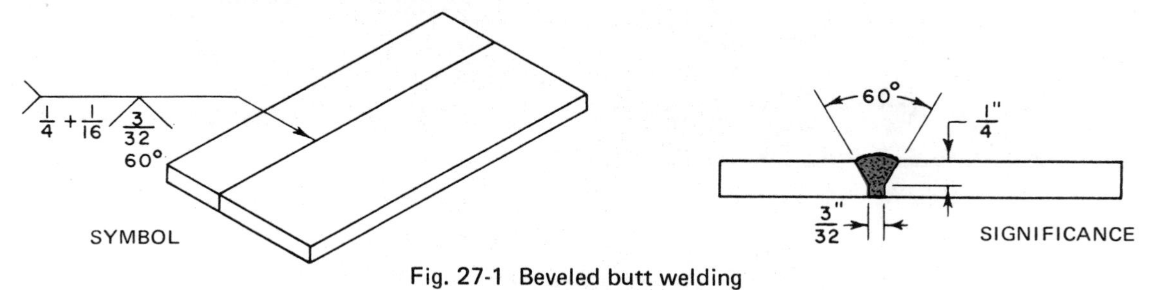

Fig. 27-1 Beveled butt welding

MATERIALS

Two pieces of 5/16- or 3/8-inch thick mild steel plate, 3 in. X 9 in. each
3/16-inch diameter steel welding rod
Airco® #7 or #8 welding tip or equivalent

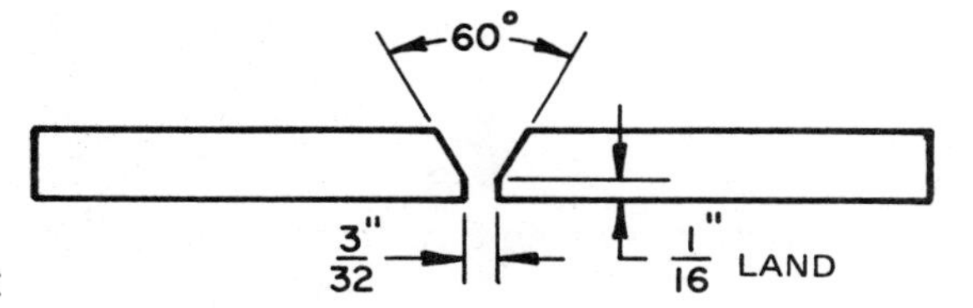

Fig. 27-2 Cross section of beveled plates

PROCEDURE

1. Bevel the plates as indicated in figure 27-1.
2. If equipment is not available to produce the *land* shown in figure 27-2, bevel the plates to a feather edge and grind them to obtain the amount of land indicated. The land is the vertical part of the opening.
3. Align and tack the plates as shown.
4. Weld the root by the backhand method so that fusion is complete. This is done by using the proper flame angle and rod application. The flame angle must be steep enough to insure good penetration. The rod may have to be dipped in and out of the puddle to allow the heat to melt the root of the base metal.
5. Make the second pass, using the motion described in unit 21. Be sure the bead is completely fused, and that it is not over 1/8 inch wider than the top of the original groove, figures 27-3 and 27-4.
6. Cut a section from this weld and examine it for penetration and fusion.
7. Weld additional plates with one pass only on each bead.
8. Cut a section from each of these joints and inspect them as in step 6. Check especially for penetration, fusion, and bead appearance.

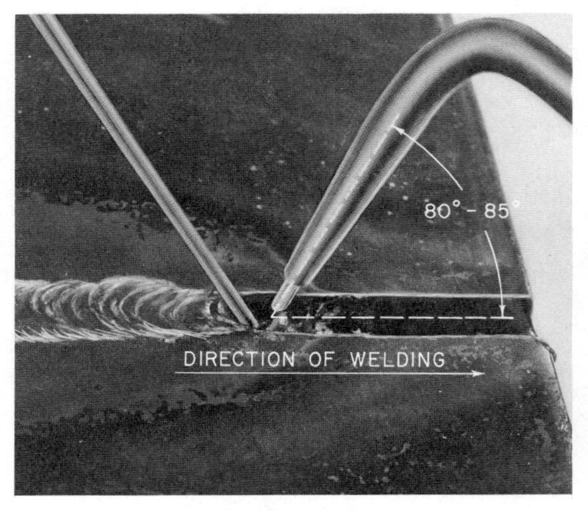

Fig. 27-3 Welder's eye view

Fig. 27-4 Beveled butt welding – backhand method

REVIEW QUESTIONS

1. What should be the shape of the surface of the root pass when making the joint at step 4?

2. Why does step 5 caution that the face of the bead must be kept narrow?

3. What is the difficulty in making this joint with a single pass?

4. Does it take longer to make the joint by the forehand method or the backhand method?

5. Will the welder feel more heat or less heat in backhand welding?

UNIT 28 BRAZING WITH BRONZE ROD

Brazing is a process in which metals are joined at a temperature greater than 800 degrees F. The base metal has a melting point at least 50 degrees F. higher than the filler rod. This indicates that:

- The base metal is not melted during this process.
- The joint is held together by the adhesion of the brazing alloy to the base metal rather than by cohesion. Cohesion takes place when the base metal and filler rod are fused.
- A brazed joint is bonded rather than welded.

This unit defines the techniques involved in oxyacetylene torch brazing using bronze filler rods. The action of flux and bronze during the brazing process will be examined.

MATERIALS

1/16- to 1/8-inch thick clean steel plates, 2 in. X 9 in. each
1/8-inch diameter bronze rod
Welding tip one size larger than for welding a similar plate
Suitable dry-type brazing flux

PROCEDURE

1. Place a piece of *clean* steel plate on the welding bench so that one end overhangs the bench.

   Note: The word clean refers to steel which has all the mill scale or iron oxide removed by either chemical or mechanical means. Mill scale and rust makes the production of strong joints either very difficult or impossible.

2. Sprinkle some flux on this plate. Apply the flame to the bottom side of the plate until the flux melts and flows over the plate.
3. Observe the color of the plate at this temperature. Also note that the flux flows freely. These two factors are the best guides to the proper brazing temperature.
4. Cool the plate and observe the metal under the flux. Compare the color of this metal to that of the unheated part of the plate. Note that the fluxed part of the plate is much whiter in color, indicating that the flux has cleaned the metal. The primary purpose of the flux is to chemically clean the surface for the brazing alloy. A secondary purpose is to protect the finished bead from the atmosphere during the brazing and cooling period.
5. Put some flux and a drop of brazing alloy on the end of an overhanging plate and heat as before. Observe that the flux melts first, then the alloy.
6. Move the flame about on the bottom of the plate. Note that the alloy flows freely in all directions as long as the flux is flowing ahead and cleaning the metal.

Note: The process of adhering a thin coating of bronze or some other metal to the surface of the base metal is called *tinning.*

7. Continue to apply heat. Note that the alloy starts to burn with a greenish flame, first in small spots and then in wider areas. At the same time, the alloy gives off a white smoke and leaves a white residue on the plate.

   Note: This residue is the zinc being overheated to the point where it evaporates and burns, to form zinc oxide. This heating is harmful to the brazed joint because the alloy is changed by the removal of the zinc which is replaced to some extent by the zinc oxide.

CAUTION: Breathing of zinc oxide may cause the operator to become violently ill.

8. Place a drop of the brazing alloy on the end of another overhanging plate and heat as before but do not use flux. Note that the alloy does not spread over the plate as before. Instead, it tends to vaporize and burn when the melting point is reached. This indicates that a brazed joint cannot be made without a dry-type brazing flux or one of the paste-type fluxes.

REVIEW QUESTIONS

1. What is the major difference between a brazed joint and a welded joint?

2. It is possible to braze brass or bronze if the proper alloy is used. What two conditions determine whether the joint is brazed or welded?

3. How does the flux act as a guide to the temperature of the joint?

4. From observations of the flowing characteristics of this alloy, what precaution should be taken to insure that the bead does not become too wide?

5. Can brazing create a toxic atmosphere for the welder? Explain.

UNIT 29 RUNNING BEADS WITH BRONZE ROD

This job provides practice in flame movement so that the operator develops skill in making good beads of a given size and shape with bronze rod.

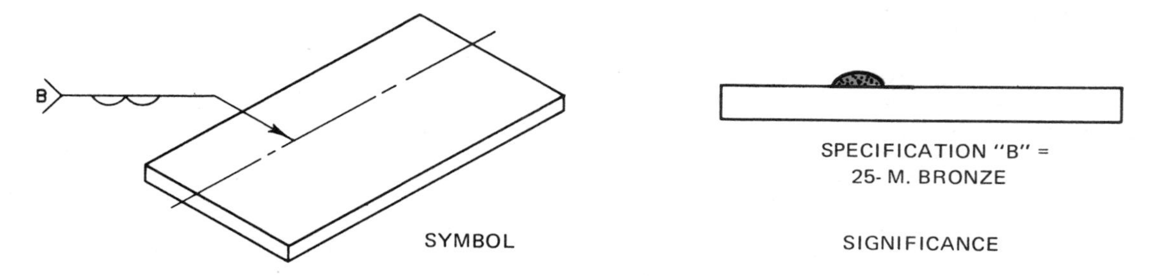

Fig. 29-1 Bead with bronze rod

MATERIALS

16- or 20-gage steel plate, 2 in. X 9 in.
1/8-inch diameter bronze rod
Suitable brazing flux
Welding tip one size larger than for welding on similar plate

PROCEDURE

1. Adjust the flame according to instructions. The flame for brazing varies with the alloy being used from slightly carburizing through neutral to slightly oxidizing.
2. Heat the rod with the flame until the flux clings to the rod when it is dipped in the flux container.
3. Apply the flame to both the rod and the work until a drop of alloy is left on the work. Remove the rod from the flame and continue to heat the plate until the drop melts and flows over an area of about 1/2-inch diameter.
4. Start a bead lengthwise on the plate. Keep the rod close to the flame and move both the rod and the flame in a spiral. Both the rod and flame are alternately close to the work and far away.

   Note: Check the flame angle. It is usually less than that used to weld a plate of similar size.
5. Bring the rod and flame in contact with the work when they are on the downswing of the spiral motion.
6. Drag the rod in the direction of the brazing before removing it for the upswing. This draws the flux ahead of the molten alloy and speeds the cleaning process.
7. Continue with the brazing and note that the flux flows ahead of the alloy. When this no longer happens, dip the still hot rod in the flux and continue with the bead.

8. Inspect the finished bead for width, height, ripples, and for the white residue that indicates overheating.
9. Make more beads, using these variations in procedure:
   a. Change the size of circle made by the flame and rod.
   b. Apply the flame to the work for longer and shorter intervals of time.
   c. Increase and decrease the amount of flux on the rod.
10. Inspect the finished beads.

REVIEW QUESTIONS

1. What effect does too much or too little heat have on the appearance of the finished bead?

2. What effect does too much or too little flux have on the ease of brazing?

3. Is temperature control in the base metal more critical in brazing than in welding? Explain.

4. Is it possible to hold the inner cone of the flame farther away from the joint than when welding in order to make a narrower braze? Explain.

5. What is the color of the plate when it is at the proper temperature for brazing?

UNIT 30 SQUARE BUTT BRAZING ON LIGHT STEEL PLATE

This job gives practice in making butt brazes in light steel plate. Although fusion welding and brazing are two different processes, they have many points in common. Thus, it is possible to use much of the experience gained in welding to produce braze joints.

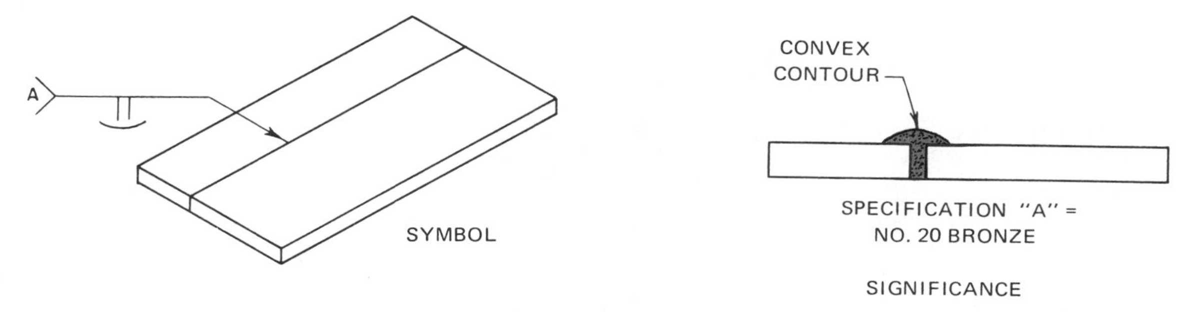

Fig. 30-1 Square butt braze

Two pieces 1/16- to 1/8-inch thick steel plate, 1 1/2 to 2 in. X 9 in. each
1/8-inch diameter bronze rod
Suitable dry brazing flux
Welding tip one size larger than for welding comparable steel plate

PROCEDURE

1. Align the plates and tack with bronze. Be sure the edges of the plates are in contact along the entire length of the joint.
2. Adjust the flame for the alloy being used.
3. Braze as in unit 25. Try to keep the bead narrow.
4. Hold the plates in a vise and bend them until they break, following the procedure for welded joints.
5. Examine the broken joint for evenness and depth of bond.
6. Tack more plates, spacing the edges slightly farther apart. Braze as before.
7. Break these plates and examine the results.

REVIEW QUESTIONS

1. What effect does plate edge spacing have on the depth of bond?

2. What is the procedure for getting a greater bond depth?

3. Is any tinning apparent on the reverse side of the joint brazed in this unit? If not, what conditions are necessary to obtain such tinning?

4. Is the square butt joint a good type of brazing joint? Why?

5. Is brazing stronger than fusion welding?

UNIT 31 BRAZED LAP JOINTS

This job develops manipulative skill in making brazed lap joints on steel plate. Some of the procedures, problems and difficulties encountered are the same as those in welding lap joints in light steel.

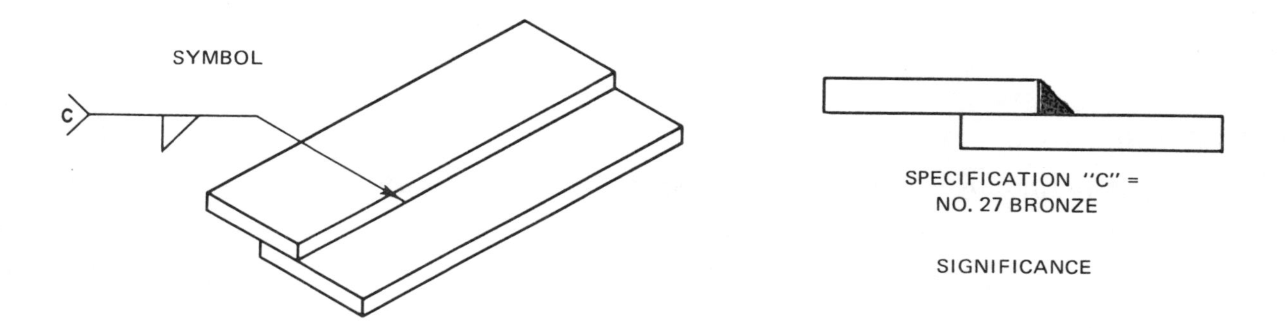

Fig. 31-1 Brazed lap joint

MATERIALS

Two pieces of 1/16- to 1/8-inch thick clean steel plate, 2 in. X 9 in. each
1/8-inch diameter bronze rod
Suitable flux
Welding tip one size larger than for welding comparable plate

PROCEDURE

1. Lap the plates following the procedure outlined for welding a lap joint, unit 18.
2. Adjust the flame for the alloy being used.
3. Proceed with brazing and observe the tendency of the alloy to flow on the top plate.
4. Change the angle the flame makes with the line of brazing to correct the tendency to over braze the top plate, figure 31-2.
5. Continue to make these joints until enough skill is acquired to make brazed lap joints with a good appearance.

   Note: Try to make all brazed beads about the same width as a weld made on material of similar thickness. The tinning action of the alloy causes the braze to become too wide if flame movement is not carefully controlled.

6. Break and examine these joints using the same procedure as that used to break welded lap joints.

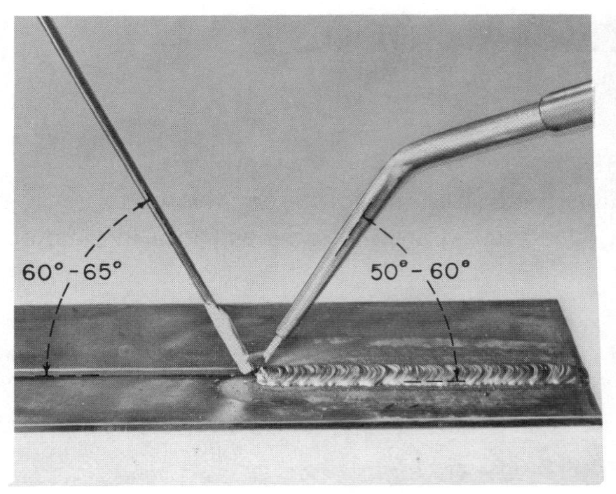

Fig. 31-2 Welder's eye view

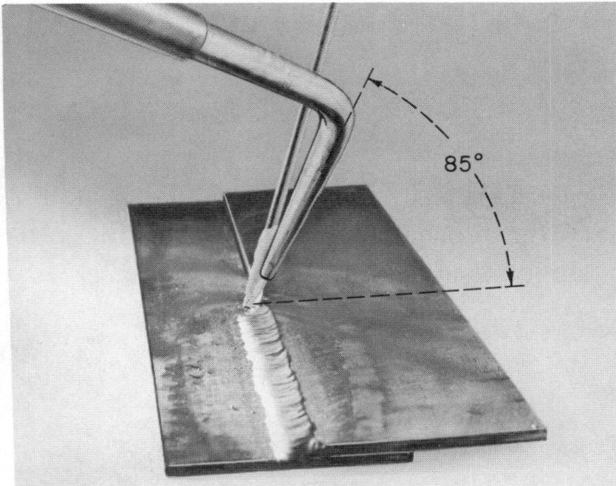

Fig. 31-3 Brazed lap joint

REVIEW QUESTIONS

1. How critical is the flame angle as compared to that used when welding a similar joint?

2. What effect does holding the inner cone too far from the work have on the width of the finished bead?

3. When brazing a lap joint, is there a tendency for the alloy to flow between the plates?

4. How does the grain size of the break in the brazed joint compare with that of a welded lap joint?

5. What is the strongest type of brazed joint?

UNIT 32 BRAZED TEE OR FILLET JOINTS

This job helps the student obtain skill in the technique of making strong joints of good appearance using bronze filler metal.

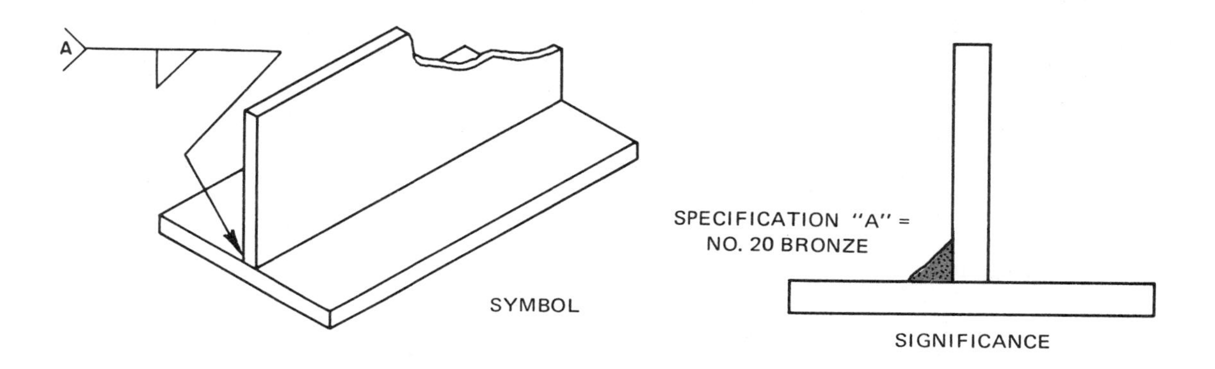

Fig. 32-1 Brazed tee or fillet joint

MATERIALS

Two pieces of 1/16- to 1/8-inch thick clean steel plate, 2 in. X 9 in. each
1/8-inch diameter bronze rod
Flux
Welding tip one size larger than for welding similar plate

PROCEDURE

1. Set up the plates as for welding, unit 19, except that the tacks are made with bronze alloy.
2. Adjust the flame.
3. Proceed with brazing in much the same manner as when welding.
4. Observe the tendency of the bronze to tin the upstanding leg of the joint over large area.
5. Correct the flame angle and the distance of the inner cone from the work until this excessive tinning is overcome. Figure 32-3 shows the correct flame angle.
6. Bend the finished joint in the manner used to test the welded fillet joint. Examine the root of the joint for uniformity of bond.
7. Braze the opposite side of the joint. Note the difficulty in achieving a good bond on the upstanding leg.

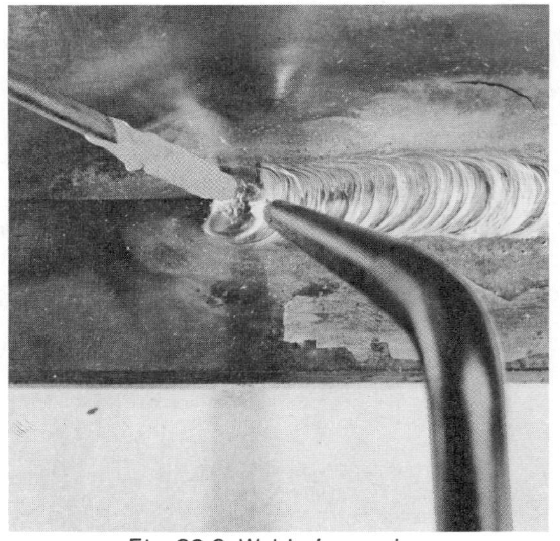

Fig. 32-2 Welder's eye view

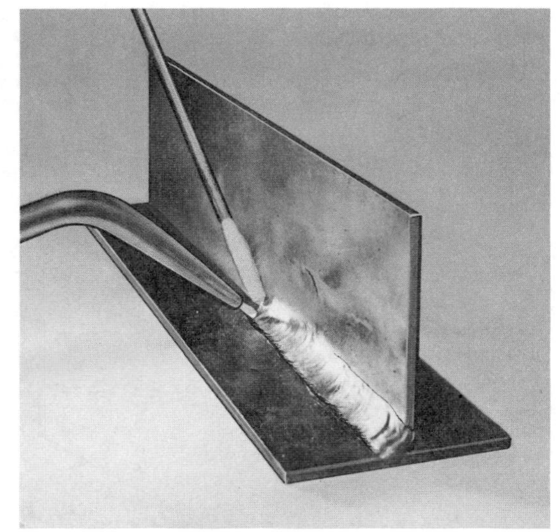

Fig. 32-3 Brazed fillet joint

REVIEW QUESTIONS

1. What factors make it difficult to obtain a good bond on the second side of the joint?

2. What preparation is necessary to obtain a good bond on the second side of the joint?

3. How does the flame for brazing the reverse side of the joint differ from that used on the first braze? Why is there a difference?

4. How should material be prepared for brazing?

5. What makes this joint more difficult to braze?

UNIT 33 BRAZING BEVELED BUTT JOINTS ON HEAVY STEEL PLATE

The lower temperatures used for brazing as contrasted with fusion welding make this process good for many jobs. Melting of the base metal is avoided, extensive preheating is not necessary, and expansion and contraction are not as severe problems as they are in fusion welding.

The student should acquire an understanding of brazing and will develop skill in this process. This unit provides an opportunity for brazing heavier steel than that used in previous units.

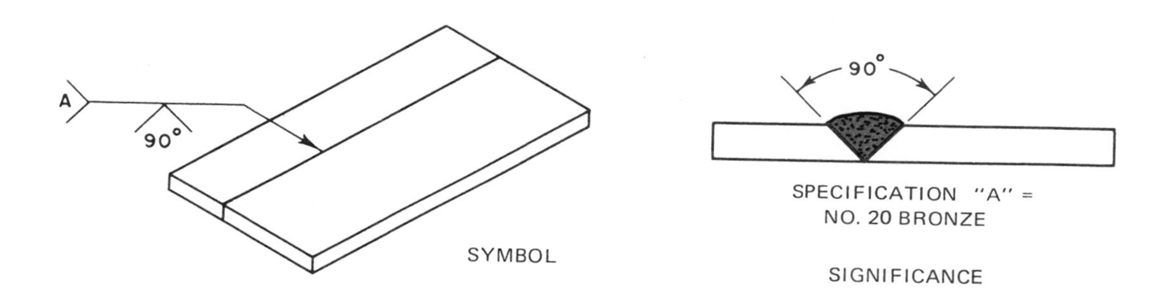

MATERIALS

Two pieces of 1/4-inch thick steel plate, 2 in. X 6 in. each
1/8-inch diameter bronze rod
Flux
Welding tip one size larger than for welding similar plate

PROCEDURE

1. Prepare the plate edges so that they make an angle of 90 degrees when they are brought together, figure 33-1.

   Note: If the plates are flame-cut, lightly grind or file the cut edges until all oxides are removed.

2. Align and tack the plates as for welding, figure 33-2.
3. Braze the joint. Be sure the plate edges are tinned to the root of the joint. Be careful not to overheat the tip edges of the plate as the brazing proceeds.
4. Make more joints of this type but apply the alloy in two layers. Apply the second layer with a weaving motion, alternating the rod and flame in the same manner as in welding heavy plate. When making a multiple-pass braze, be sure that each bead, except the finished bead, is slightly concave and that the alloy tins well up on the sides of the joint, figure 33-3.
5. Vary the angle of the flame and observe the results.

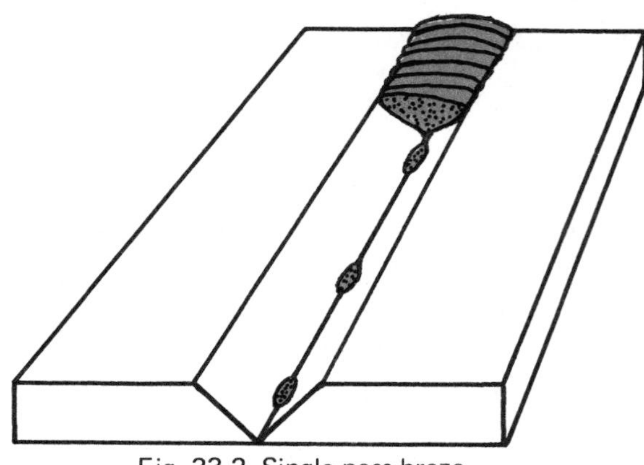
Fig. 33-2 Single-pass braze

Fig. 33-3 Multiple-pass braze

REVIEW QUESTIONS

1. What effect does the flame angle have on the penetration into the root of the groove?

2. What effect does the flame angle have on the appearance of the finished bead?

3. Why is the included angle for this joint 90 degrees instead of the 60 degrees used when welding a beveled butt joint?

4. Should the surface of the beads be flat, convex or concave?

5. Should the bottom sides of the material being brazed be cleaned? Why?

UNIT 34 BUILDING-UP ON CAST IRON

Worn areas of cast iron machine parts may be built up with bronze by the brazing process. The new surface may be machined if necessary. The resulting job is often as good as a new part.

Brazing on cast iron has advantages over fusion welding, largely because of the lower temperatures used. Brazing saves time, uses less gas, and involves less expansion and contraction of the metal. This unit gives practice in this process which is often called *bronze surfacing.*

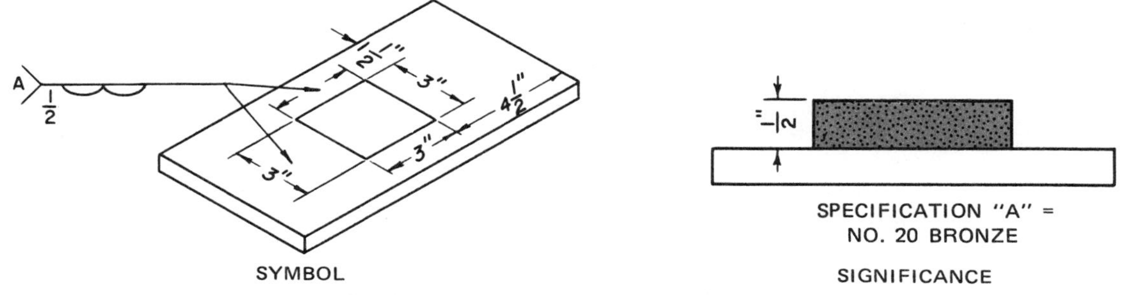

Fig. 34-1 Built-up cast iron using bronze

MATERIALS

Cast iron
1/8-inch diameter bronze rod
Suitable flux
Welding tip one size larger than that indicated for welding metal of similar thickness

PROCEDURE

1. Clean the surface to be brazed. Remove all rust, scale, and oil.
2. Preheat the cast iron with the flame before attempting to braze.
3. Build up an area about 3 inches square with the bronze rod using beads 3/4 inch wide, figure 34-1. Make sure that each bead is fused into the preceding bead. Use a standard dry flux for this operation.

   Note: If a Hi-bond® cast iron brazing flux or its equivalent is available, use this in with the standard flux. Note the ease of tinning on the cast iron surface.
4. Make the first layer about 1/8 inch high. Apply a second bead over the first as soon as the work has been inspected and before the plate has a chance to cool.
5. Apply a third and fourth bead. Inspect each bead and correct any faults when making the next bead.
6. Grind one edge of a piece of cast iron 1/4 inch thick and 6 inches long until it is square and clean. Stand this plate on edge so that a bead can be applied to the ground edge, figure 34-2.

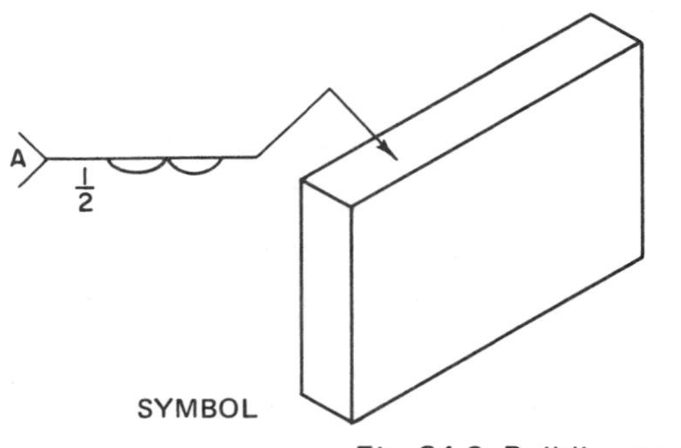

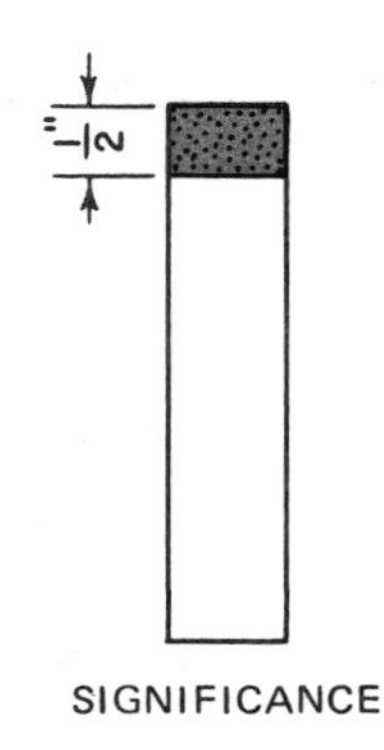

Fig. 34-2 Building-up a ground edge

7. Apply a bead along this edge, moving the flame and rod so that the bronze flows to each side of the plate but does not overhang.
8. Inspect the finished bead for uniformity of thickness and ripples. Note the presence or absence of overhang.
9. Deposit additional beads over the first until the edge is built up to a height of 1/2 inch above the original plate, figure 34-2. Check for uniformity.

REVIEW QUESTIONS

1. How do the tinning characteristics of cast iron compare with those of the same thickness of steel?
2. Why is preheating of cast iron necessary?
3. Draw a sketch of the cross section of the bead and casting in step 8.
4. Can cast iron be too hot to braze? What happens?
5. Is there an advantage to using stringer beads in brazing instead of wide beads? Why?

UNIT 35 BRAZING BEVELED JOINTS ON CAST IRON

The quality of a brazed joint on cast iron is greatly affected by the surface preparation. The surfaces must be clean to allow good tinning and to get a strong bond. The two surfaces to be joined must have enough area to provide a strong joint.

Testing the brazed joint to destruction demonstrates some of the qualities of a good joint. This unit provides an opportunity to make and test brazed joints on cast iron.

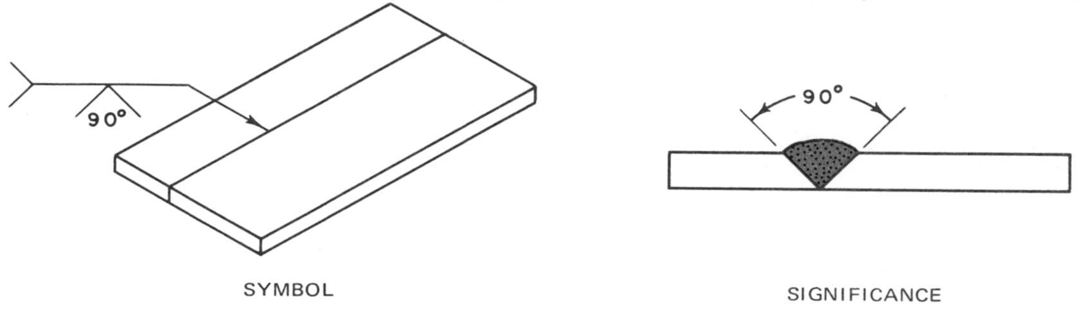

MATERIALS

Two pieces of cast iron 1/4 inch or thicker, 2 in. X 6 in. each
1/8-inch diameter bronze rod
Suitable flux

PROCEDURE

1. Grind or machine one edge of each of the two pieces of cast iron to 45 degrees. When the two pieces are aligned, the included angle of this opening should be 90 degrees, figure 35-1.

   Note: The included angle for brazing is normally made larger than that for welding the same size material. The reason is that in a bonding process such as brazing, the wider V presents more bonding surface for the brazing alloy; as a result a stronger joint is formed.

2. Align the pieces and preheat them by playing the flame over the joint until the work becomes dull red in color.
3. Sprinkle some Hi-Bond® flux along the joint and tack each end of the assembly.
4. Braze in a manner similar to that in unit 33. Be sure that good tinning action takes place along the sides of the V. Do not try to complete the braze in one pass, figure 35-2. Attempts to build up too great a thickness in one step usually result in poor tinning and a weak joint.
5. After inspecting the braze for appearance, cut the piece in two and place each piece in a vise with the brazed joint slightly above the vise jaws. Bend one piece toward the root of the braze and one toward the face of the braze, figure 35-3.

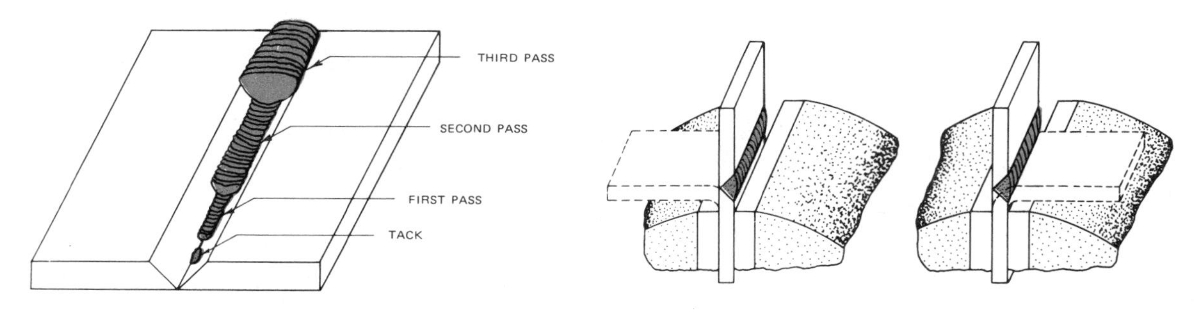

Fig. 35-2 Multilayer brazed joint

Fig. 35-3 Testing the braze

6. After these pieces break, inspect the break for a good bond. This is indicated by a coating of bronze on the cast iron and small particles of cast iron clinging to the bronze.
7. Prepare two more pieces. After beveling, draw a coarse file over the beveled surfaces to roughen them. This presents a surface of greater area to the brazing alloy. In addition, any free graphite which may be left on the beveled surface by the grinding operation is removed.
8. Braze this joint and test as before. Compare the bond strength of these plates with the first set tested.

REVIEW QUESTIONS

1. What prebrazing operation is most important to insure success in making a brazed joint?

2. What is the effect if brazing is attempted on a surface coated with free graphite?

3. When the joints made in steps 4 and 7 are tested, is there any difference in the amount of bending necessary to break the joints? Which joint requires the most bending? Why?

4. What color should the base metal have for good brazing?

5. Why is the included angle of the brazed bevel joint wider than for welding?

UNIT 36 SILVER SOLDERING NONFERROUS METALS

Silver soldering or *silver brazing* is, in reality, low-temperature brazing. A typical alloy, such as the one used in this unit, contains 80 percent copper, 15 percent silver, and 5 percent phosphorus. This type of alloy is effective at temperatures well below the melting point of brass and copper, the metals most commonly silver-soldered.

A number of different alloys with various melting points and fluxes is available.

The development of skill in silver soldering greatly increases the scope of the work which a welder can undertake.

MATERIALS

Two pieces of strip brass or copper 1/16-inch thick, 1 in. X 6 in.
Two pieces of brass or copper tubing, one of which just slips inside the other.
Handy-Flux® or equivalent
Sil-Fos® brazing alloy or equivalent, 1/16 in. X 1/8 in. X 14 in.
Airco® #3 welding tip or equivalent

PROCEDURE

1. Prepare the strips to be brazed by wire brushing, rubbing with emery cloth or steel wool, or by dipping in an acid bath to insure absolute cleanliness.

   Note: All welding and brazing operations are more successful if attention is given to surface cleanliness. In the silver brazing operations, failure to observe the proper precautions results in joints of low strength and poor appearance. A bright surface does not necessarily mean a clean surface from a welding or brazing viewpoint. Surface oxides sometimes appear very bright.

2. Apply a thin layer of flux to both surfaces of the strips that are to make contact in the braze. Allow this fluxed area to extend somewhat beyond the area to be brazed.
3. Set up the work so that the lapped surfaces to be brazed are in close contact, with the ends lapped 1 inch.

   Note: Proper joint spacing is very important in silver brazing and silver soldering. The ideal clearance between the surfaces to be brazed is 1 1/2 thousandths of an inch (.0015 inch). If this clearance is kept, the joint develops its maximum strength. The strength of the finished joint falls off rapidly as the clearance is increased beyond .0015 inch. At the same time, the amount of expensive silver soldering alloy used rises at a rapid rate.

4. Apply a 2X or 3X flame to the work with a back-and-forth motion to heat the brazing area. Do not apply the flame directly to the brazing rod.
5. Heat the work until the alloy can be applied to the joint a short distance from the flame. If the work is hot enough, the heat is conducted into the alloy causing it to melt and flow into the joint by capillary action.

6. During the brazing operation, the flux serves as a temperature guide, as well as a cleaning agent.

   Note: When the heat is first applied, the flux dries and turns white. As the amount of heat in the work increases, this white powder starts to melt, forming small beads of molten flux on the surface. Further heating causes the flux to become more fluid and flow out over the work surface in a thin, even coating. When this condition is noted, the work is at the proper brazing temperature, 1,300 degrees F. Temperatures beyond this point tend to make the molten flux *crawl* or leave bare areas on the work surfaces which oxidize. The result is a poor bond and unacceptable appearance.

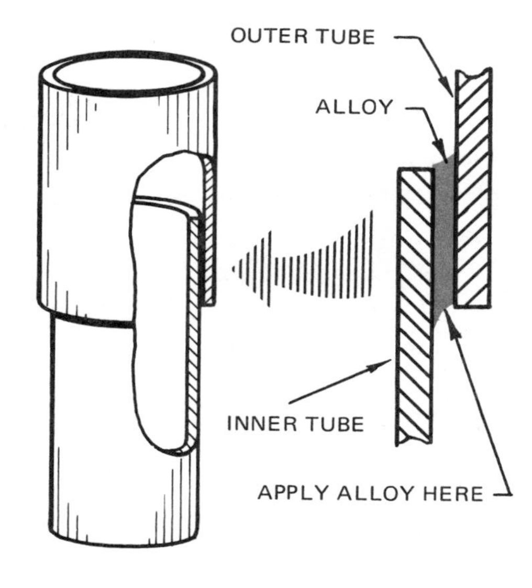

Fig. 36-1 Silver brazing tubes

7. Test the finished joint by trying to peel one of the lapped pieces from the other. If the braze has been properly made, the metal tears before the joint breaks.
8. Clean two pieces of copper or brass tubing inside and out. Make sure that one piece just slips inside the other, as in figure 36-1.
9. Place flux on both pieces and slide one into the other for a distance of 1 inch. Set this assembly on the bench or in a vise in a vertical position with the larger tube at the top.
10. Proceed with the brazing, making sure that all precautions outlined above are observed. Apply the alloy only on the outside of the joint. Make sure that the heating is complete.
11. Cool the finished work and note the thin, even line of alloy around the tube. Look into the tube and note that the alloy flowed by capillary action up between the two tubes and formed a small bead at the end of the inner tube. A further check can be made by hacksawing the joint diagonally and noting the white silvery appearance of the alloy between the entire lapped surfaces.

REVIEW QUESTIONS

1. If the base metal appears highly oxidized after the braze is completed, what factors must be checked before this condition can be corrected?

2. What are the factors that must be observed to produce good silver brazed joints?

3. What is a 3X flame?

4. What alloys are contained in typical silver solder?

5. How can material be prepared for silver soldering?

UNIT 37 SILVER SOLDERING FERROUS AND NONFERROUS METALS

Silver soldering, silver brazing, and low-temperature brazing are terms used for the same process in industry. To be technically correct, the process done at a temperature greater than 800 degrees F. is called *brazing.* The melting points for the different brazing alloys are:

Easy-Flo®	1,175° F.
Sil-Fos®	1,300° F.
Phos-Copper®	1,600° F.
Bronze	1,750° - 1,800° F.

The operations in this unit are done at a lower temperature than in the previous unit. The Easy-Flo® alloy is used with either ferrous or nonferrous metals.

All silver soldering or silver brazing operations require very small amounts of the alloy. Any attempt to use these alloys as welding rods or high-temperature brazing rods results in a weak and costly joint.

MATERIALS

Strips of steel, stainless steel, copper, and brass about 1/16-inch thick, 1 in. X 4 in. each
Handy-Flux®
Regular Easy-Flo® and #3 Easy-Flo® 1/16-inch diameter alloy
Airco® #3 welding tip

PROCEDURE

1. Prepare steel plates following the procedure of unit 36.
2. Set up two of the plates, lapping the ends about 1 inch to make an assembly 1 inch wide X 7 inches long. Be sure that the opposite end of the top plate is supported so that the two plates are kept parallel.
3. Braze the joint, using the same flame adjustment and technique as in unit 36. Apply the alloy (regular Easy-Flo®) by wiping the rod slowly on the one-inch face of the lap on the top side only. Apply enough heat to cause the alloy to flow between the lapped surfaces completely.
4. Cool the brazed joint and observe the opposite face of the lap and both edges. The alloy should be visible all around.
5. Try peeling one lapped plate from the other. This is impossible if the joint is properly made.
6. Set up two stainless steel plates and repeat the steps above.

   Note: The upper limit of temperature is very critical when brazing stainless steel. Overheating causes the formation of chromium oxide which can be removed only by filing or grinding. Flux does not remove chromium oxide.

7. Experiment with stainless steel joints by heating two of the plates until the color indicates that oxide has formed. Then proceed with the brazing as in previous joints and observe the results.
8. Check the joint of step 7 visually for complete bonding; peel one plate from the other.

   Note: Tensile strength tests made on properly brazed joints on stainless steel break outside the brazed area at a load of 100,000 to 120,000 pounds per square inch. The actual strength of the braze is somewhat higher than the above figures.
9. Set up two brass plates to make a fillet- or tee-type joint. Make sure both sides of the joint are covered with flux.
10. Braze this joint using #3 Easy-Flo®. Note that this alloy does not flow as freely as other types, but allows a slight fillet to build up. This is desirable in certain applications.
11. Cool and inspect the finished braze. In particular, check the draw through of the alloy on the reverse side of the joint.
12. Hammer the upstanding leg flat against the bottom plate. Note that the color of the alloy is very close to the color of the work. This is desirable in many jobs where color match is important.
13. Prepare two more plates but, in this case, make a butt joint. Test in the same manner as for a butt weld and observe the results.

REVIEW QUESTIONS

1. If the joint is overheated when low-temperature brazing stainless steel, what procedure must be followed to obtain a satifactory braze?
2. Is it possible to make fillets when using silver brazing alloys? Explain.
3. What is the proper clearance between the lapping surfaces for maximum joint strength?
4. How do butt-brazed joints compare with lapped joints?
5. What can be said about the amount of filler metal required for a given joint?

ACKNOWLEDGMENTS

The authors wish to express their appreciation and acknowledge the contributions of the following organizations for their assistance in the development of this text:

- The American Welding Society for permission to use and adapt the Chart of Standard Welding Symbols; and the illustration of Common Faults that Occur in Hand Cutting.
- Air Reduction Sales Company (AIRCO) for permission to use and adapt certain illustrations.
- Linde Air Products Division of the Union Carbide Corporation for permission to use and adapt certain illustrations.

The following members of the staff at Delmar Publishers assisted in the preparation of this edition:

Vice President-Editorial – Alan N. Knofla

Source Editor – Mark W. Huth

Director of Manufacturing/Production – Frederick Sharer

Production Specialists – Debbie Monty, Patti Manuli, Betty Michelfelder, Sharon Lynch, Jean LeMorta, Margaret Mutka, Lee St. Onge

Illustrators – Mike Kokernak, George Dowse, Tony Canabush

This material has been used in the classroom by the Oswego County Board of Cooperative Educational Services, Mexico, New York. Improvements in the text have in part been a result of feedback from these classes.

BASIC TIG & MIG WELDING

PREFACE

Tungsten inert-gas (usually called TIG) welding has become an important process, especially when it is necessary to do unusual jobs or work with difficult materials. The TIG process is actually an extension of electric arc welding. An intense arc is drawn between the work and a tungsten electrode, and an inert gas shields the weld zone.

In the metallic inert-gas (MIG) process, a consumable wire electrode is used to maintain the arc and to provide filler metal. As in TIG welding, an inert gas is fed into the weld zone to prevent contamination. *Basic TIG and MIG Welding* emphasizes basic metallurgy required for the thorough understanding of TIG and MIG processes, and stresses skill in the use of these processes.

The pattern of instruction employed is to teach students to master the techniques of TIG and MIG welding by having them perform specific welds on a variety of materials. The necessary fundamentals of the processes are covered, but the primary emphasis is on performance.

This text is divided into two sections: the first deals with TIG welding; the second concentrates on the MIG process. Each section is divided into units which furnish related information (use of equipment, variables, safety, data for welding metals) or provide procedures for welding. The theory parts of the text aid the students in completing the procedural units. Each procedural unit contains a list of materials and steps for performing the weld. Upon completion of the weld, students are instructed to make additional welds to strive for uniformity and consistency in appearance. A method for testing the weld is given in some units as a means to judge its quality. Since TIG welding is especially useful in welding aluminum, most of the procedural units in the TIG section concentrate on welds made on aluminum. Several units also provide the opportunity to weld less common metals such as stainless steel, magnesium and its alloys, copper and its alloys, and nickel and its alloys. At the end of every unit are review questions designed to help students draw conclusions and to check their progress.

CONTENTS

SECTION 1 TIG WELDING

SECTION 2 MIG WELDING

CHARTS

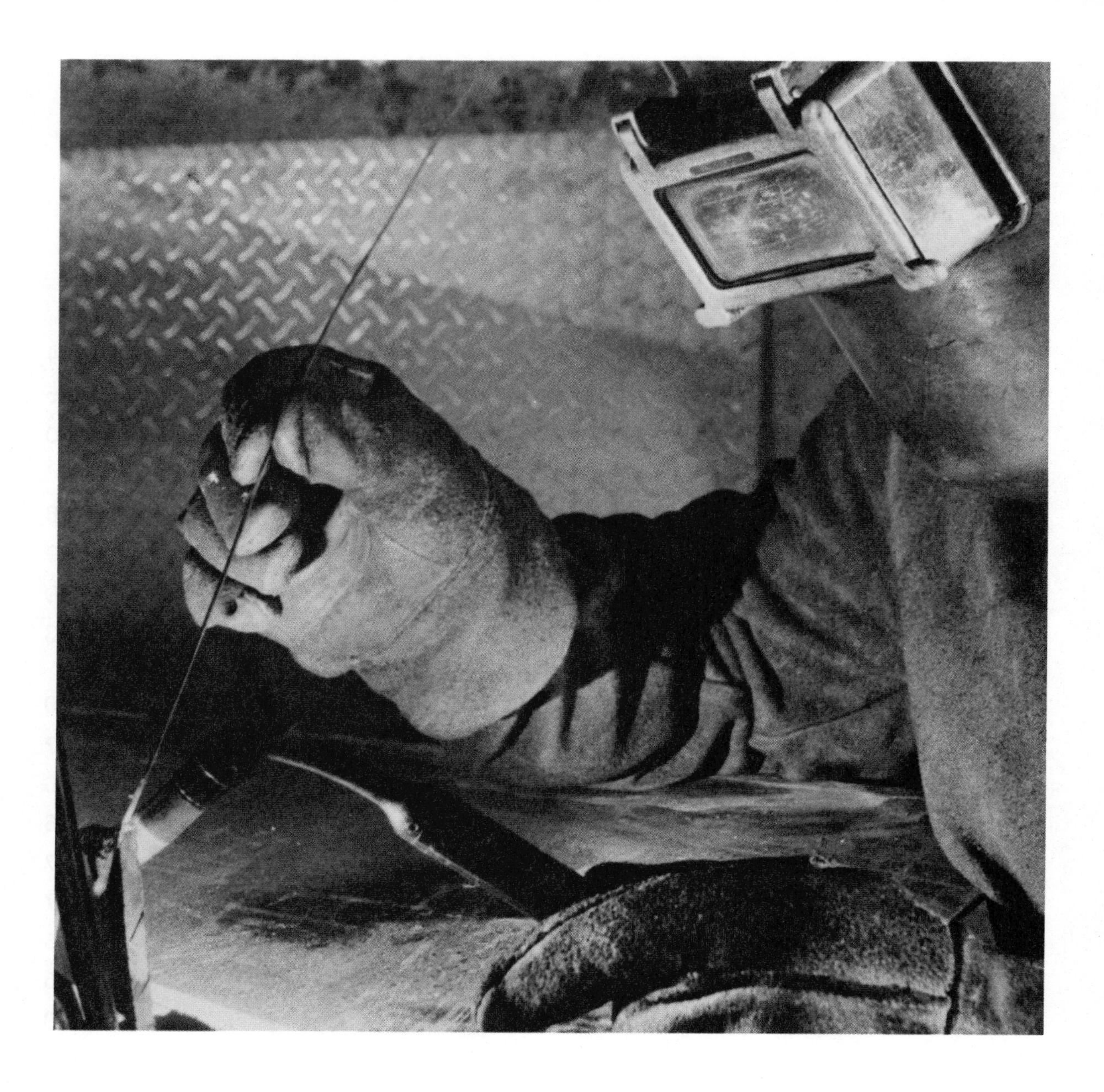

SECTION **1**

TIG WELDING

TIG is basically a form of arc welding. It is especially useful in welding aluminum. Developed in the period of 1940 to 1960, it has rapidly become one of the indispensable welding methods.

The equipment used is more complex and expensive than that used for arc welding because electricity, water and gas must all be provided and controlled. MIG (metallic inert-gas) welding is closely related to TIG welding.

UNIT 1 THE TUNGSTEN INERT-GAS SHIELDED-ARC WELDING PROCESS

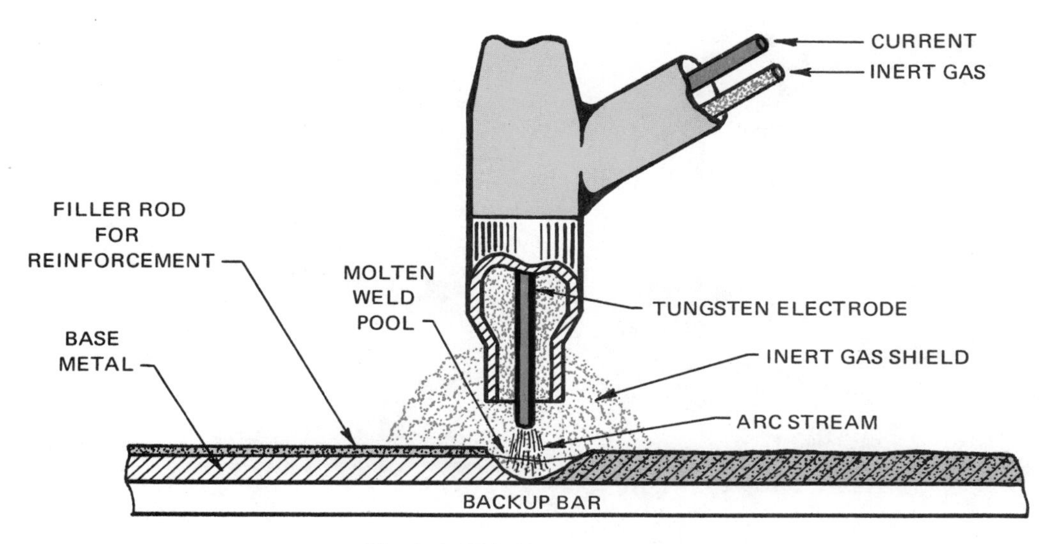

Fig. 1-1 TIG Welding process

The tungsten inert-gas shielded-arc welding process, figure 1-1, is an extension, refinement, and improvement of the basic electric arc welding process.

In the complete name of this process:

Tungsten refers to the electrode which conducts electric current to the arc.

Inert refers to a gas which will not combine chemically with other elements.

Gas refers to the material which blankets the molten puddle and arc.

Shielded describes the action of the gas in excluding the air from the area surrounding the weld.

Arc indicates that the welding is done by an electric arc rather than by the combustion of a gas.

The process is commonly referred to as *TIG welding* which is obtained from the first letter of each of the words, tungsten, inert, and gas. This type of welding is often referred to as Heliarc®, which is the trade name of a particular manufacturer. The TIG process generally produces welds which are far superior to those made by metallic arc welding electrodes.

ELEMENTS OF THE PROCESS

As shown in figure 1-2, the basic process uses an intense arc drawn between the work and a tungsten electrode. The arc, the electrode, and the weld zone are surrounded by an inert gas which displaces the air to eliminate the possibility of contamination of the weld by oxygen and nitrogen in the atmosphere. The tungsten electrode has a very high melting

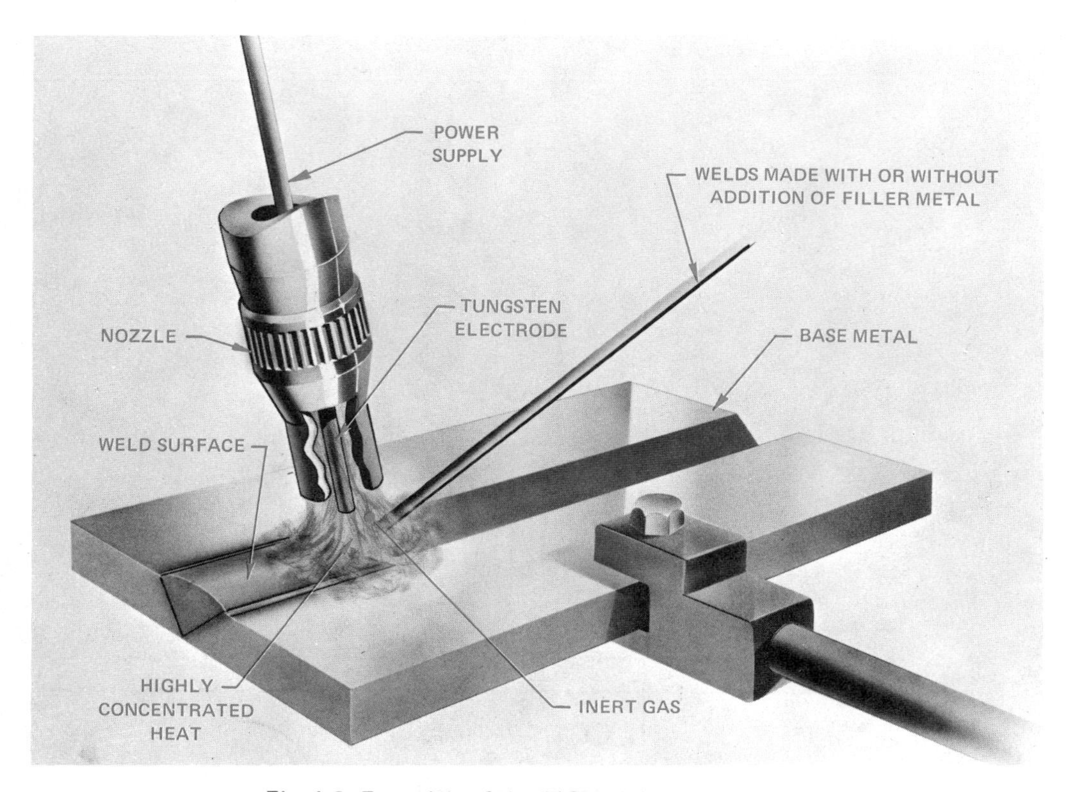

Fig. 1-2 Essentials of the TIG welding process

point (6,900 degrees F.) and is almost totally nonconsumable when used within the limits of its current-carrying capacity.

The inert gas supplied to the weld zone is usually either helium or argon, neither of which will combine with other elements to form chemical compounds. Argon gas is usually recommended because it is more generally available and better suited for use in the welding of a wide variety of metals and alloys. The basic components for a water-cooled TIG welding outfit are indicated in figure 1-3.

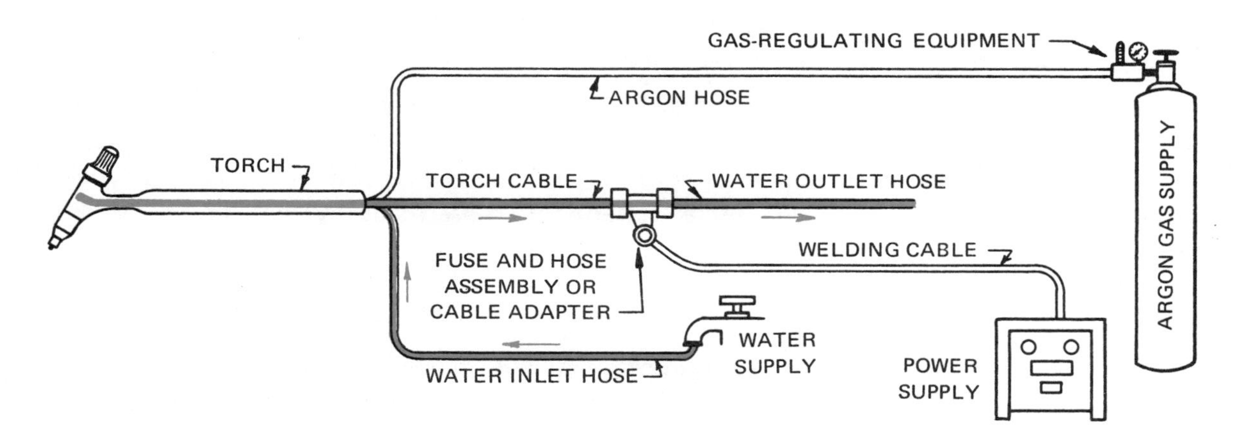

Fig. 1-3 Essentials for water-cooled TIG welding

ADVANTAGES OF TIG WELDING

Examples of the beads welded by arc, oxyacetylene and TIG processes are shown in figure 1-4.

- No flux is required and finished welds do not have to be cleaned of corrosive residue. The flow of inert gas keeps air away from the molten metal and prevents contamination by oxygen and nitrogen.
- In the chemical composition, the weld itself is usually equal to the base metal being welded. It is usually stronger, more resistant to corrosion, and more *ductile* (ability of a metal to deform without fracturing) than welds made by other processes. The inert gas will not combine with other elements or permit contamination by such elements, thus keeping the metal pure.
- Welding can be easily done in all positions. There is no *slag* (waste material entrapped in weld) to be worked out of the weld.
- The welding process can be easily observed. No smoke or fumes are present to block vision, and the welding puddle is clean.
- There is minimum distortion of the metal near the weld. The heat is concentrated in a small area and thus tends to minimize stresses.
- There is no splatter to cause metal-cleaning problems. Since no metal is transferred across the arc, this problem is avoided.

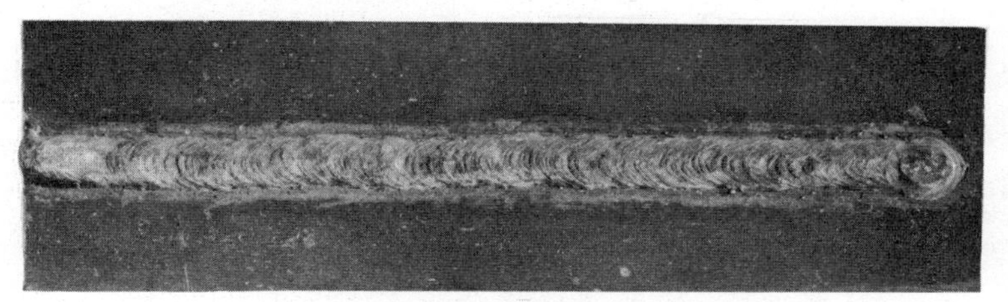

ARC WELDING

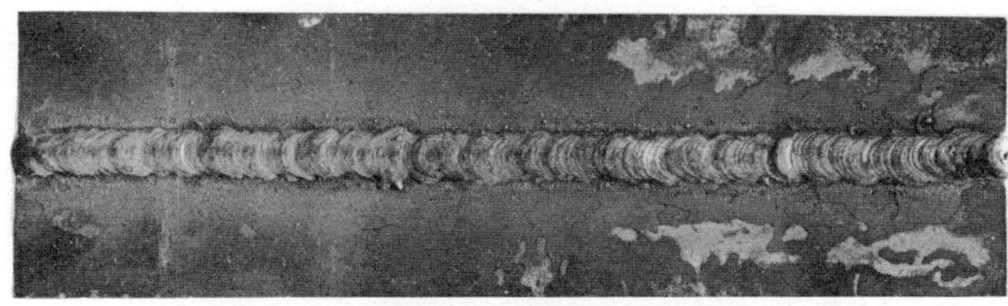

OXYACETYLENE WELDING

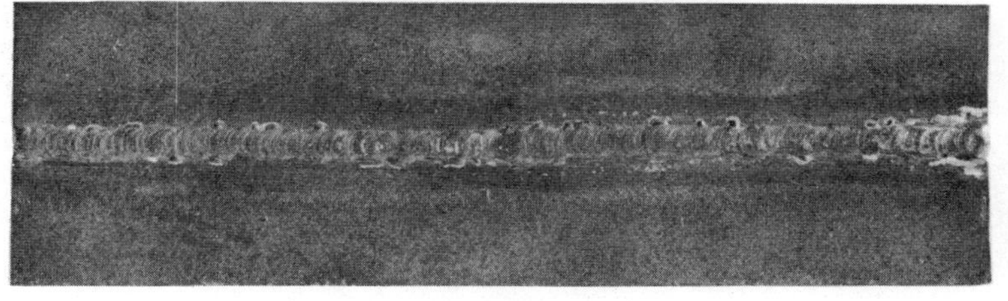

TIG WELDING

Fig. 1-4 Comparison of beads as welded

- Practically all the metals and alloys used industrially can be fusion-welded by the TIG process in a wide variety of thicknesses and types of joints.
- TIG welding is used particularly for aluminum and its alloys (even in very thick sections), magnesium and its weldable alloys, stainless steel, nickel and nickel-base alloys, copper and copper alloys, some brasses, low alloy and plain carbon steel, and the application of hard-facing alloys to steel.

REVIEW QUESTIONS

1. What does the term TIG welding refer to?

2. What are the essentials of TIG welding?

3. How does a TIG weld compare chemically with a metallic arc weld?

4. How do the mechanical properties of TIG welds compare with those of welds made by other manual processes and with the base metal?

5. What is meant by the term slag entrapment? Why is it harmful?

6. How do the uses of TIG welding compare with other manual processes?

7. Why is argon recommended as the shielding gas for most TIG welding?

UNIT 2 EQUIPMENT FOR MANUAL TIG WELDING

The equipment and material required for TIG welding consist of an electrode holder, or torch, containing gas passages and a nozzle for directing the shielding gas around the arc; nonconsumable tungsten electrodes; a supply of shielding gas; a pressure-reducing regulator and flowmeter; an electric power unit; and on some machines a supply of cooling water.

THE TORCH

A specially designed torch is used for TIG welding. It is so constructed that various sizes of tungsten electrodes can be easily interchanged and adjusted. The torch is equipped with a series of interchangeable gas cups to direct the flow of the shielding gas. Some of the torches are air-cooled, but water-cooled torches are more widely used.

SOURCE OF ELECTRIC CURRENT

The source of the electric current used in modern TIG welding is a specially designed welding machine, figures 2-1 and 2-2. It is possible to adapt standard alternating-current (AC) and direct-current (DC) welding machines such as are used in arc welding operations to TIG welding operations. However, a unit of this type is bulky and hard to manage when compared with the modern machines that are designed for TIG welding. An AC arc is best

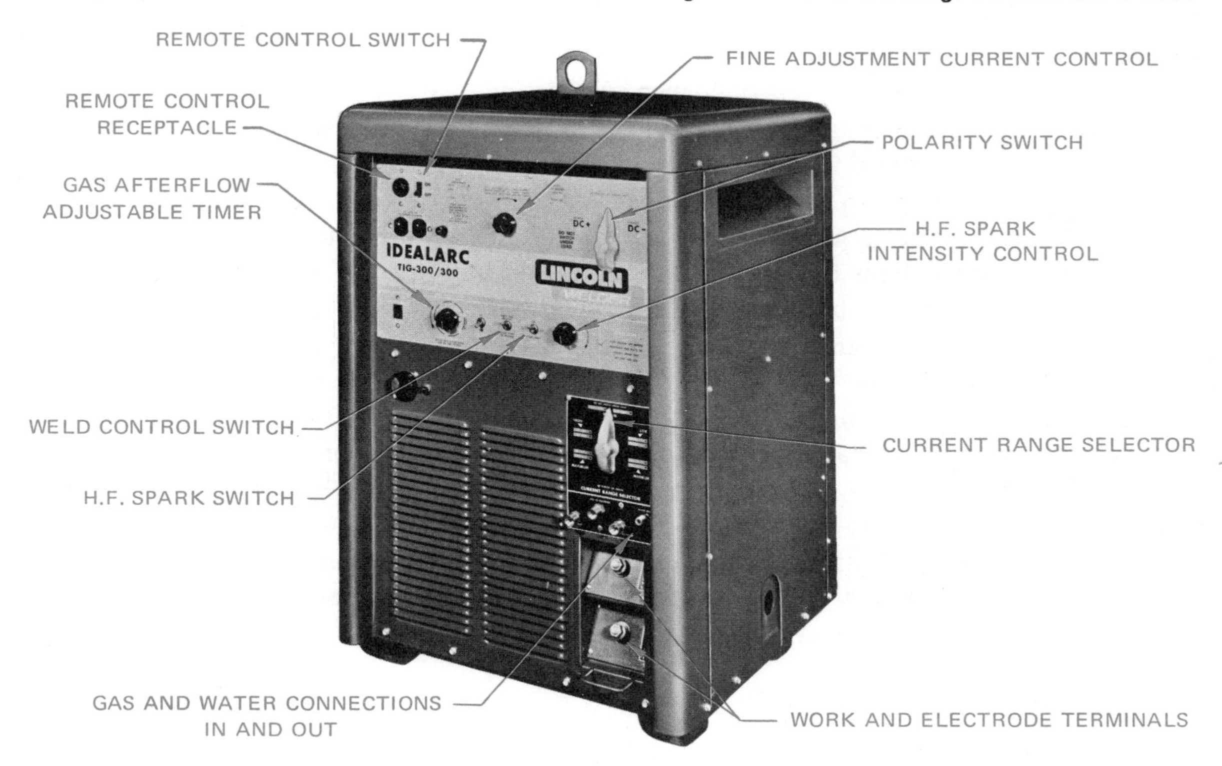

Fig. 2-1 AC - DC TIG welding machine

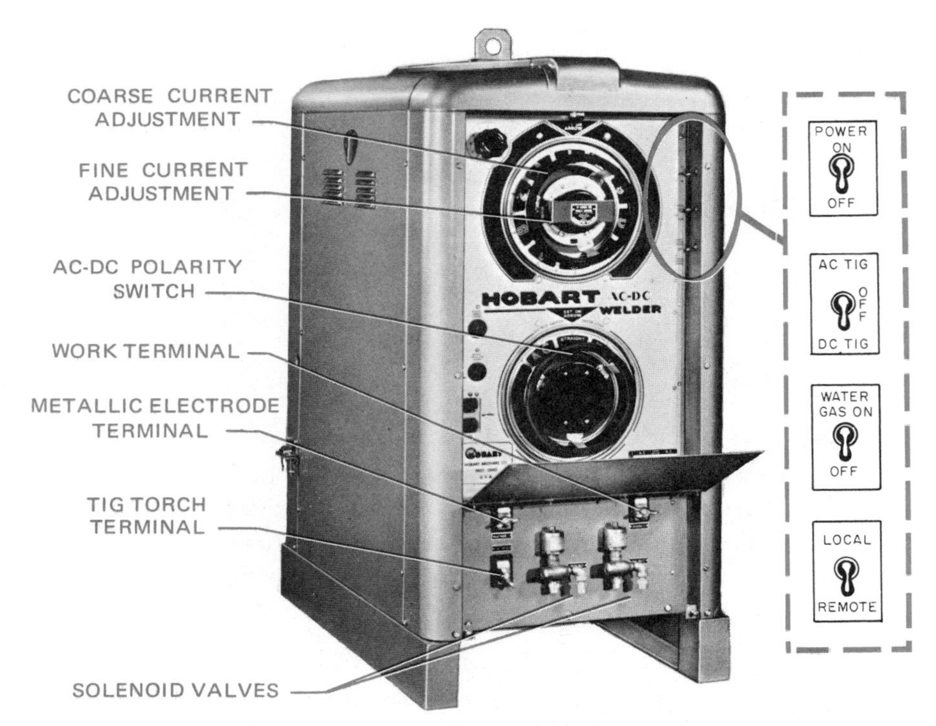

Fig. 2-2 AC - DC TIG welding machine

suited for aluminum and some other metals and alloys. But a standard 60-cycle alternating current, which changes its direction of flow 120 times a second, is unsuited for welding because the electrical characteristics of the oxides on these metals cause the arc to extinguish (go out) at every half cycle or change of direction.

If, however, an *igniter arc current* is added to the standard 60-cycle current, the tendency to extinguish will be overcome because the igniter current will maintain a path for the standard 60-cycle current to follow. This igniter current is usually generated within the machine by a spark-gap oscillator which causes the current to change direction, not 120 times a second, but millions of times each second. Because this frequency of change is very high, the term high frequency is used in describing this current. Since standard 60-cycle alternating current actually does the welding, it is called AC welding; hence the term, *high-frequency alternating current* welding or, as it is commonly referred to, HFAC.

TIG welding power units, in most cases, can supply either AC or DC power to the electrode. These welding machines are equipped with a high-frequency oscillator which injects the high-frequency igniter current into the welding circuit. This high-frequency current causes a spark to jump from the electrode to the work without contact between the two. This ionizes the gap and allows the welding current to flow across the arc. Some manufacturers provide an external means of varying the frequency of this alternating current. In other machines the housing must be opened to make adjustments.

CONTROLS

Figure 2-3 shows the control panel for an AC-DC welding machine. TIG power units are usually equipped with solenoid valves to turn the flow of shielding gas and cooling water

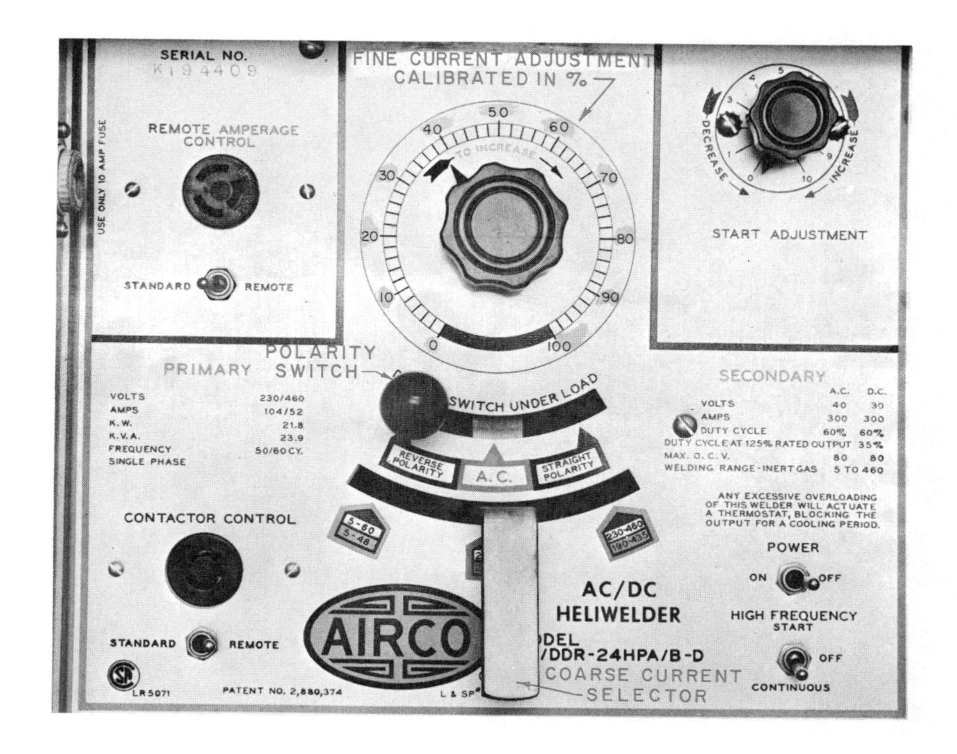

Fig. 2-3 Control panel for AC - DC welding machine

on and off. They are also provided with a remote-control switch, either hand- or foot-operated, to turn the water and gas on and off. Some of these remote-control devices also turn the main welding current on or off at the same time.

Most manufacturers equip the solenoid valves with a delayed-action device which allows the cooling water and shielding gas to continue to flow after the remote-control switch has been set at the stop or open position. This delay allows the tungsten electrode to cool to the point that it will not oxidize when the air comes in contact with it.

Some machines have an external means of varying the time of this afterflow to correspond to the electrode which is being used. In other types of machines, the housing must be opened to make this adjustment, figure 2-4. In any case, the shielding gas must be allowed to flow long enough so that the tungsten electrode cools until it has a bright, shiny surface.

SHIELDING GAS

The shielding gases are distributed in standard cylinders, which contain 330 cubic feet at 3,000 p.s.i.

As with all compressed gases, a regulator must be provided to reduce the high cylinder pressure to a safe, usable working pressure.

The main difference between the regulators used for oxyacetylene welding and those used for TIG welding is that the working pressure on the oxyacetylene regulators is indicated in pounds per square inch while the regulators used for TIG welding indicate the flow of shielding gas in cubic feet per hour. The latter are generally referred to as *flowmeters.* A combination regulator and flowmeter is shown in figure 2-5.

Fig. 2-4 Lower portion of machine: cover removed

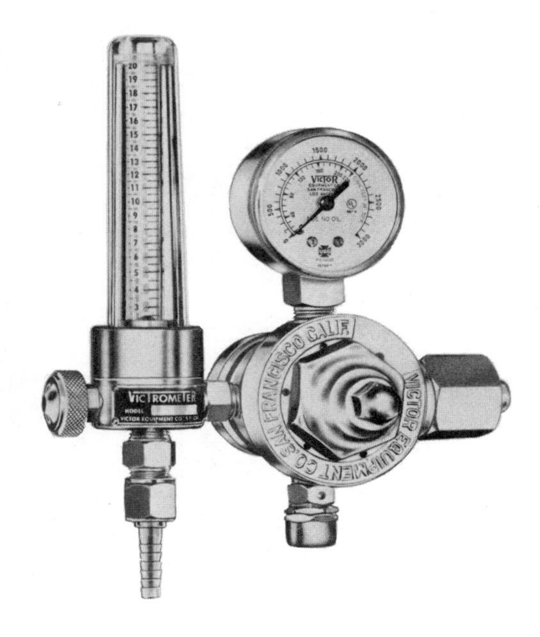

Fig. 2-5 Combination regulator and flowmeter

Another significant difference between a standard regulator and the flowmeter is that the regulator will indicate the working pressure to the torch regardless of the regulator's position, while the tube on the flowmeter must be in a vertical position if an accurate reading is to be obtained.

REVIEW QUESTIONS

1. How long should the shielding gas and cooling water be allowed to flow after the welding arc is broken?

2. How do modern TIG welding machines compare with earlier models?

3. What makes the modern TIG torch adaptable to a wide range of welding operations?

4. What precaution is used to install a flowmeter on a gas cylinder?

5. In the air-cooled TIG torch, what does the cooling?

UNIT 3 THE WATER-COOLED TIG WELDING TORCH

The TIG torch, figure 3-1, is a multipurpose tool. It serves as:

- A handle.
- An electrode holder.
- A means of conveying shielding gas to the arc.
- A conductor of electricity to the arc.
- A method of carrying cooling water to the torch head.

COOLING WATER FLOW

Figure 3-2 shows a cross section of a torch with the cooling water flow indicated. The water cools the torch head, the collet and the electrode; it also cools the relatively light welding current cable which will overheat and burn if it is not surrounded by cooling water at

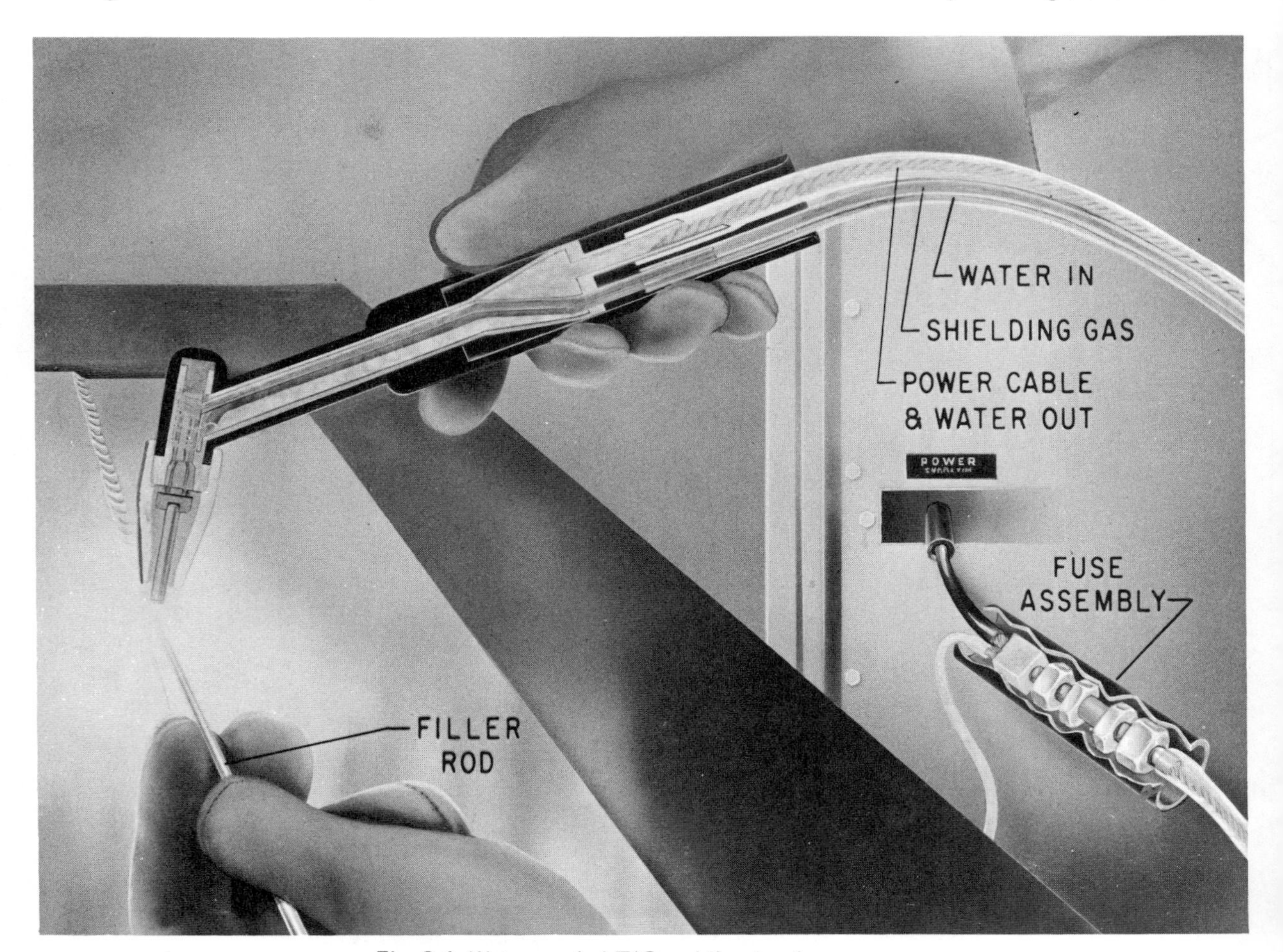

Fig. 3-1 Water-cooled TIG welding torch

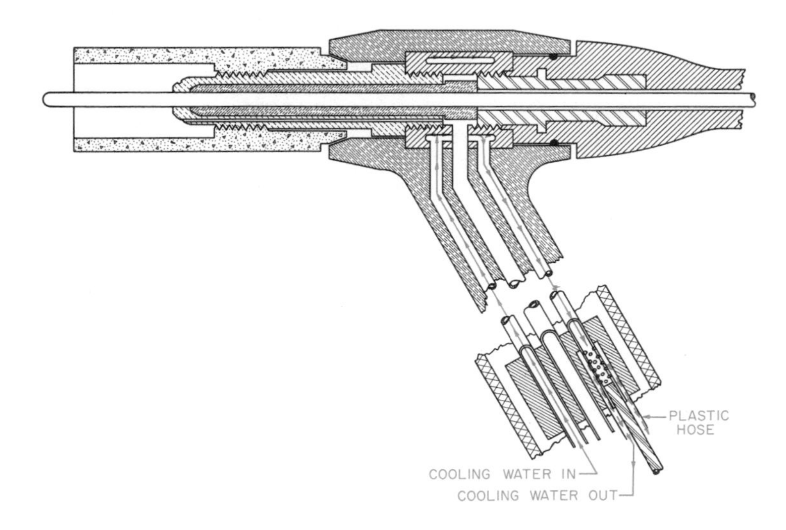

Fig. 3-2 Cooling water circuit

all times when current is flowing. All equipment manufacturers will supply cooling water requirements. A typical recommendation for cooling a medium-duty (300-amp.) torch is one quart of water per minute at 75 degrees F. or less, at not over 50 pounds per square inch pressure.

THE FLOW OF GAS

Figure 3-3 indicates the path of the regulated shielding gas through a hose to the torch head and through the collet holder. The gas then flows through a series of holes around the collet holder which direct the flow around the tungsten electrode through the ceramic nozzle to the work zone. The diameter and length of this nozzle vary with the size of the electrode used, the type of current being used, the material being welded, and the shielding gas being used.

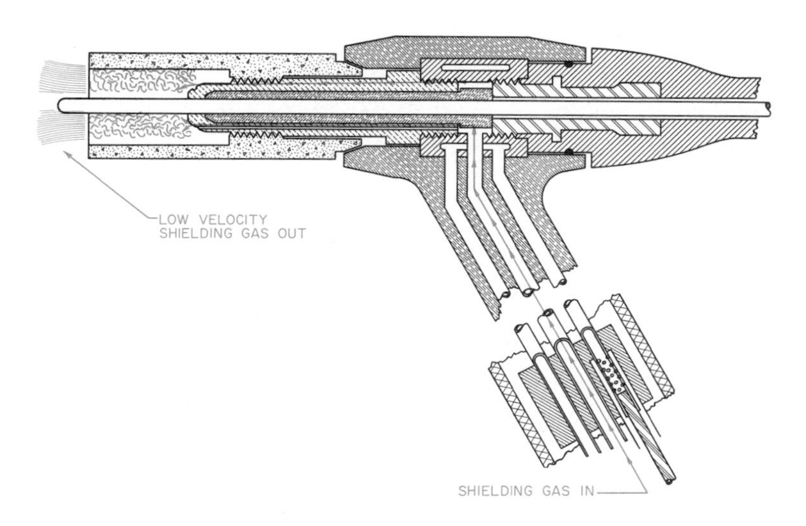

Fig. 3-3 Shielding gas flow

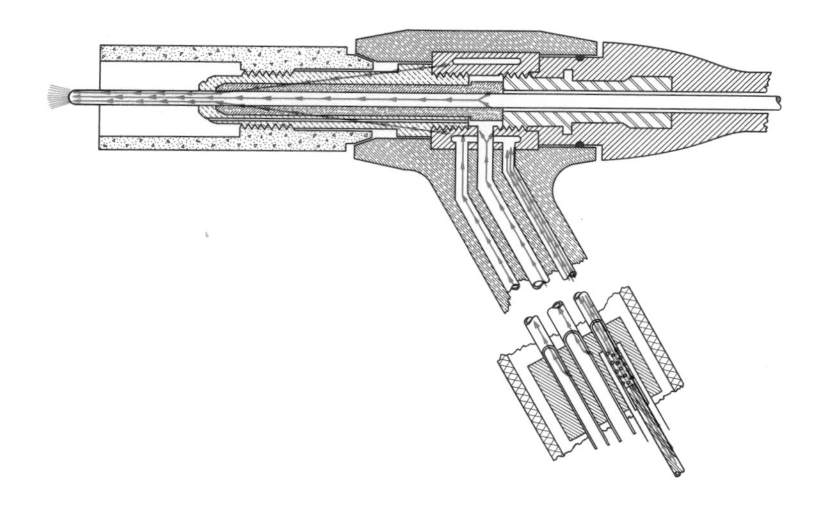

Fig. 3-4 Electric circuit

THE FLOW OF ELECTRICITY

As shown in figure 3-4, the electric current flows through the water-cooled welding cable, through the torch head to the collet holder and collet, to the tungsten electrode which forms one terminal of the arc, then through the work, and back through the ground cable to the power source. The student should examine a torch and compare the size of this ground cable with that of the cable leading into the torch.

The description of the electric circuit through the torch and work indicates that the electrode is negative or on straight polarity. If reverse polarity is used, the current flow is in the opposite direction.

If 60-cycle alternating current is used, the direction of flow changes 120 times each second so the electrode is positive (+) 60 times each second and negative (-) 60 times each second. This alternating of welding current is found to be very advantageous in many TIG welding operations. Direct-current straight polarity is referred to as DCSP; direct-current reversed polarity, as DCRP; and alternating current as AC, or in the case of AC with a superimposed high-frequency igniter current, as HFAC. The student should learn and remember these abbreviations since they are used in nearly every technical manual and paper written on the subject of TIG welding.

THE ASSEMBLY AND OPERATION OF THE TIG TORCH

To prepare the torch, figure 3-5, for welding operations, first choose the proper size electrode, matching collet and gas cup or nozzle. Some torches also require that the collet holder be changed with each different collet. Figure 3-6 shows a TIG torch with interchangeable collets.

1. Remove the collet cap or gas cap.
2. Remove the nozzle or gas cup by turning it counterclockwise.
3. If the collet holder is removable, take it off by turning it in a counterclockwise direction.

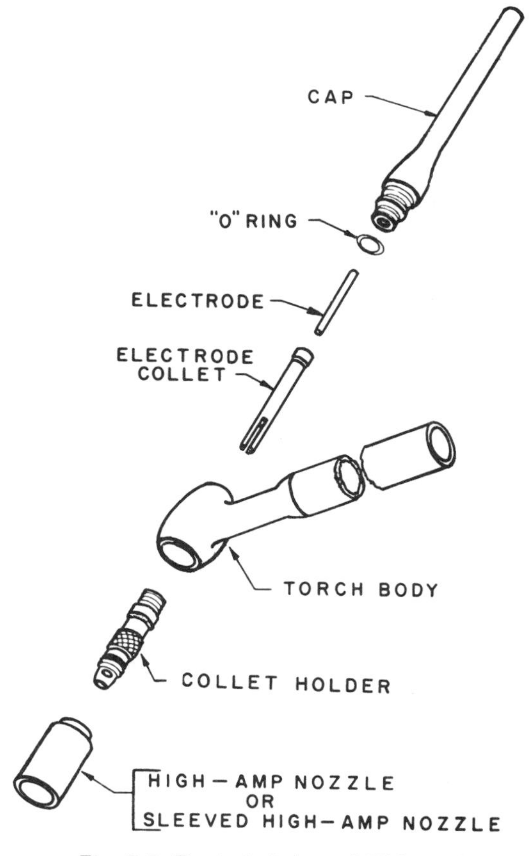

Fig. 3-5 Exploded view of TIG torch

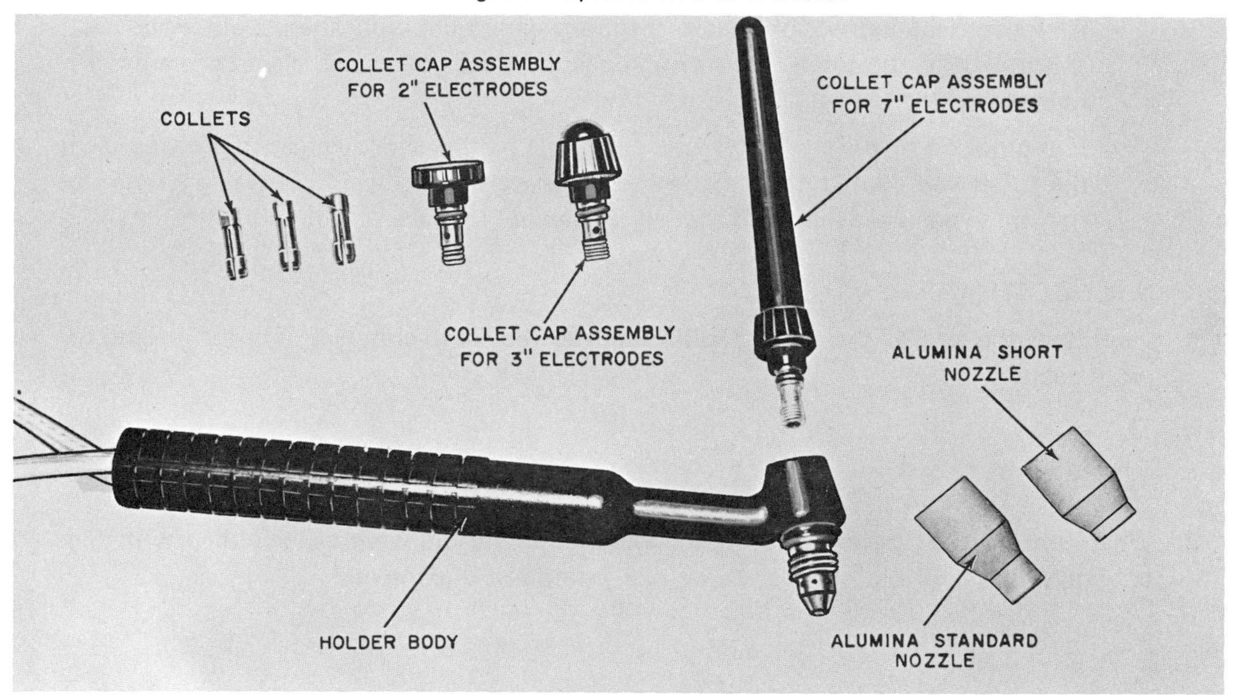

Fig. 3-6 TIG torch with interchangeable collets

Note: In the case of a torch with a single collet holder, step 3 can be ignored and the collet and electrode can be removed from the gas cup side of the torch.

4. Remove the collet and the electrode from the collet holder.
5. Choose the proper size electrode, collet holder, collet and gas cup or nozzle.
6. Screw the collet holder firmly into the torch.
7. Screw the nozzle into the collet holder firmly against the O ring on the torch body.
8. Place the proper collet in the collet holder and replace the gas cap or collet cap in the torch, leaving it loose by one-half to one turn.
9. Insert the electrode through the gas nozzle into the collet.

   Note: Never insert the electrode in the collet before inserting the collet in the torch. This guards against inserting a used electrode in the collet and, after use, finding that each end of the electrode has a ball formed on the end thus preventing the removal of the electrode from the collet.
10. Adjust the electrode for the recommended extension beyond the nozzle and tighten the gas cap or collet cap until the electrode is firmly fixed in the torch.
11. When the electrode extension needs to be adjusted to compensate for the slow burn-off, loosen the gas cap and adjust the electrode, and then firmly tighten the cap. Check the electrode to be sure it is firmly seated in the collet.

    CAUTIONS:

    - The ceramic nozzles are brittle, expensive, and easily broken. Always handle them with great care.
    - Any electrical connection that is not thoroughly tight will generate extensive heat and may ruin the torch. Be sure all collet holders, collets and electrodes are tight to avoid costly damage.
    - If the nozzle or gas cap is loose, it is possible for the shielding gas to draw air into the torch and contaminate the electrode as well as the weld. Always make sure that these parts are tight and that all O rings are in place.

REVIEW QUESTIONS

1. How does the size of the water-cooled cable to the torch compare with the ground or work cable?

2. What result would be expected if the water-cooled cable were not supplied with the cooling water at all times while the welding operation is going on?

3. What effect will a loose electrode or collet have on the torch?

4. What effect does a loose gas cap or ceramic nozzle have on the electrode and work zone?

5. If the electrode collet is put in the collet holder upside down, what happens?

UNIT 4 THE WELDING OF ALUMINUM

Most of the basic research and development in the use of the tungsten inert-gas shielded-arc process has been concentrated on the welding of aluminum and its weldable alloys. A brief analysis of the properties of aluminum will help to account for its widespread popularity.

ADVANTAGES OF ALUMINUM

- It is one of the earth's most abundant metals, making up about eight percent of the earth's crust.
- It has great strength in comparison to its weight.
- It is generally highly resistant to most forms of corrosion.
- It gives a very attractive appearance.
- It is very ductile and malleable.
- It has very good electrical and thermal conductivity characteristics.
- When its other qualities are considered, it is reasonably inexpensive.

As shown, aluminum has many advantages, but certain other facts must be considered before aluminum can be sucessfully joined by any of the welding processes, including TIG.

CONSIDERATIONS

- Chemically, aluminum is a very active metal.
- It combines with oxygen from the atmosphere even at room temperature to form a very hard oxide film on the surface. The hardness is illustrated in abrasive wheels composed of aluminum oxide.
- While aluminum melts at 1218 degrees F., its oxide melts at 3600-3900 degrees F. Even in the molten condition, it has a large amount of surface strength when compared to the oxides of many other metals.
- Aluminum oxide tends to absorb moisture. Under the extreme heat of the welding arc, this moisture breaks down to free hydrogen which often leads to porosity in the weld.
- The oxide can be removed mechanically by filing, scraping. or wirebrushing. It can be removed chemically with some liquid cleaners or by the use of flux. Most important in this discussion, it can be vaporized by the intense heat of the electric arc under the proper circumstances.
- Regardless of how the oxide is removed, it starts to re-form immediately. Too long a time lapse between cleaning and welding leaves the surface in about the same condition as it was before it was originally cleaned.
- Chemical fluxes must be immediately and thoroughly removed to prevent highly corrosive action on the metal. This is no problem in TIG welding because no flux is required.

- Compared to most metals, aluminum has a very high thermal conductivity. Thus the heat is conducted away from the weld zone at a fast rate. This means that a very high heat input must be maintained in the weld zone to balance the heat loss to the adjacent metal. TIG welding, with its intensely hot arc, is an excellent method of maintaining this high heat input.
- Aluminum is easily welded in overhead or other positions by the TIG process. When molten, it has a high surface tension for such a light metal. This surface tension tends to hold the molten puddle in position. The high rate of thermal conductivity causes the molten pool to solidify rapidly.
- Aluminum does not change color when it nears the melting point as do most other metals. When chemical fluxes are used to clean the surface, they cause considerable glare which makes accurate observation of the molten puddle difficult. With TIG welding there is no glare or smoke. The welder has a clear view of the size, shape, and condition of the molten pool at all times.

WELDABILITY OF ALUMINUM ALLOYS

The basic factor governing the ability of an aluminum alloy to be successfully welded is its chemical composition. Whether the alloy is wrought, die cast or sand cast makes little difference in the welding procedure.

According to the Air Reduction Company,

> *Weldability is the capacity of a metal to be fabricated by welding under the imposed conditions into a structure adequate for the intended purpose.*

Figure 4-1 shows a section of aluminum sheet. The alloy, its degree of hardness or temper, and its thickness in thousandths of an inch are indicated.

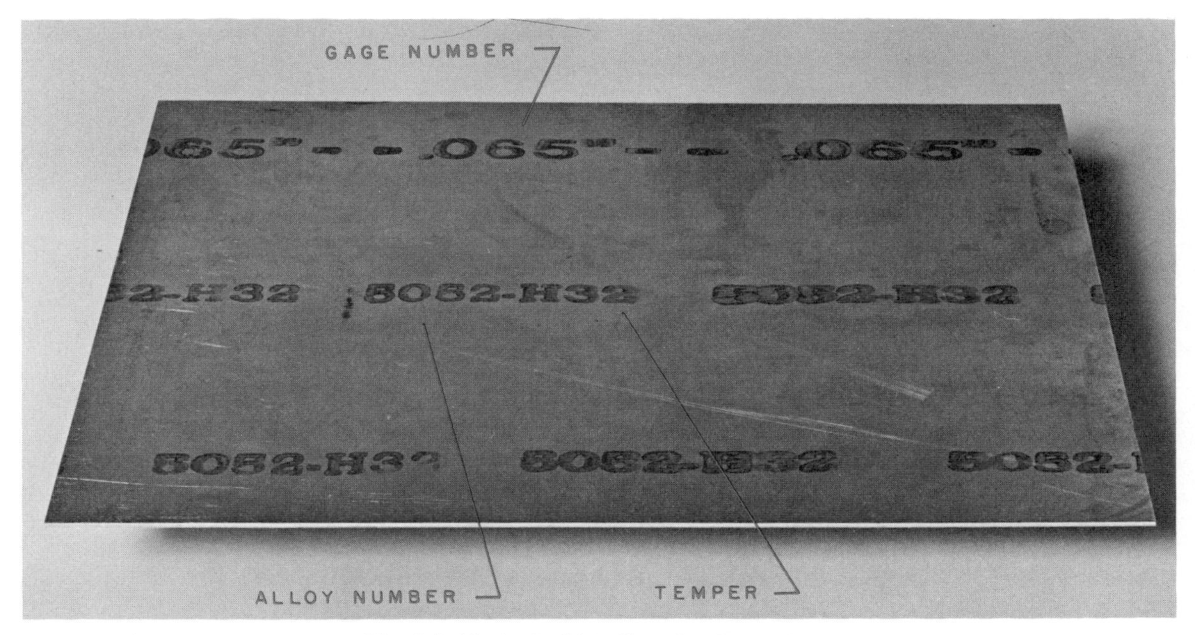

Fig. 4-1 Method of labeling aluminum sheets

The chemical composition, physical characteristics and heat treatment of the large array of alloys produced by manufacturers of aluminum and its alloys are out of the scope of this book. However, these manufacturers can supply a large amount of information about their products. A study of material of this type will increase the welder's knowledge and make him a more valuable employee.

REVIEW QUESTIONS

1. Would the use of aluminum be expected to increase or decrease in relation to the other commonly used metals? Why?

2. What is meant by the term thermal and electrical conductivity?

3. Where can information be found about the chemical composition or physical characteristics of the various aluminum alloys?

4. What major condition causes aluminum to be considered difficult to weld?

5. What other characteristics cause aluminum to be considered difficult to weld?

UNIT 5 THE ACTION IN AND AROUND THE WELDING ARC

The experiments in this unit will provide an opportunity to investigate some of the phenomena that take place in and around the welding arc. This provides a basis on which to judge the conditions necessary to produce high-quality TIG welds.

Note: As the following experiments are carried out in the unshielded atmosphere, aluminum oxide is formed rapidly. Also, when welding with DCRP the weld zone is covered with black residue. Remember, arc phenomena and not welding techniques are being investigated. Some aluminum has been welded in the past using the carbon arc process DCSP. TIG welding has made this method almost obsolete.

EXPERIMENT 1

Materials

DC welding machine

2 carbon electrodes 1/4-inch diameter

Procedure

1. Grind one end of each electrode to a pencil point.
2. Place one electrode in the holder and the other in a vise or other clamping device which is attached to the ground cable.
3. Set the controls on the machine to produce 150 amps. or more with the electrode holder negative – DCSP.
4. Draw an arc between the pointed ends of the carbon electrode and carbon work. This makes one pole of the arc negative and the other positive, which is a normal condition in all DC welding arcs.
5. Observe the arc and the electrodes to find out how great an area at each pole becomes incandescent (gives off bright light).
6. After thirty to sixty seconds of arc action, break the arc and observe both electrodes closely to see what effect the arc action has had on both the negative and positive carbons.

Observations

1. At which pole of the DC arc was the greatest amount of heat released?
2. About how much heat was released at each pole? On what is this conclusion based?
3. What was the conditon of the electrodes after completing the experiment?
4. If the poles had been reversed what would the result be?

EXPERIMENT 2

Materials

DC welding machine
Sharpened carbon electrode 1/4-inch diameter
Mild steel plate about 4 inches square and 1/4 inch to 1/2 inch thick

Procedure

1. Set the machine to deliver 150 or more amps., DCSP.
2. Draw an arc and melt the metal to form a bead about 1 1/2 inches long. Observe the arc action during the welding and inspect the electrode after welding.
3. Reverse the welding current using DCRP.
4. Run a second bead some distance from the first and parallel to it, observing the arc action and the condition of the electrode after completing this weld.
5. Center punch the following using a hammer and center punch.
   a. the plate
   b. the weld made with DCSP
   c. the weld made with DCRP

   CAUTION: Be sure to wrap the pointed end of the punch with cloth, a leather glove or other material that will guard against flying metal should the punch break. Because of the hazard involved, the test described in step 5 should be conducted by the instructor.

Observations

1. What was observed about the ease or difficulty of welding in this experiment?
2. How did the finished beads compare?
3. What was the condition around the welds in regard to residue?
4. What observations and conclusions were made when the plate and the DCSP weld were center punched?
5. When the DCRP weld was center punched, what conclusions were made?
6. Is resistance to penetration a good test of hardness? Explain.
7. From observations made in this experiment, is the greatest mass action in the carbon arc from negative to positive or from positive to negative? Explain.

EXPERIMENT 3

Materials

DC welding machine
Sharpened 1/4-inch diameter carbon electrode
Aluminum plate about 2 in. x 6 in. x 1/8 in.

Procedure

1. Strike an arc on the plate and run a bead using about 150 amps., DCSP. Observe the arc action and, in particular, the molten aluminum. A wrinkled, dull-appearing skin will form over the molten pool.
2. Examine the finished bead for appearance and note the condition of the electrode.
3. Repeat step 1 but use DCRP. Note the size of the pool of molten metal directly under the arc as compared with that found with DCSP. Also compare the brightness of this pool.

Observations

1. What was observed about the amount of penetration when using DCSP in this experiment? Is this consistent with the findings in Experiment 1?
2. What surface condition was observed directly under the arc when using DCRP?
3. What conclusion can be drawn from the above?
4. What accounts for the black residue on the aluminum when using DCRP, as this condition was not observed when welding the steel with the same polarity?
5. What was observed about the amount of penetration and the size of the bead in this experiment? What significance is attached to this observation?

THE ALTERNATING CURRENT ARC

In the direct-current welding circuit the flow of current across the arc is always from (+) to (-). In the alternating current arc, the current reverses itself many times a second, which means that alternating current is a combination of DCSP and DCRP. The most common welding frequency is 60-cycle AC. In 60-cycle AC, the electrode is positive (+) 60 times each second and negative (-) 60 times each second. It is also momentarily zero 120 times each second, a fact which makes the AC arc difficult to maintain. In this case, a high-frequency AC carrier or igniter arc is superimposed on the standard 60-cycle AC welding current to produce HFAC which eliminates the problem of maintaining an arc.

EXPERIMENT 4

Materials

AC welding machine
Sharpened 1/4-inch diameter carbon electrode
Aluminum plate about 2 in. x 6 in. x 1/8 in.

Procedure

1. Strike the arc, using about 150 amps., and again observe the arc. Note the size, shape and degree of brightness of the molten metal.
2. Try varying the arc length, making continuous observations. Note the ease or difficulty of maintaining the arc.

Observation

1. What conclusions can be drawn from this experiment?

REVIEW QUESTIONS

This unit has led the students through a series of experiments and questions designed to help them form conclusions that are important to the study of TIG welding. What can be concluded regarding:

1. Oxidation of aluminum?

2. Effect of straight polarity?

3. Effect of reverse polarity?

4. Comparison of AC with DC with regard to penetration and cleaning?

5. Carbon transfer?

UNIT 6 FUNDAMENTALS OF TIG WELDING

Although it can produce outstanding results, the TIG welding process may be unnecessarily expensive. A careless operator can cause a major expense by damaging the equipment. In particular, the tungsten electrode and the ceramic nozzle are subject to misadjustment. Care must be taken with these parts, therefore, since they are important to the production of high-quality work at a moderate cost.

COMPARISON OF METALLIC ARC WELDING WITH TIG WELDING

The variables found in TIG welding are almost identical to those in metallic arc welding. These variables are

- Length of arc
- Amount of current
- Rate of arc travel
- Angle of electrodes

The student who is skilled in the metallic arc welding process is already familiar with these variables. This prior knowledge is helpful in the study of TIG welding.

The following differences in costs must be considered:

- The overall value of the equipment used in TIG welding is much higher. This includes the welding machine, the cable and hoses, as well as the torch, regulator and nozzles.
- The shielding gas used is much more expensive than gases used in most other welding processes. For example, helium and argon gases are far more expensive than acetylene.
- The electrodes used in TIG welding are much more expensive. Actually, these electrodes are consumed so slowly that the cost of electrodes per foot of weld is very slight. However, the student should realize that any waste of electrodes due to bending, breaking or the use of excessive current is a very expensive error.
- The material being welded is generally much more expensive and sometimes hard to obtain, especially in the sizes and alloys desired.

In general, the student of TIG welding should be aware of the costs involved and should follow an intelligent, rigid procedure to protect the equipment from costly damage and to avoid the waste of expensive materials.

ELECTRODES

Tungsten and tungsten alloys are supplied in diameters of .010 inch, .020 inch, .040 inch, 1/16 inch, 3/32 inch, 1/8 inch, 5/32 inch, 3/16 inch, and 1/4 inch. They are manufactured in lengths of 3 inches, 6 inches, 7 inches, 18 inches, and, in some instances, 24 inches. The electrodes are made with a cleaned surface, either chemically cleaned and etched, or with a ground finish which holds the diameter to a closer tolerance. Electrodes are supplied in pure tungsten and in three alloys: 1 percent thorium, 2 percent thorium alloy and zirconium alloy.

CHART 6-1

ELECTRODES AVAILABLE

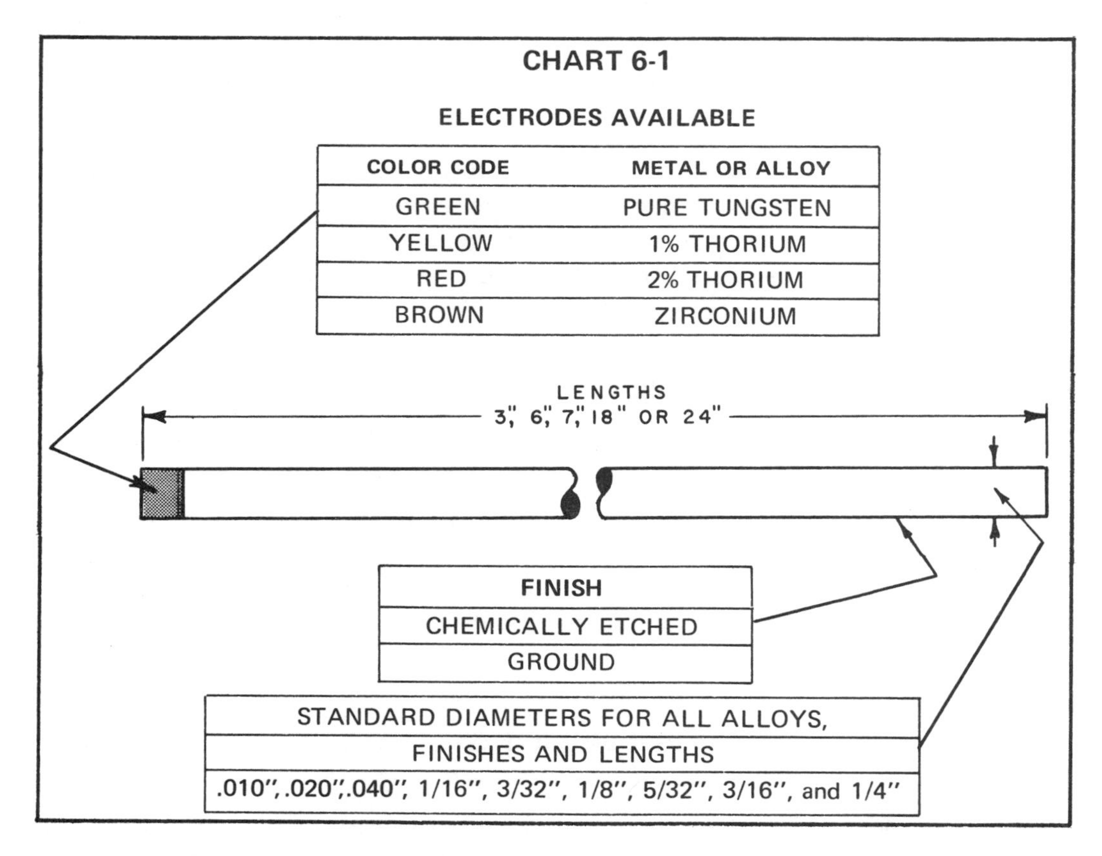

COLOR CODE	METAL OR ALLOY
GREEN	PURE TUNGSTEN
YELLOW	1% THORIUM
RED	2% THORIUM
BROWN	ZIRCONIUM

FINISH
CHEMICALLY ETCHED
GROUND

STANDARD DIAMETERS FOR ALL ALLOYS, FINISHES AND LENGTHS
.010", .020", .040", 1/16", 3/32", 1/8", 5/32", 3/16", and 1/4"

Pure tungsten is generally used with AC welding. The thoriated types are mostly used for DCSP welding and give slightly better penetration and arc starting characteristics over a wider range of current values. The zirconium alloy is excellent for AC welding and has high resistance to contamination. Its chief advantage is that it can be used in those instances when contamination of the weld by even very small quantities of the electrode is absolutely intolerable.

Chart 6-1 condenses information on electrodes available for TIG welding. The color code shown is being used by the major producers and distributors of tungsten electrodes.

The recommended amperage for any given size electrode varies with the type of joint being welded and the type of current used. A general recommendation when welding with AC is that the current be equal to the diameter of the electrode in thousandths of an inch multiplied by 1.25. For example, an electrode with a diameter of .040 requires a current of 40 x 1.25 or 50 amperes. Of course, the size of the electrode is a function of the thickness of the metal being welded. Chart 6-2 gives current ratings and electrode sizes for butt welding the various thicknesses of aluminum using HFAC arc with argon shielding and pure tungsten electrodes.

While charts are valuable as guides, a degree of sound judgment on the part of the operator is also desirable. Electrodes operated at a current value which is too low cause an erratic arc just as with metallic arc welding. If the current is correct, the end of the electrode appears as in figure 6-1.

CHART 6-2

DATA FOR STRINGER BEADS IN ALUMINUM

Thickness in Inches	HFAC Welding Current Flat* Amperes	Tungsten Electrode Diameter	Welding Speed Inches per Min.	Filler Rod Diameter	Recommended Argon Flow Cu. Ft. per Hr. ***	Gas Nozzle Size
1/16	60-80	1/16	12	1/16	15 to 20	4, 5, 6
1/8	125-145	3/32	12	3/32 or 1/8	17 to 25	6, 7
3/16	190-220	1/8	11	1/8	21 to 30	7, 8
1/4	260-300**	3/16	10	1/8 or 3/16	25 to 35	8, 10
3/8	330-380**	3/16, 1/4	5	3/16 or 1/4	29 to 40	10
1/2	400-450**	3/16, 1/4	3	3/16 or 1/4	31 to 40	10

* – Current values are for flat position only. Reduce the above figures by 10% – 20% for vertical and overhead welds.
** – For current values over 250 amps., use a torch with a water-cooled nozzle.
*** – Use lower argon flow for flat welds. Use higher argon flow for overhead welds.

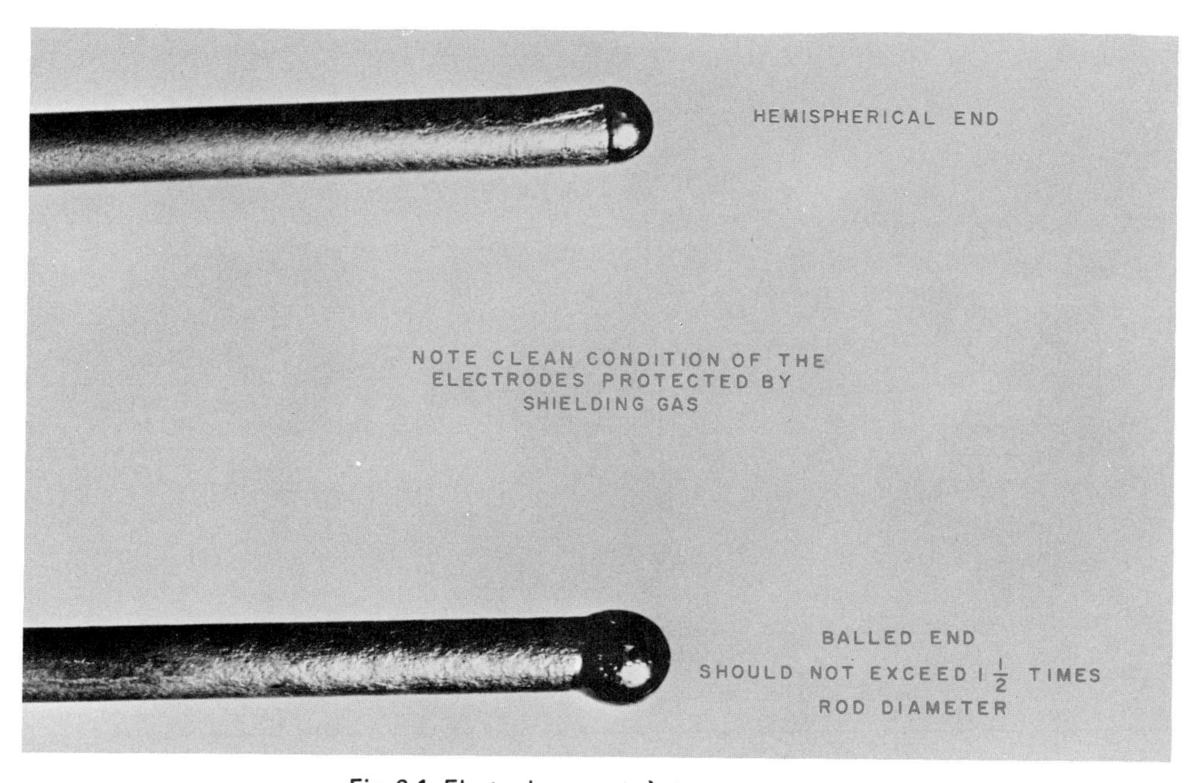

Fig. 6-1 Electrodes operated at proper current

Note that this figure shows a round, shiny end in one case and an end which forms a ball in the other. If this ball is over one and one-half times the diameter of the electrode, the current is too high and the electrode is consumed at an excessively high rate.

CERAMIC NOZZLES

The ceramic nozzles in chart 6-2 are indicated in fractions of an inch to describe the recommended inside diameter of these nozzles. There is a trend in the industry to indicate the nozzles by numbers such as 4, 5, 6, and 7. These numbers give the nozzle size in sixteenths of an inch. For example, a #6 nozzle indicates 6/16-inch or 3/8-inch inside diameter.

In general, the inside diameter or orifice of the nozzle should be from four to six times the diameter of the electrode. Nozzles with an orifice which is too small tend to overheat and either break or deteriorate rapidly. Smaller-diameter nozzles are also more subject to contamination. Ceramic nozzles are usually recommended for currents up to 250-275 amps. Above this point, special torches with water-cooled metal nozzles are generally used.

One other important factor is the amount the electrode extends beyond the nozzle, figure 6-2.

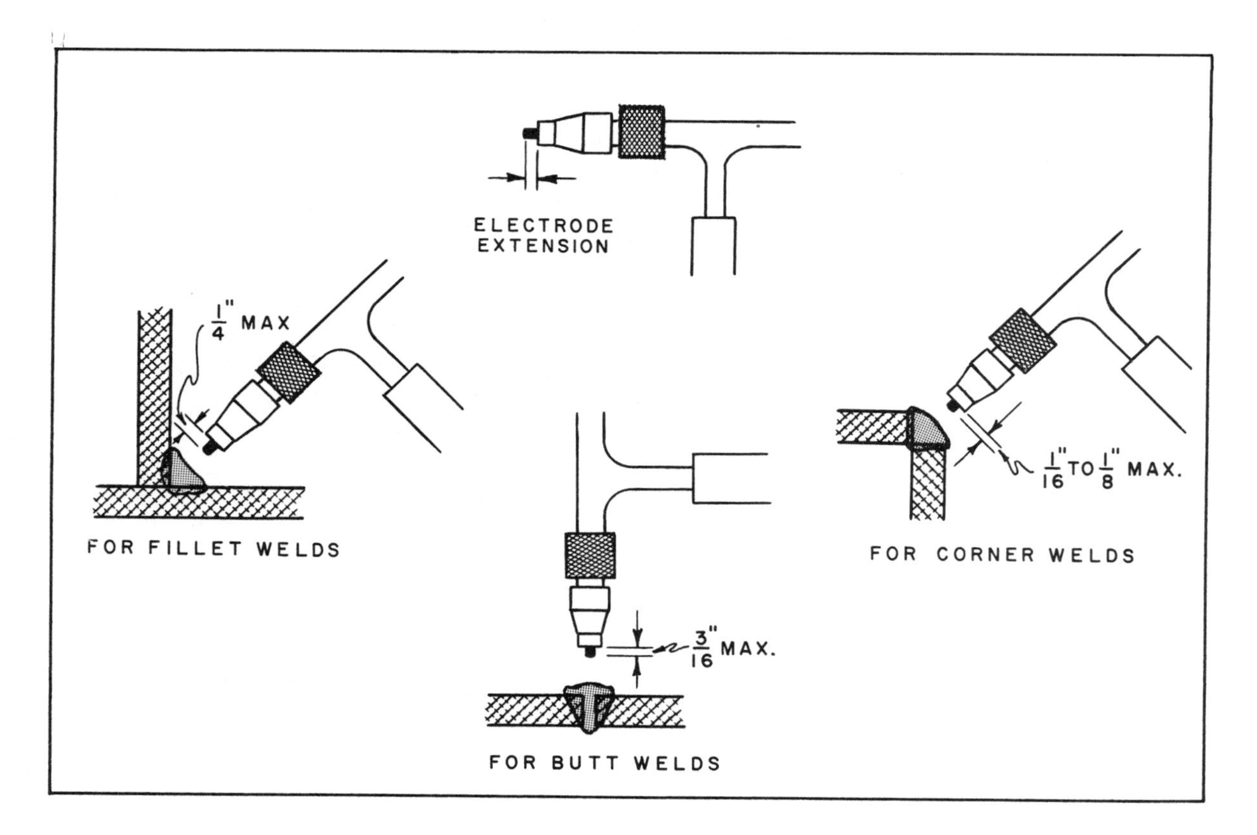

Fig. 6-2 Electrode extension

REVIEW QUESTIONS

1. Zirconium electrodes cost fifty percent more than tungsten but have excellent characteristics for HFAC welding. Would it be justifiable to use them?

2. Chart 6-2 shows the shielding gas flow for various sizes of electrodes. When welding aluminum would it be justifiable to experiment with this factor? On what basis?

3. Why would it be desirable to use seven-inch tungsten electrodes whenever possible instead of three-inch electrodes?

4. When would it be justifiable to use three-inch tungsten electrodes?

5. What is the objection to a long electrode extension?

6. What current in amperes is required for a 1/16-inch electrode?

UNIT 7 STARTING AN ARC AND RUNNING STRINGER BEADS ON ALUMINUM

Rigid attention to detail and procedure is of extreme importance in TIG welding. Errors due to carelessness may prove to be very expensive. For instance, failure to turn on the cooling water usually results in destruction of the torch as well as the cable and hose assembly. Striking an arc with the machine set for normal amperage but with the polarity selector on DCRP will result in destruction of the electrode and usually the collet holder and collet.

Materials

Clean aluminum plate 1/8 in. thick x 4 in. x 6 in.
AC welding machine equipped with high-frequency oscillator
Cylinder of argon gas equipped with flowmeter
TIG torch fitted with 1/8-inch pure tungstem electrode and a #7 or #8 nozzle, (7/16 inch or 1/2 inch.)

Preweld Procedure

1. Make sure the torch is well away from the ground or work cable.
2. Turn on the cooling water.
3. Set the high-frequency switch to AC TIG.
4. Set water and gas switch to ON position.
5. If remote control is used, set switch to ON or remote. Otherwise leave it at LOCAL or OFF.
6. If the machine is equipped with a balanced wave filter or batteries, set this switch to ON.
7. Set gas and water afterflow timer for 1/8-inch electrodes.
8. Turn on the gas from the argon cylinder and adjust the flowmeter to supply 17 to 21 cubic feet per hour.

   Note: The flow of argon and the flow of water cannot be checked unless the remote control switch is ON or, if local control is used, the machine power switch is in the ON position.
9. Set the polarity switch to AC.
10. Adjust the curent as indicated in chart 6-2.
11. Check the electrode for the proper extension. Refer to figure 6-2.
12. With the power OFF, check the electrode to be sure it is firmly held in the collet. To do this, place the exposed end of the electrode against a solid surface and push the torch down gently but firmly. If the electrode tends to move back into the nozzle, either the collet holder or gas cap needs to be tightened. Be sure to set the electrode extension and then tighten the collet holder or gas cap.

13. Turn the power switch ON. Turn the remote switch to ON. Note that the flowmeter indicates the proper flow of shielding gas. Check the waste line to be sure the cooling water is flowing.
14. Strike an arc by bringing the electrode close to a grounded workpiece, preferably copper. If the arc fails to jump from the electrode to the work without actual contact, set the high-frequency intensity control to a higher setting.
15. Again strike an arc on the copper and allow the current to flow until the electrode becomes incandescent. Break the arc and check the afterflow by watching the electrode cool. The instant the electrode becomes bright, look at the small ball in the flowmeter tube. It should drop to the bottom of the tube in two or three seconds. If it indicates a flow of a longer duration, adjust the afterflow timer to a slightly lower setting and repeat the above procedure until the timing is right.

   Note: If the afterflow timing was originally set for too short duration, the electrode would cool in the atmosphere and oxidize. This is indicated by the electrode becoming blue or black in color. In this event, adjust the afterflow timer higher until the ideal afterflow is reached.

The above procedure should be strictly followed each time the TIG welding process is used, even if someone else has just completed a weld with the equipment. It is the personal responsibility of each operator to be sure that all controls are properly adjusted at all times. Carelessness can lead to serious damage to the equipment.

Welding Procedure

1. Use all standard safety precautions. In addition, use a welding helmet filter plate one or two shades darker than the one for metallic arc welding.
2. Place the clean aluminum workpiece in firm contact with a clean worktable surface. The radio-frequency high-voltage (2000-4000 volts) igniter arc will jump a wide gap. If it jumps from the worktable to the reverse side of the workpiece, it may cause damage to the surface finish.
3. There are times when backing bars are advantageous. Either steel or stainless steel is recommended for backing bars; stainless steel is the better choice. Copper is not good for a backing bar when welding aluminum, and carbon should never be used for this purpose.
4. Turn on the remote switch and draw an arc with the electrode held as nearly vertical as possible while observing the molten puddle. Use an arc about equal in length to the electrode diameter.
5. Run a straight bead about 1/2 inch from, and parallel to, the edge of the plate, with the rate of the arc travel adjusted to maintain a pool of molten metal about 3/8 inch in diameter.
6. Examine the finished bead for uniformity and surface appearance. The weld should have a shiny appearance along its entire length. Note that there is an area about 1/8 inch on each side of the weld which is quite white but dull in lustre, figure 7-1. This

Fig. 7-1 Partially vaporized oxide, no preweld cleaning

is aluminum oxide which has been partially vaporized by the high-frequency igniter arc. Also examine the electrode for brightness and shape of the arc. Examine the reverse side of the plate for penetration.

7. Run a second bead 1/2 inch from the first and parallel to it. While making this bead, pay close attention to the area just outside the molten pool. A great number of pin-point-size arcs should be seen. This is the high-frequency arc partially breaking up the aluminum-oxide film.
8. Examine the finished bead for surface appearance, uniformity, penetration and for the amount of white residue along the edges of the weld. Check the bead for cracking.
9. Make a third and fourth bead, observing the arc action and adjusting the rate of arc travel to obtain a uniform weld with complete penetration.
10. Examine these beads critically as in step 8. Examine the electrode after each bead for evidence of excessive burn-off or electrode contamination.

Figure 7-2, view A, shows an electrode contaminated from the atmosphere because of too short a duration of afterflow of the shielding gas plus excessive electrode extension. In view B, an electrode is shown that was contaminated because of too short a duration of afterflow of shielding gas. View C shows an electrode contaminated by allowing it to come in contact with the molten pool.

11. Make a fifth bead but this time allow the electrode to contact the molten pool two or three times as the weld progresses. Observe the arc action and the area around the weld, especially after contact has been made.
12. Observe the finished bead as in step 7 and note the difference. Examine the electrode and gas nozzle for evidence of contamination as in figure 7-2, view C.

Note: In the interest of electrode economy, the aluminum contamination of the electrode can be burned off by allowing the arc to dwell for several seconds on a copper plate. This is acceptable for practice welding. However, for high-production, high-quality welds, the accepted practice is to remove the electrode and either grind away the contaminated area or notch the electrode just back of the contamination and break it off. If this is done, be sure to grasp the electrode close to the notch to avoid bending.

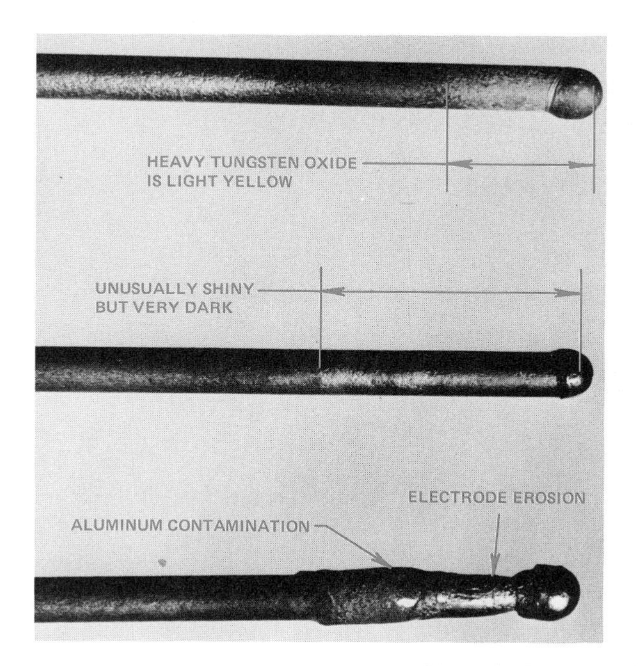

A. TOO LONG AN ELECTRODE EXTENSION PLUS TOO SHORT A DURATION OF AFTERFLOW

B. SLIGHTLY TOO SHORT A DURATION OF AFTERFLOW

C. CONTAMINATION DUE TO CONTACT OF ELECTRODE AND MOLTEN PUDDLE

Fig. 7-2 Electrode Contamination

13. Do the experiment shown in figure 7-3. After the aluminum has broken cool the pieces and examine them. Observe that there is no evidence of melting. This indicates that the metal has broken rather than melted. This phenomenon, termed *hot-short,* is not uncommon. Many metals and alloys, such as copper, brass, and even cast iron display this characteristic at a temperature slightly below their melting point. Consider what would happen if the aluminum strip were laid on a flat steel plate and heated. Would the piece have broken or simply melted? What conclusion can be drawn from the above?

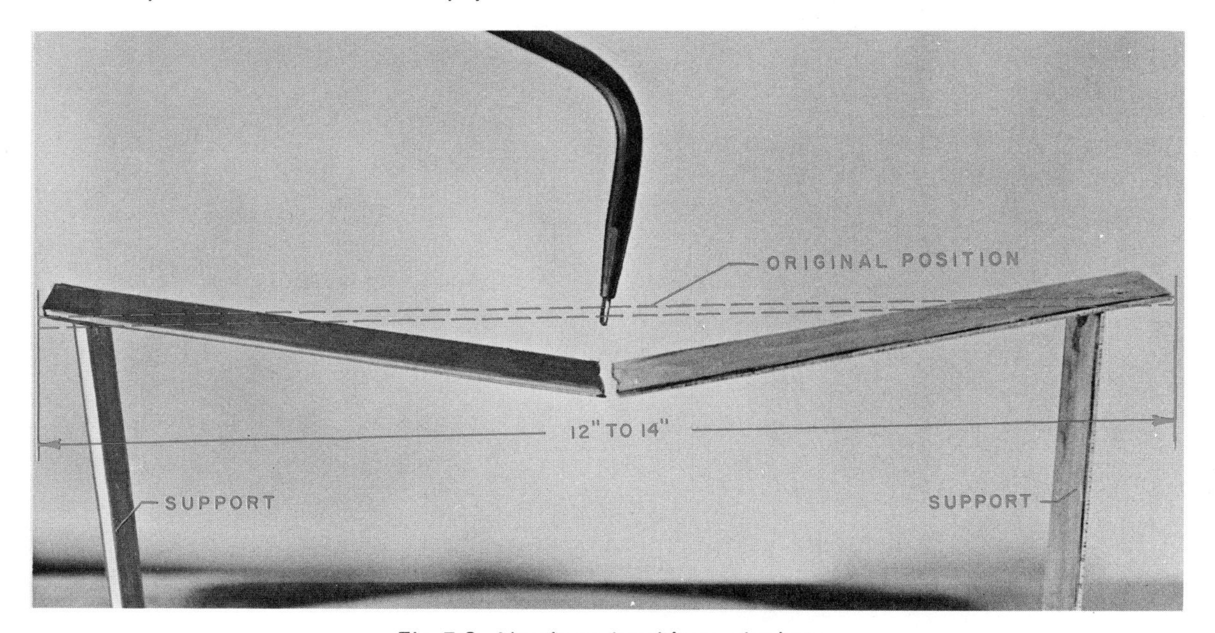

Fig. 7-3 Aluminum breaking under heat

REVIEW QUESTIONS

1. What four steps in the preweld procedure, in order of their importance, are most essential in avoiding damage to the equipment or electrodes?

2. For what two reasons would stainless steel be preferred over steel for a backing bar when welding aluminum?

3. What two reasons make copper a poor choice for a backing bar when welding aluminum?

4. Why would carbon be unacceptable for a backing bar when welding with tungsten electrodes?

5. Compare the amount of the white or light residue observed adjacent to the first weld with that observed in the succeeding welds. How do the welds differ?

6. What conclusion can be drawn from the observation in Question 5?

7. What practical use is there for the phenomenon observed in Questions 5 and 6?

8. What observations were made when welding with an electrode contaminated with aluminum?

9. What conclusions can be drawn from the answer to Question 8?

10. What conclusions can be drawn from the experiment in step 13?

UNIT 8 RUNNING PARALLEL STRINGER BEADS ON ALUMINUM

This unit provides the first opportunity to use TIG welding in which additional metal is added to the bead by using a filler rod. Those who have welded with a carbon arc using a filler rod will find much similarity between TIG welding and the processes and techniques with which they are familiar.

Materials

Aluminum plate 1/8 in. x 4 in. x 6 in. to 9 in. long, free from oil, grease and lint
1/8-inch aluminum or 5% silicon-alloy rod

Procedure

1. Choose the proper size tungsten electrode, nozzle and gas flow from chart 6-2.
2. Follow the preweld procedure in unit 7.
3. Make a second check of the four most important factors.
   a. Water flow
   b. Polarity of electrode
   c. Flow of shielding gas
   d. Firmness of contact between the electrode and collet
4. Strike an arc and run a bead about 3/8 inch wide parallel to the edge of the aluminum plate and about 1/2 inch from the edge, holding the torch and filler rod as shown in figure 8-1.
5. Examine the finished bead for fusion, uniformity and completeness of penetration, and white residue.

   Note: Since there is no requirement that either the plate or filler rod be cleaned prior to welding, some difficulty may have been found in fusing the filler rod and plate. This is due to the tough skin or film of aluminum oxide on the work.

   Oxyacetylene welders of aluminum overcome this difficulty by alternately jabbing the filler rod into the molten pool and then withdrawing it in a regular, rhythmic motion. This allows the rod to break through the oxide film at each down cycle. TIG welding uses much the same procedure except as the rod is moved into the molten pool, the electrode is moved slightly back into the finished weld to avoid contacting the electrode with the filler rod. This would cause electrode contamination and the difficulties found in unit 7.
6. Run beads until the technique shown in figure 8-2 is perfected.
7. Examine each bead for appearance, penetration and residue.
8. Check the electrode after each bead for electrode extension and contamination.
9. Obtain more aluminum plates and clean them prior to welding, either with steel wool or abrasive cloth.

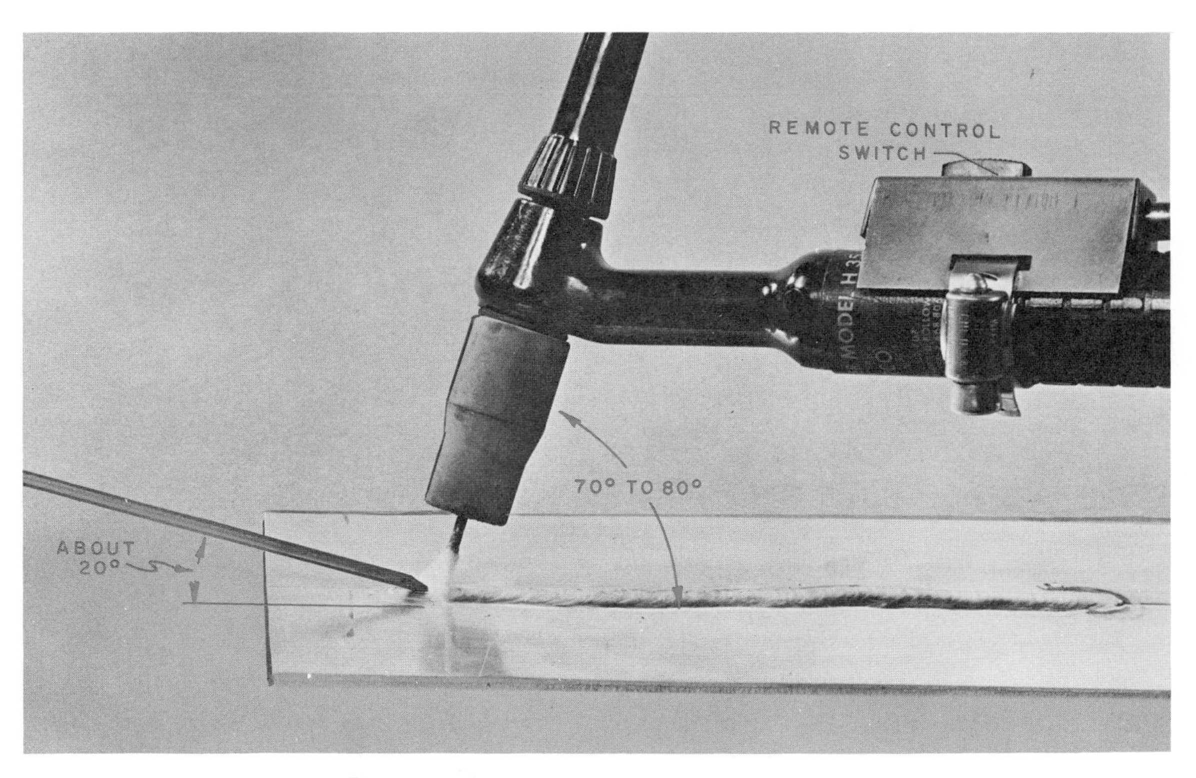

Fig. 8-1 Relative position of torch and filler rod

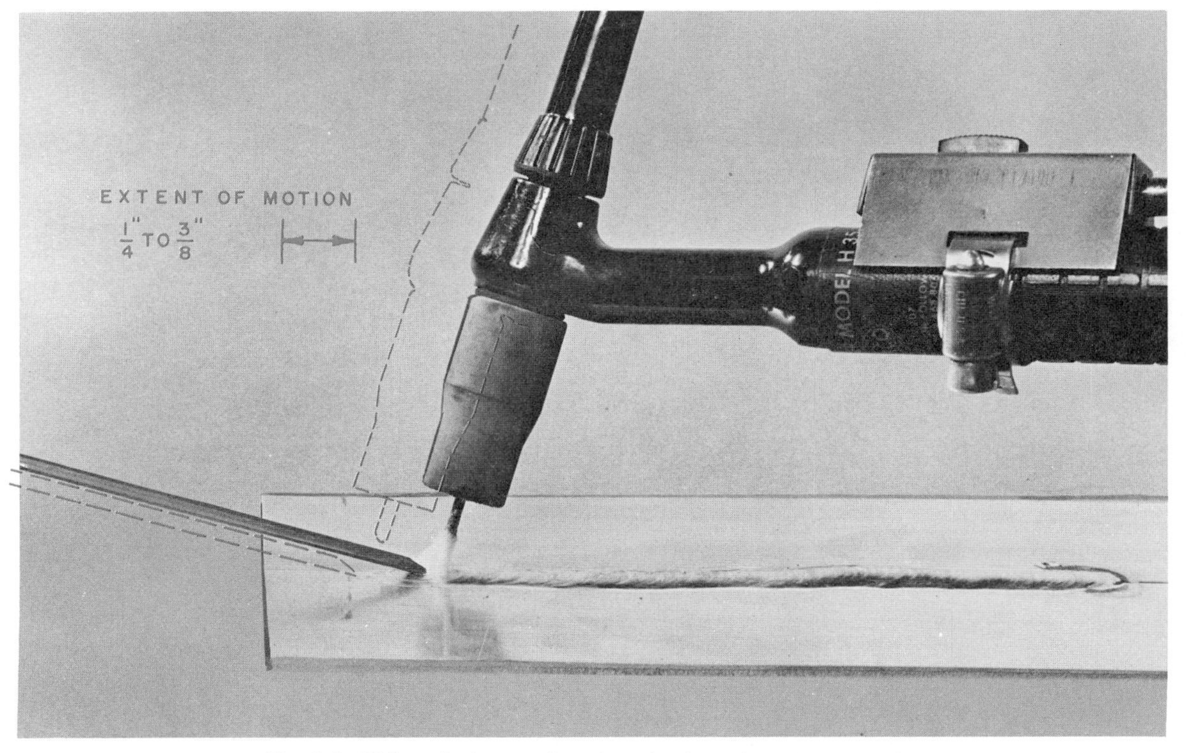

Fig. 8-2 TIG technique – Synchronized torch and rod motion

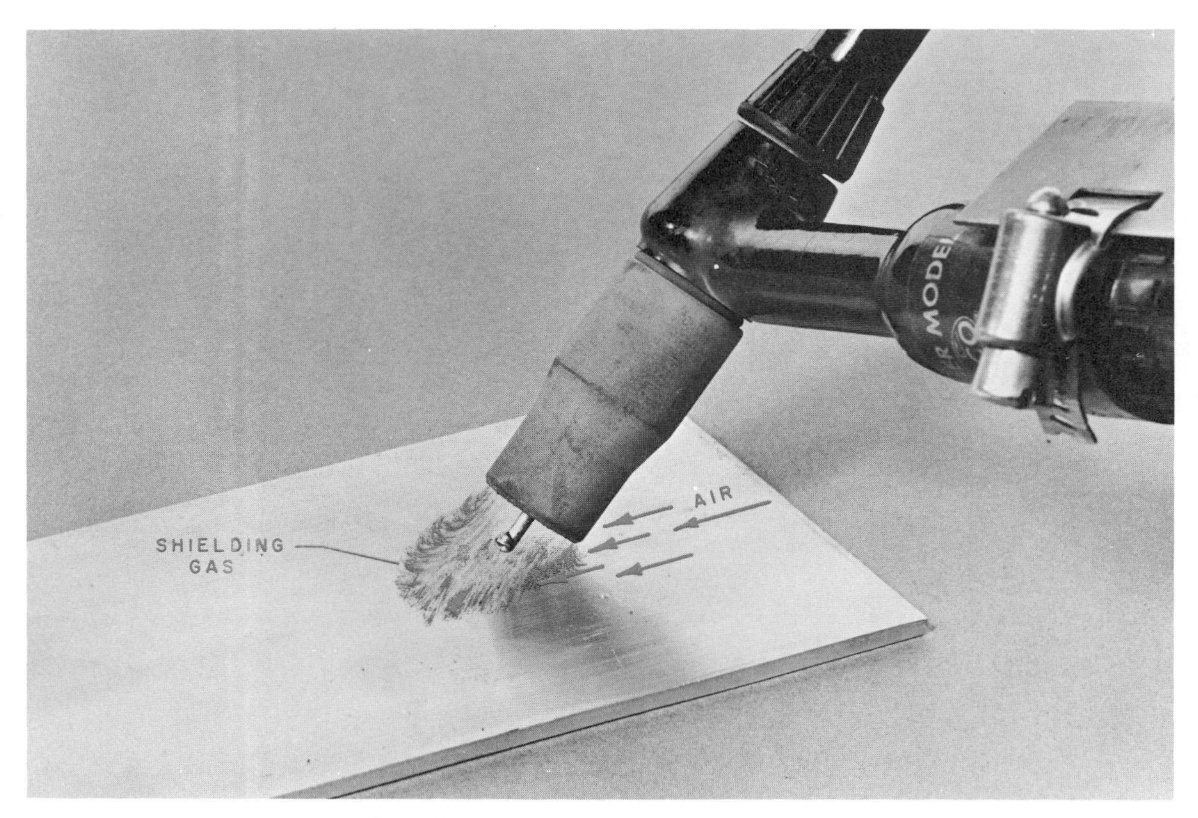

Fig. 8-3 Torch operated at too low an angle

10. Run a bead on this cleaned plate as in steps 4 and 6. Pay close attention to the ease or difficulty of fusing the filler rod and work.

11. Examine the finished bead as before. Be sure to compare the amount of white residue along the edges of this weld with that observed in steps 5 and 7.

12. Cool the plate and again clean it in the area where the next bead is to be made. Also clean the filler rod thoroughly and run another bead. Note the ease or difficulty of fusing the filler rod and workpiece.

13. Examine this bead, paying close attention to the amount of white residue along the edges of the weld as compared to that along the edges of beads observed in steps 5,7, and 11.

14. Make more beads with both the plate and filler rod. Cool and clean the plate and rod before each bead. Try varying the angle of the filler rod. Remember that the height of the bead is a function of the angle between the filler rod and the work as in oxyacetylene welding. Also try varying the angle of the work and electrode slightly. If the torch angle is too low the argon gas draws some air along with it and breaks down the shielding qualities, figure 8-3.

REVIEW QUESTIONS

1. What are the proper filler rod and torch angles for TIG welding?

2. When using chart 6-2 to find the amperage setting and flow of shielding gas, would the higher current setting and gas flow be used or the lower? Why?

3. What are the objections to using aluminum plate which has not been cleaned of oil, grease, and dust?

4. What harmful effects does failure to clean the oxide from the work and filler rod have on the finished weld?

5. In observing the reverse side of the welded plates, it is found that the surface has a dull, wrinkled appearance. How can this be overcome?

6. Is abrasive cleaning of the work and filler rod worthwhile? Why?

7. When would abrasive cleaning be undesirable? What cleaning method could be used?

UNIT 9 OUTSIDE CORNER WELDS ON ALUMINUM

This unit provides the opportunity to make high-quality welds of excellent appearance without the responsibility of manipulating a filler rod. It also gives the added opportunity to construct and use a simple jig or welding fixture. In all following units, material designated as aluminum is understood to mean either pure aluminum, if available, or any of its weldable alloys.

Outside corner welds on aluminum (see figure 9-3) require the construction of a jig, either of steel or stainless steel, as shown in figure 9-1. Be sure that all oxides and dirt are cleaned from this jig before using. A better jig can be made by adding clamps and clamping bars as shown in figure 9-2.

Materials

2 pieces 1/8 in. x 1 1/2 in. x 6 in. to 9 in. long aluminum or weldable aluminum alloy

Procedure

1. Place the two workpieces in the jig in such a way that their edges are brought in close contact along their entire length as shown in figure 9-4.
2. Before welding make a complete check on all welding factors as in unit 7. Be sure to refer to figure 6-2 to find the proper electrode extension.

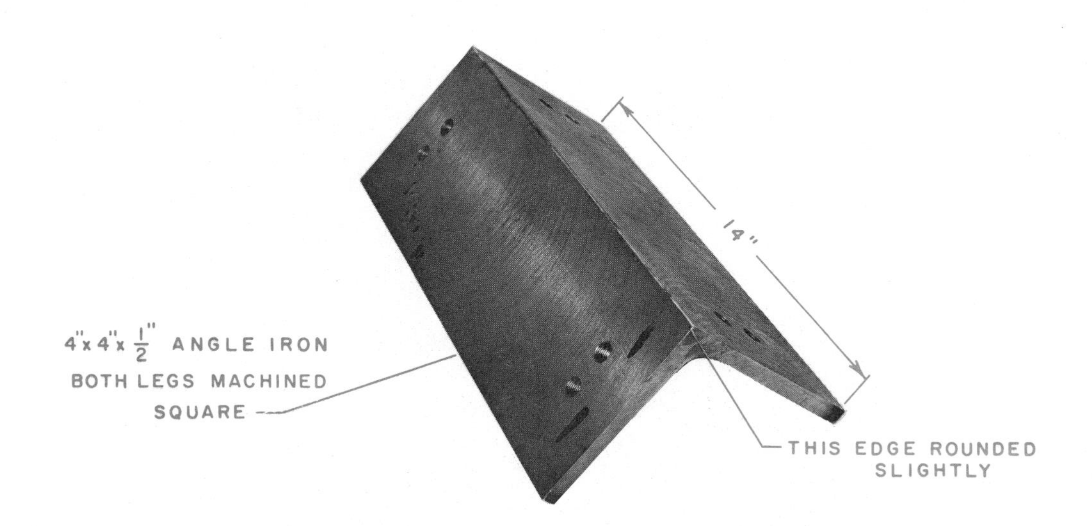

Fig. 9-1 Simple jig

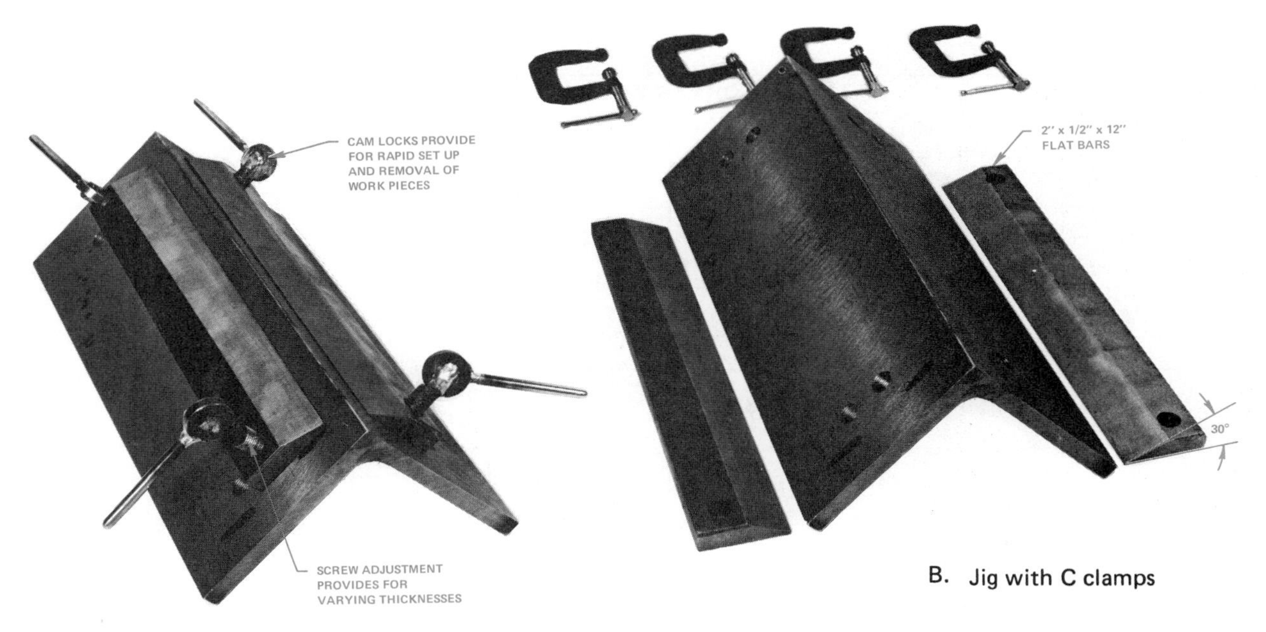

A. Jig with cam locks

B. Jig with C clamps

Fig. 9-2 Jigs with clamping bars

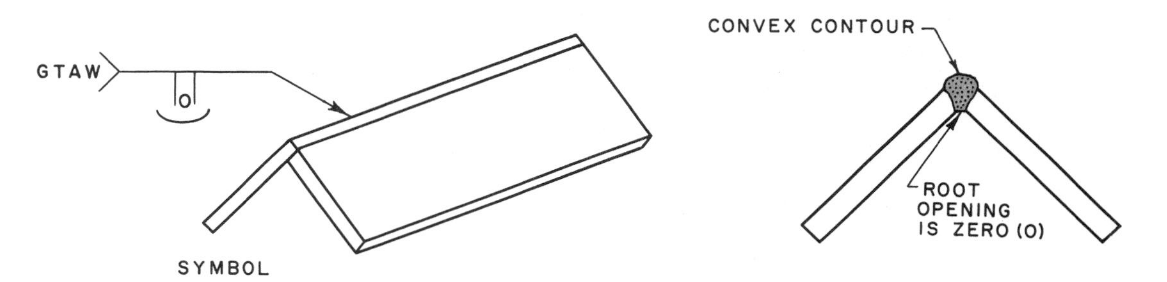

Fig. 9-3 Outside corner weld

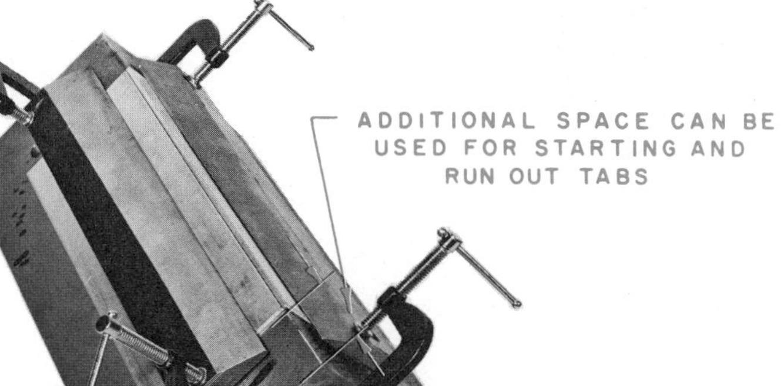

Fig. 9-4 Plates in position for an outside corner weld

CHART 9-1					
DATA FOR CORNER WELDS IN ALUMINUM					
Thickness in Inches	**HFAC Welding Current Flat* Amperes**	**Tungsten Electrode Diameter**	**Welding Speed Inches per Min.**	**Filler Rod Diameter**	**Recommended Argon Flow Cu. Ft. per Hr. *** **
1/16	60-80	1/16	12	1/16	15 to 20
1/8	125-145	3/32	12	3/32	17 to 25
3/16	190-220	1/8	11	1/8	21 to 30
1/4	280-320**	3/16	10	1/8 or 3/16	25 to 35
3/8	350-400**	3/16, 1/4	5	3/16 or 1/4	29 to 40
1/2	420-470**	3/16, 1/4	3	3/16 or 1/4	31 to 40

* — Current values are for flat position only. Reduce the above figures by 10% — 20% for vertical and overhead welds.

** — For current values over 250 amps., use a torch with a water-cooled nozzle.

*** — Use lower argon flow for flat welds. Use higher argon flow for overhead welds.

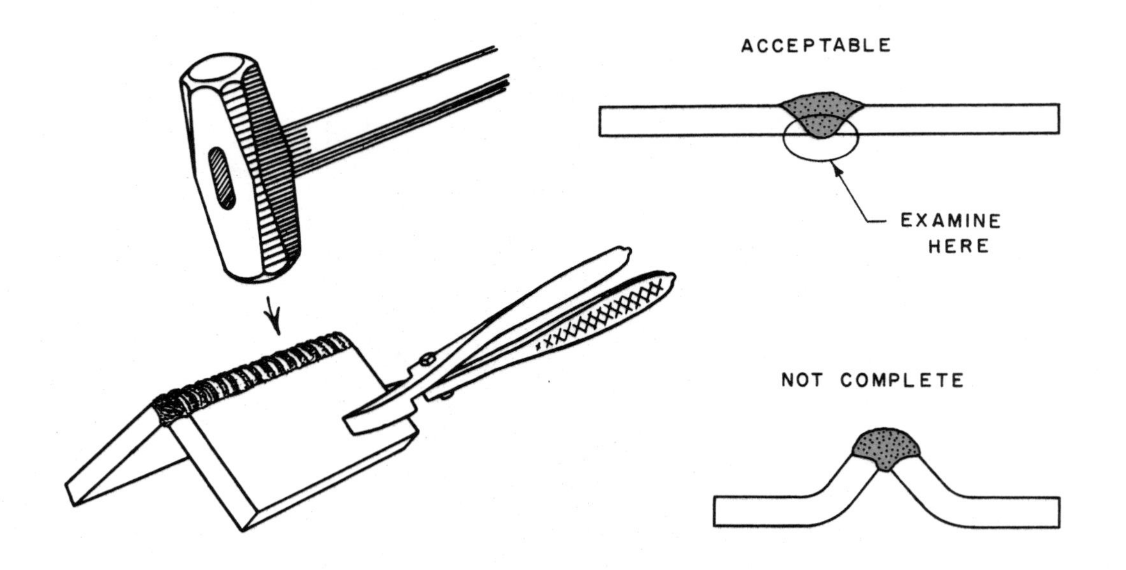

Fig. 9-5 Testing an outside corner weld

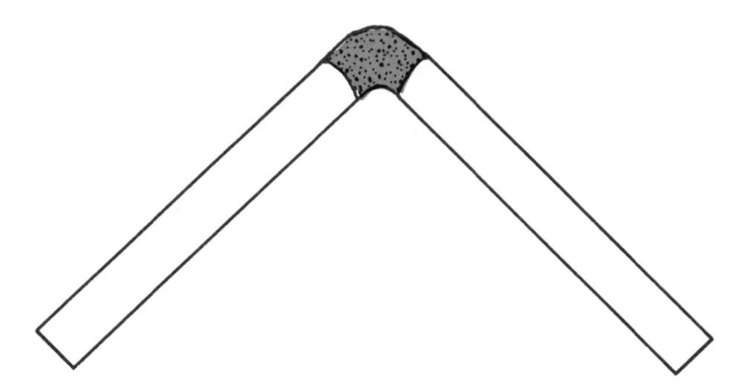

Fig. 9-6 Root of weld with slight fillet

3. Refer to chart 9-1 for electrode size, amperage, and argon flow.
4. Strike an arc and weld the corner; try for a smooth joint with 100 percent fusion.
5. Cool and examine the weld and the reverse side of the joint for uniformity and fusion. Also note any residue.
6. To test this weld place it on an anvil and hammer it flat as shown in figure 9-5.
7. Clean two more plates in the manner used in unit 8 (with steel wool or abrasive cloth) and place them in the jig.
8. Make this weld and be sure to note any change in the ease or difficulty of welding as compared to step 4.
9. Cool and inspect this weld for appearance, fusion, residue and evidence of wrinkled aluminum oxide on the root side of the assembly. Compare with the joint previously made without cleaning the material.
10. Make more corner welds and test them as in step 6. If it is found that there is a tendency for the joint to break, remove a little material from the corner of the jig by filing or grinding so that the root of the weld can take the shape indicated in figure 9-6.
11. If polishing and buffing equipment are available, try buffing a joint and examine it to find out how a rough weld, or a weld with deep ripples, tends to increase the buffing time and cost of finishing.

REVIEW QUESTIONS

1. What care should be taken of the jig once it is put in use? Why?

2. What condition leaves the root side of the joint in the best appearance?

3. What effect does a slight fillet have on the ability of the joint to withstand hammer testing?

4. If it is necessary to provide a joint with the root side absolutely free from oxide, how would this be accomplished?

5. What is the best material to use when making the jig? Why?

UNIT 10 BUTT WELDS ON ALUMINUM

This unit provides the opportunity to gain skill and knowledge in making what is probably the most common joint used in welding, the butt joint. The butt weld is shown in figure 10-1.

Materials

2 pieces 1/8 in. x 2 in. x 6 in. to 9 in. long clean aluminum
1/16 in. or 3/32 in. x 36 in. long 4043 welding rod

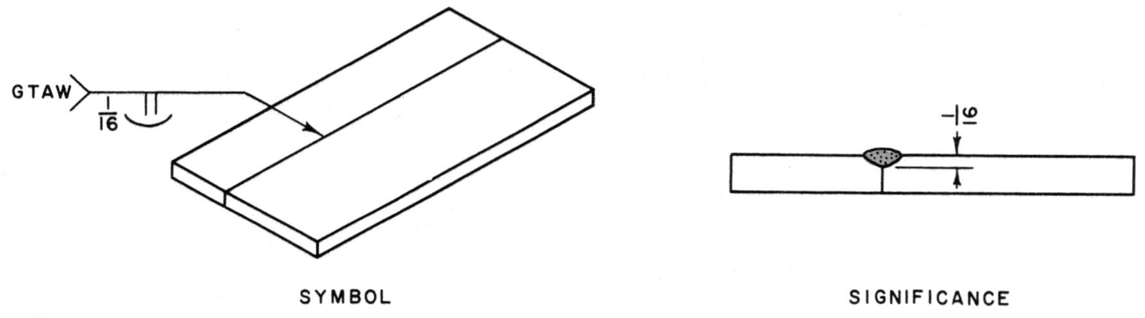

Fig. 10-1 Butt weld

Procedure

1. To get a better job, use a backing bar as shown in figure 10-2, rather than tacking and welding the aluminum on a welding bench. A steel bar is acceptable but stainless steel is better.

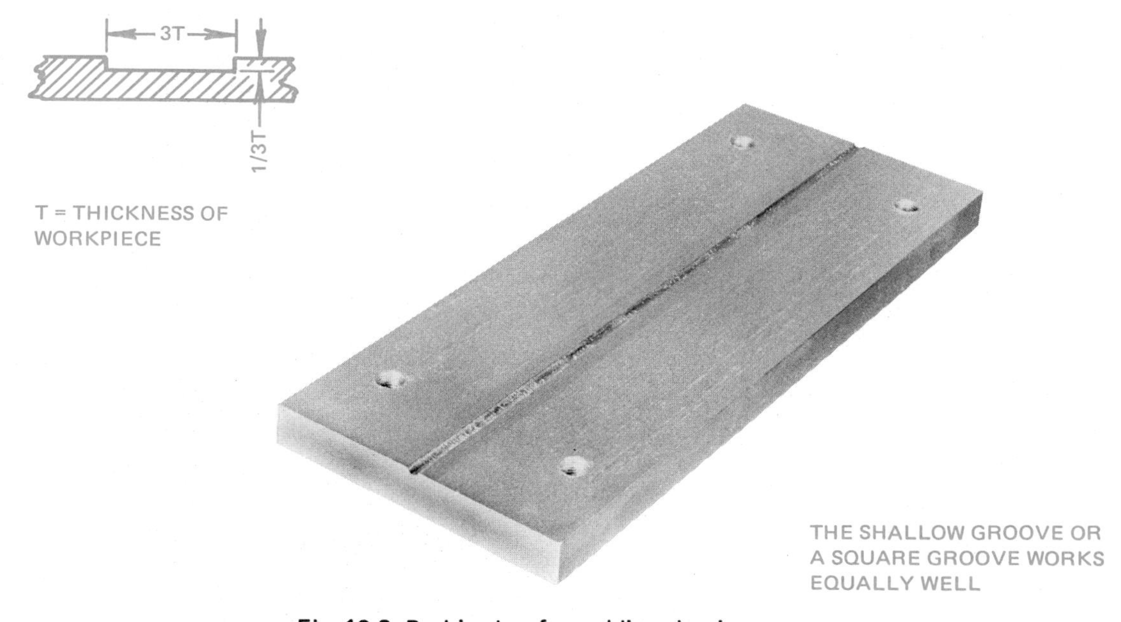

Fig. 10-2 Backing bar for welding aluminum

CHART 10-1					
DATA FOR BUTT WELDS IN ALUMINUM					
Thickness In Inches	**HFAC Welding Current Flat* Amperes**	**Tungsten Electrode Diameter**	**Welding Speed Inches per Min.**	**Filler Rod Diameter**	**Recommended Argon Flow Cu. Ft. per Hr. *** **
1/16	60-80	1/16	12	1/16	15 to 20
1/8	125-145	3/32	12	3/32 or 1/8	17 to 25
3/16	190-220	1/8	11	1/8	21 to 30
1/4	260-300**	3/16	10	1/8 or 3/16	25 to 35
3/8	330-380**	3/16, 1/4	5	3/16 or 1/4	29 to 40
1/2	400-450**	3/16, 1/4	3	3/16 or 1/4	31 to 40

* – Current values are for flat position only. Reduce the above figures by 10% – 20% for vertical and overhead welds.

** – For current values over 250 amps., use a torch with a water-cooled nozzle.

*** – Use lower argon flow for flat welds. Use higher argon flow for overhead welds.

2. Use the standard preweld procedure before starting to weld.
3. Refer to chart 10-1 for electrode size, current setting, and argon gas flow.
4. Check the electrode extension and the seating of the electrode in the collet.
5. Make a bead using procedure and technique similar to those described for making stringer beads in unit 8.

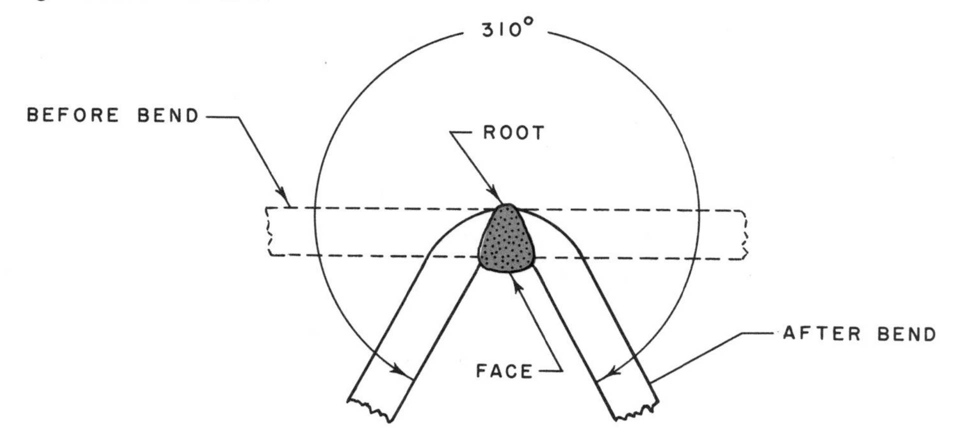

Fig. 10-3 Bend test for butt weld

6. Cool and examine the finished bead for appearance and complete fusion.
7. Test the finished weld by bending it in a brake or vise as shown in figure 10-3.
8. Examine the root side of the bent piece. There should be no indication of cracking or failure.
9. Continue to make tests and examine butt welds until the welds made are consistently of excellent quality and appearance with a cross section similar to that shown in figure 10-4.

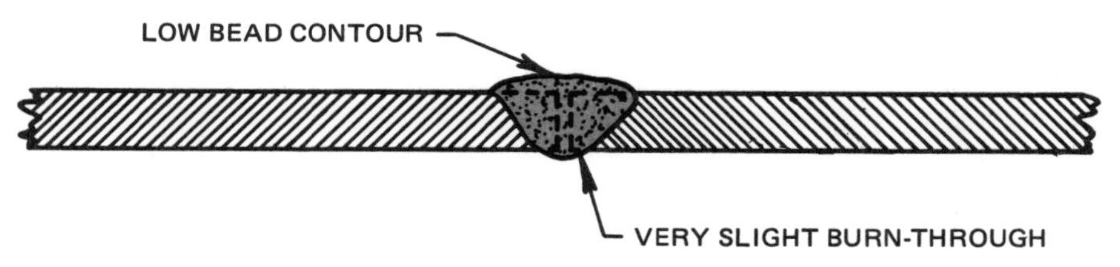

Fig. 10-4 Cross section of ideal butt weld

10. It is not always practical to make butt welds using a backing bar. Try tacking some plates and supporting them so that there is an air space between the workbench and the root of the joint. Make butt welds using this procedure until the welded joints are of uniform appearance with good fusion at the root, but not too great an amount of burn-through.

    Note: If there is some difficulty in fusing this type of joint, try the technique shown in figure 8-2.
11. Test the fusion at the root of the weld by bending these welds as in step 7.

REVIEW QUESTIONS

1. What dimensions should the groove in a backing bar have?

2. What will the result be at step 10 if the filler rod is allowed to move far enough to be outside the shielding gas zone?

3. What are the recommendations for joint spacing for a square butt joint in 1/8-inch plate?

4. If the surface of this bead is white and dull what does it indicate?

5. What is the thickest material that can obtain 100 percent penetration from one side?

UNIT 11 LAP WELDS ON ALUMINUM

While lap welds, figure 11-1, and fillet welds (which will be studied in unit 12) have many similarities, there are enough differences to make the separate study of each joint worthwhile.

Materials

2 pieces clean aluminum, 1/8 in. x 2 in. x 6 in. to 9 in. long
1/16 in. or 3/32 in. x 36 in. long 4043 welding rod

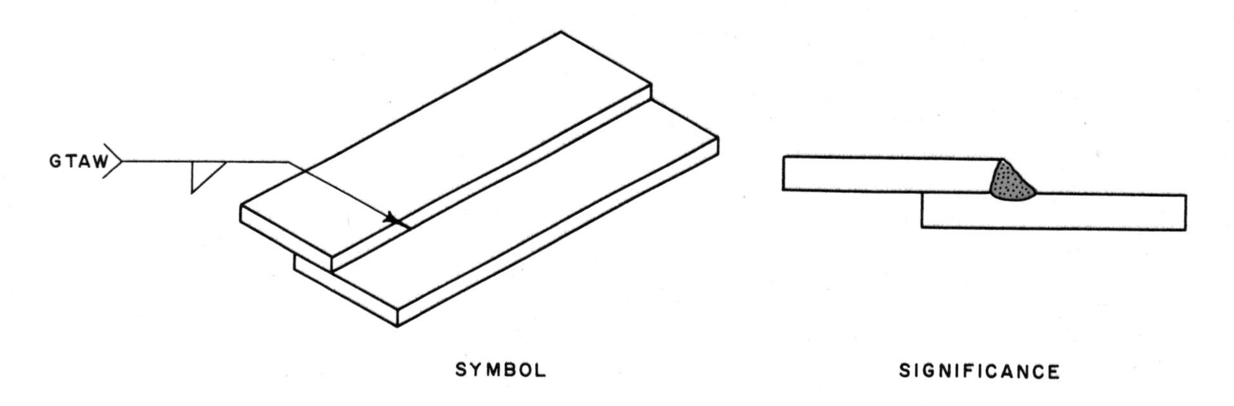

Fig. 11-1 Lap weld

Procedure

1. Follow the standard preweld procedure for TIG welding equipment.
2. Set up the cleaned plates as shown in figure 11-2.
3. Refer to chart 11-1 for electrode size, amperage setting, argon gas flow and nozzle size.

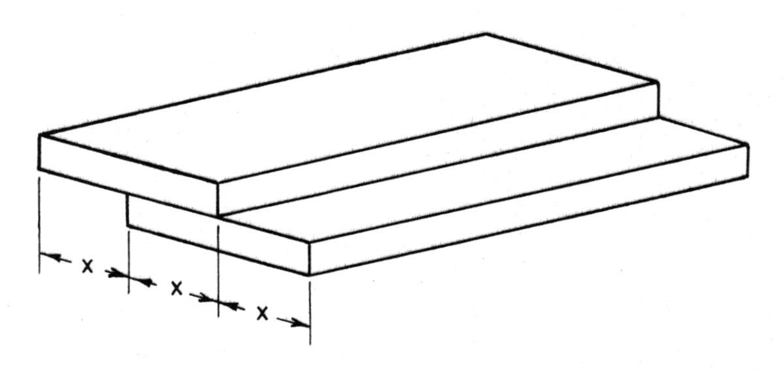

Fig. 11-2 Setup for lap joint

CHART 11-1

DATA FOR LAP AND FILLET WELDS IN ALUMINUM					
Thickness in Inches	HFAC Welding Current Flat* Amperes	Tungsten Electrode Diameter	Welding Speed Inches per Min.	Filler Rod Diameter	Recommended Argon Flow Cu. Ft. per Hr. ***
1/16	70-90	1/16	10	1/16	15 to 20
1/8	140-160	3/32	10	1/16 or 3/32	17 to 25
3/16	210-240	1/8	9	1/8	21 to 30
1/4	280-320**	3/16	8	1/8 or 3/16	25 to 35
3/8	330-380**	3/16, 1/4	5	3/16 or 1/4	29 to 40
1/2	400-450**	3/16, 1/4	3	3/16 or 1/4	31 to 40

* – Current values are for flat position only. Reduce the above figures by 10% – 20% for vertical and overhead welds.

** – For current values over 250 amps., use a torch with a water-cooled nozzle.

*** – Use lower argon flow for flat welds. Use higher argon flow for overhead welds.

Note: Use an electrode extension which is just slightly longer than for butt welds.

4. Make a lap weld, applying the filler rod along the edge of the top plate. Pay close attention to the molten pool. A notch effect is usually observable in the joint as in figure 11-3. Avoid melting the edge of the top plate more rapidly than the center of the bottom plate.

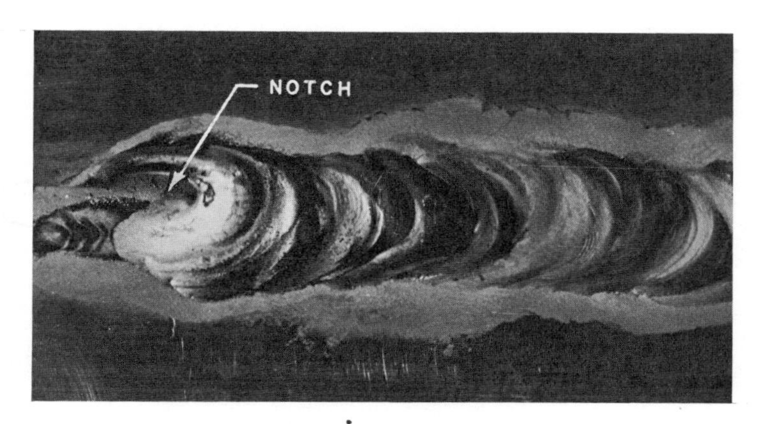

Fig. 11-3 Notch effect

5. Each edge of the weld tends to melt some distance ahead of the center of the pool, which results in a loss of the continuity of the weld. To overcome this, apply the technique used in making stringer beads and butt welds.
6. Make more lap welds, using both sides of the plates, except in those instances where it is desired to test the finished weld. Note that the electrode angle in figure 11-4 is much greater than that used for metallic-arc welding.

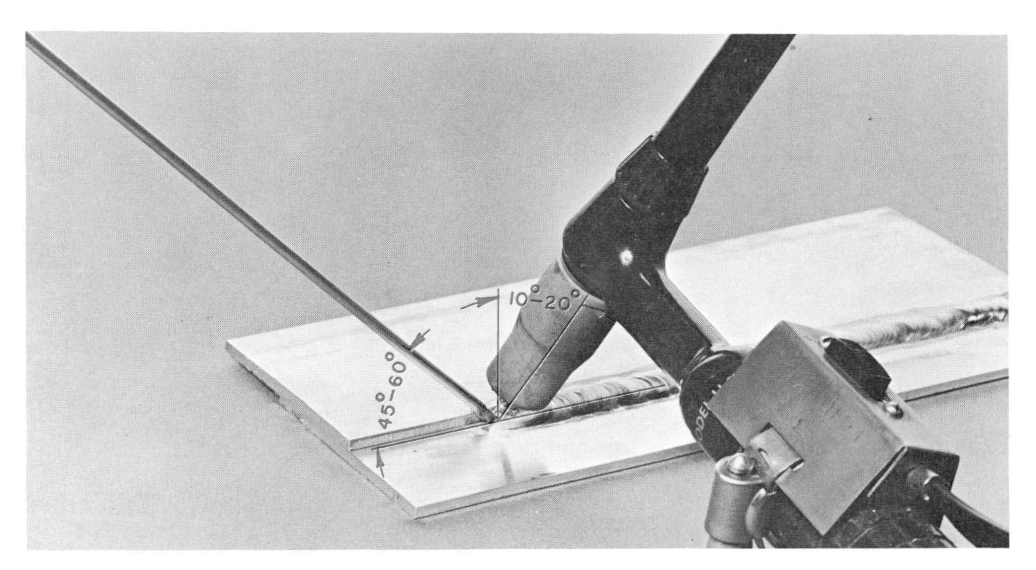

Welder's view

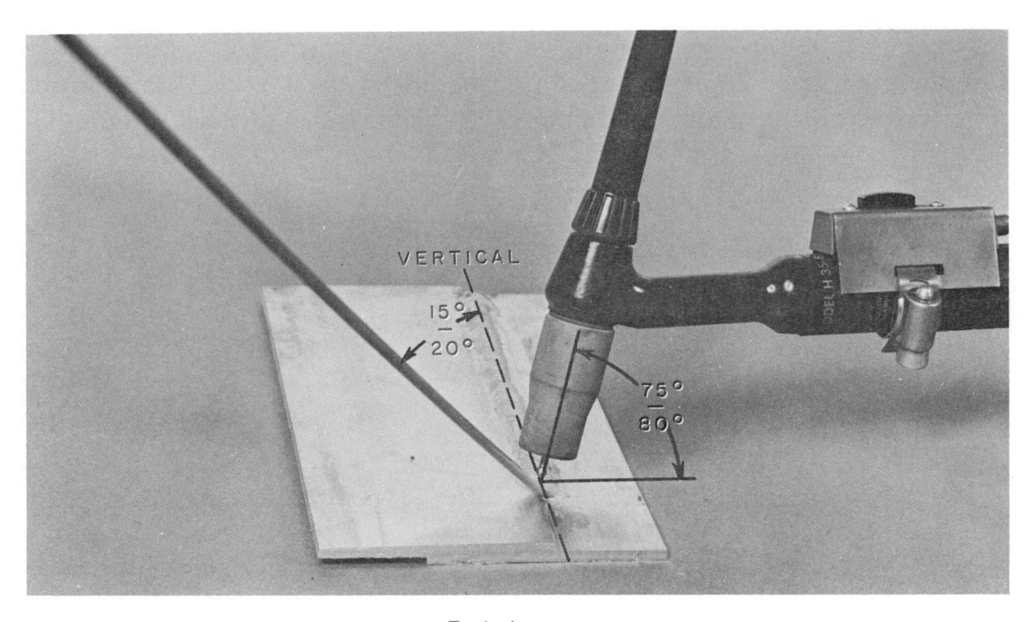

End view

Fig. 11-4 Rod and torch angles

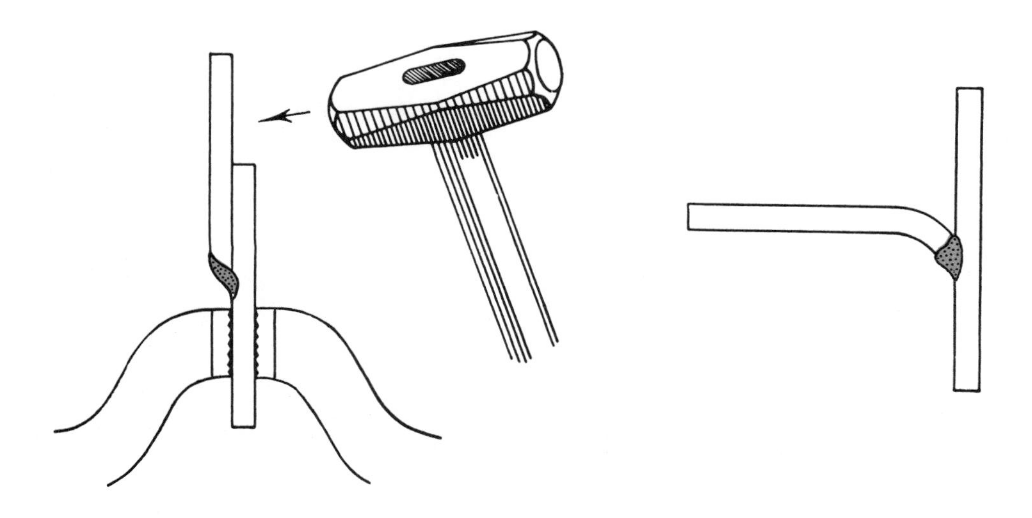

Fig. 11-5 Testing a lap weld

7. Test the finished welds as shown in figure 11-5.
8. Examine the root of the weld for evenness and uniformity of fusion.
9. Make more lap welds, but vary the technique to produce welds with a cross section similar to that in figure 11-6.

   Note: It has been proven by tests that a lap weld of this contour transfers stresses from one plate to the other more efficiently than a joint made with the face of the weld at a 45-degree angle.

10. Test these welds as in step 7, after first observing the finished bead for uniformity of fusion with both plates, as well as general appearance of the contour and ripples.

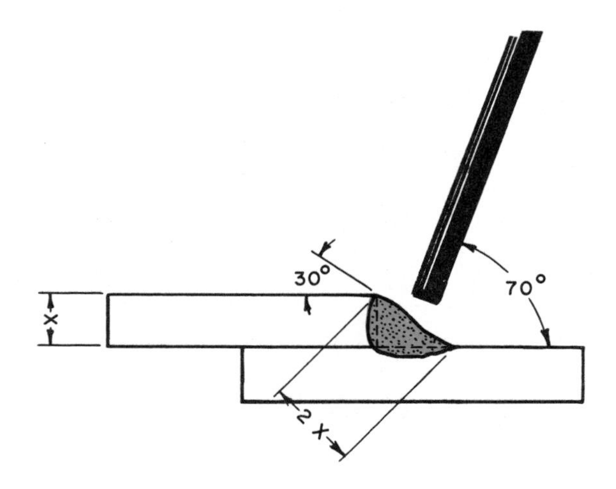

Fig. 11-6 Lap joint weld

REVIEW QUESTIONS

1. Should the current range for lap welds on aluminum be low, medium, or high? Why?

2. Where is the welding rod held in relation to the lap joint?

3. What rod angle affects melt-off rate?

4. What does the notch indicate in the weld crater?

5. Regarding penetration, what must be considered on this joint?

UNIT 12 FILLET WELDS ON ALUMINUM

Fillet welds, figure 12-1, using the TIG torch may be slightly awkward to make. The gas nozzle gives some interference not found in either oxyacetylene or metallic-arc welding. However, attention to detail makes it possible to produce welds of excellent appearance and quality.

In making fillet welds care must be taken to avoid *undercutting,* figure 12-2, an absence of metal along the top edge of the weld. It is caused by faulty manipulation of the electrode and filler rod.

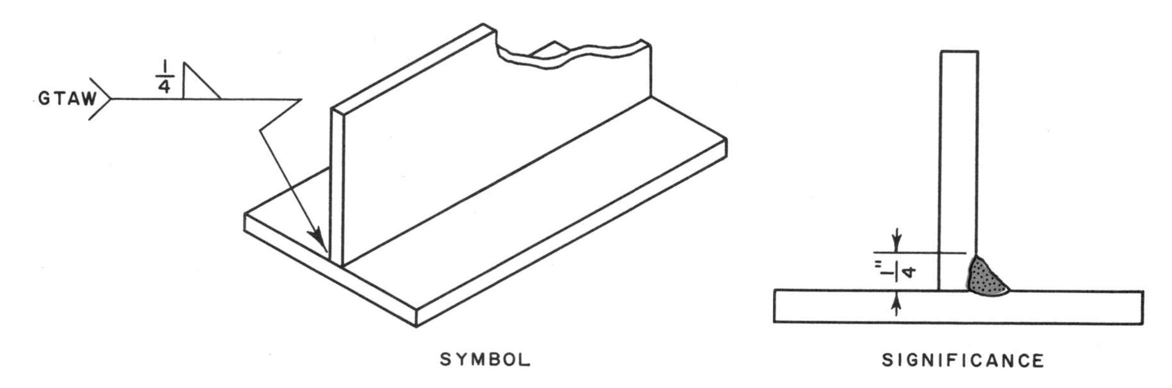

Fig. 12-1 Fillet Weld

Materials

2 pieces clean aluminum 1/8 in. x 2 in. x 6 in. to 9 in. long
1/16 in. or 3/32 in. x 36 in. long 4043 aluminum welding rod

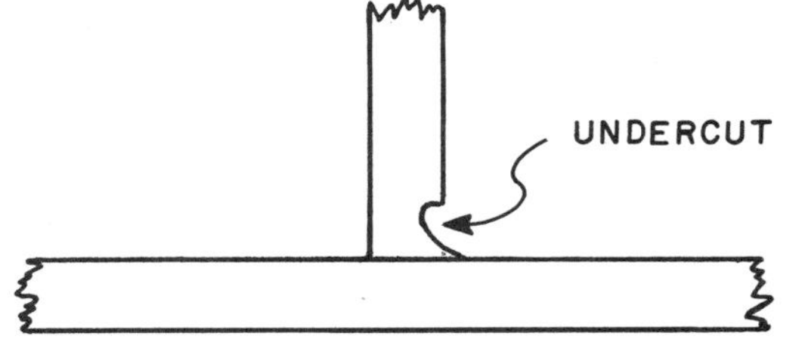

Fig. 12-2 Undercutting

Procedure

1. Check all factors of the equipment.
2. Set up the plates as in figure 12-1.
3. Refer to chart 11-1 for electrode size, nozzle size, amperage, and argon flow.
4. Refer to figure 6-2 and adjust the electrode extension.
5. Make a fillet weld with the torch angle and rod angle as shown in figure 12-3, view A.
6. Operate the electrode and filler rod as shown in figure 12-3, view B. Proper manipulation avoids the condition of undercutting which can happen as easily in TIG welding as in other processes.
7. Observe the finished weld for surface appearance and uniformity.
8. Test the weld as shown in figure 12-4. If the penetration and fusion are not deep enough, this joint will probably break. Examine the break for uniformity of the line of fusion.

(A) Welder's view

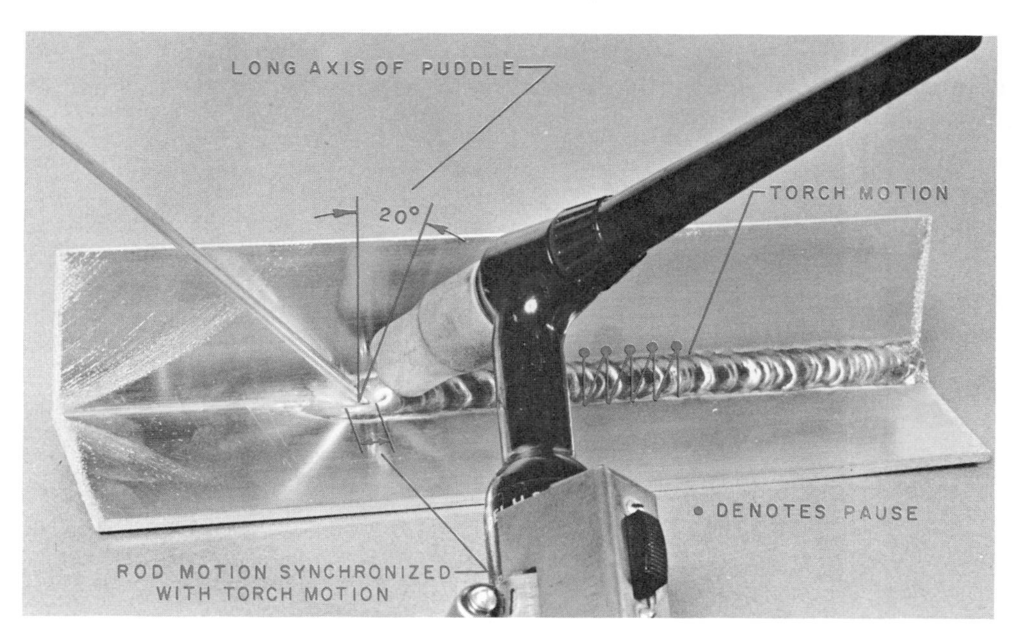

(B) Torch and rod motion

Fig. 12-3 Fillet weld

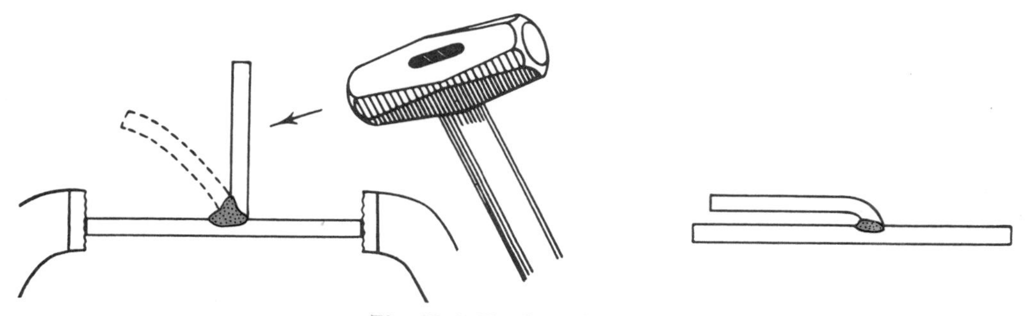

Fig. 12-4 Testing a fillet weld

9. Make more fillet welds. Try to get deeper fusion and better bead appearance.
10. Continue to make and test this type of joint until the result is a high-quality weld every time.

REVIEW QUESTIONS

1. Why is undercutting considered such a serious error in the welding industry?

2. Is it possible to use a lower argon gas flow when making a fillet weld?

3. Would the afterflow time be the same duration as for other types of joints? Shorter? Longer?

4. Could it be possible to use a smaller ceramic cup on this joint? Why?

5. What shape does the surface of the bead present?

UNIT 13 FLANGED BUTT WELDS ON ALUMINUM

Light-gage aluminum, as well as other thin metals and alloys, requires the maintenance of a rather short arc at a comparatively high rate of travel. The addition of filler rod to the joint further complicates the problem. For this reason, the flanged butt joint, figure 13-1, is often used when welding the lighter gage materials. This relieves the welder of the responsibility of manipulating the small-diameter, very flexible filler rod necessary to make a weld of acceptable appearance. The flanged butt weld is shown in figure 13-2.

Fig. 13-1 Flanged butt joint

Materials

2 pieces .030-inch to .050-inch clean aluminum, 2 in. wide x 9 in. long, flanged as shown in figure 13-1

Pure tungsten electrodes, .040 inch or 1/16 inch

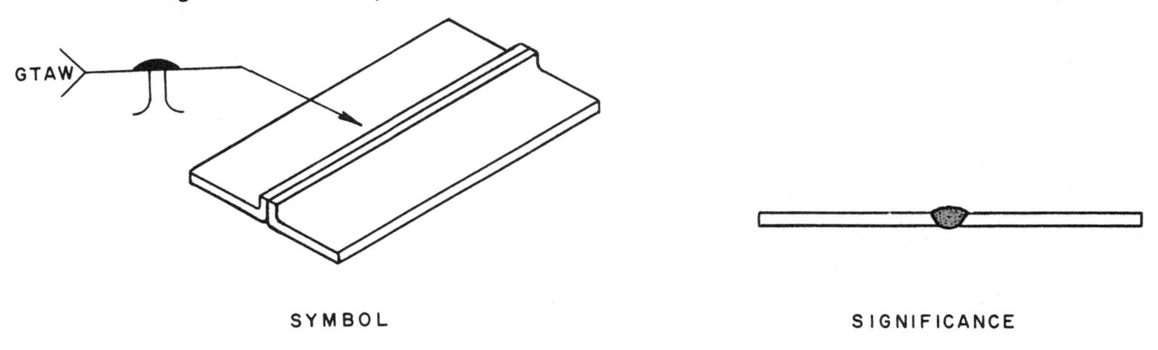

Fig 13-2 Flanged butt weld

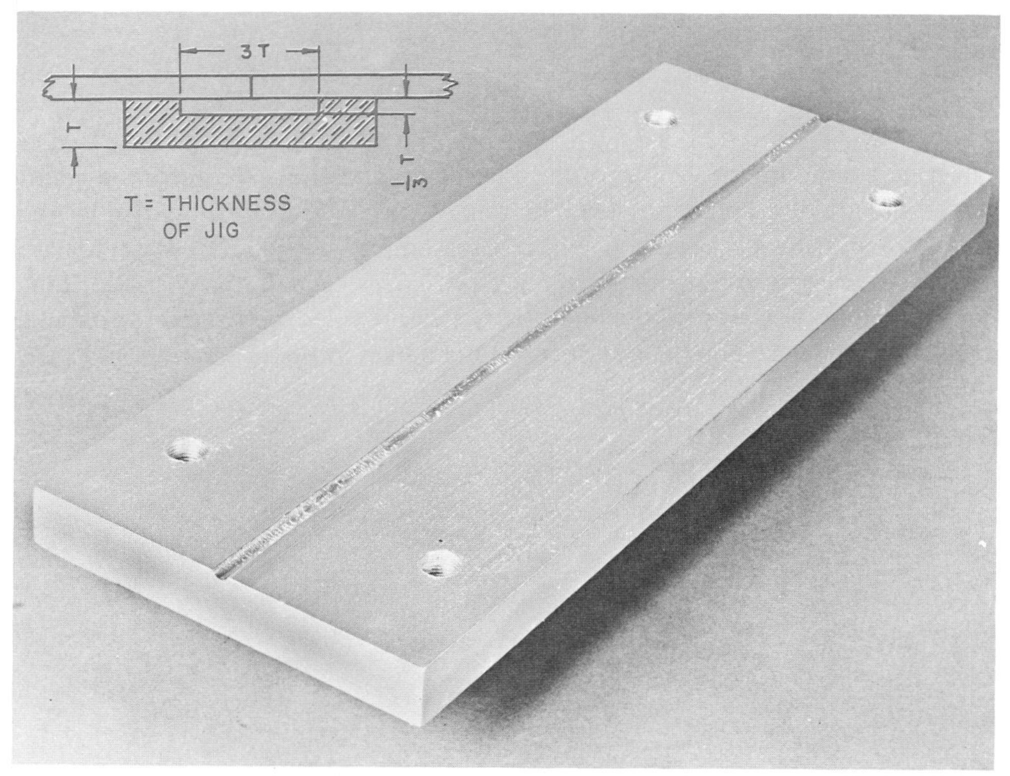

Fig. 13-3 Jig for welding light sheets

Procedure

1. Follow the standard preweld procedure for checking prior to HFAC TIG welding.
2. Use a .040-inch or 1/16-inch pure tungsten electrode with the machine set for the proper current value.
3. Set up the flanged sheets in a jig which holds the parts in proper alignment.

   Note: A steel or stainless steel jig such as that shown in figure 13-3 provides a quick and easy method of making the setup.
4. Strike the arc and make a weld, adjusting the rate of travel to provide good fusion and penetration.
5. Remove the finished work, cool, and examine both sides carefully for uniformity of the bead as well as for good fusion and uniform penetration.
6. Obtain more sheets and make more welds until each one made is of acceptable uniform appearance.

   Note: At this point, it is advisable to add starting and run-out tabs as shown in figure 13-4. Tabs are often used in high-speed welding to provide for a more uniform bead, with easier starting and stopping.
7. Set up more flanged plates, but change the electrode to one that is four times as large as previously used. Set the machine for DCRP electrode positive. Leave the current setting as it was for the small electrode used with the previous joints.

Fig. 13-4 Flanged butt joint with starting and run-out tabs

Note: If the torch is equipped to handle electrodes only 5/32 inch in diameter, do not exceed the current setting used for a .040-inch diameter electrode when using HFAC.

8. Make a weld using DCRP and note the ease or difficulty of producing the weld. Compare the rate of travel necessary with that required when using HFAC. Use a slightly longer arc when welding this joint.
9. Remove the work and examine it as in step 5.
10. Obtain some 1/16-inch clean aluminum plates and make square butt joints with DCRP.
11. Remove and examine these joints. Pay close attention to the width of the beads and the penetration at the root of the weld. Also, examine the electrode and compare its condition with the electrode used for HFAC welding.
12. Tack two flanged sheets of aluminum and weld them without the use of a jig. Compare the ease or difficulty of making this joint with the previous joints. Compare the finished assembly with those made before.

REVIEW QUESTIONS

1. How does the weld made with DCRP compare with that made with HFAC?

2. How does the end of the electrode used with DCRP compare with that used with HFAC?

3. How do the observations made in this unit compare with those made in unit 5?

4. What disadvantage is there to welding light-gage aluminum without a jig?

5. What happens when the tungsten comes in contact with the work?

UNIT 14 WELDING MAGNESIUM AND MAGNESIUM ALLOYS

The welding of magnesium and its alloys started laboratory experimentation and investigation of the use of inert-gas shielded-arc welding in the early 1940s.

CHARACTERISTICS OF MAGNESIUM

The metal is silvery white in appearance, relatively light in weight and comparatively strong. The specific gravity of magnesium is 1.74 as compared with 2.70 for aluminum. This is only 65 percent or approximately two-thirds of the weight of aluminum. Comparing this 1.74 with steel which has a specific gravity of 7.86, it is found that magnesium weighs only 22 percent or roughly one-quarter as much as steel. The melting point of magnesium is almost identical with that of pure aluminum: magnesium melts at 1204 degrees F. and aluminum, at 1218 degrees F. While magnesium itself is not too strong, its alloys possess excellent strength, as noted in chart 14-1. The thermal conductivity of magnesium is reasonably high when compared to other metals.

WELDING MAGNESIUM

Welding magnesium is no more difficult than welding aluminum. The same equipment and techniques are used and the same preparation and precautions must be carried out. The rate at which magnesium expands when heated is higher than aluminum. More severe warpage results if proper precautions are not taken, especially in the lighter thicknesses.

CHART 14-1

MAGNESIUM ALLOY WELDABILITY			
Wrought Alloy	**TIG Weld Rating**	**Cast Alloy**	**TIG Weld Rating**
AZ10A	A	AM100A	B+
AZ31B	A	AZ63A	C
		AZ81A	B+
AZ80A	B	AZ91C	B+
HK31A	A	AZ92A	B
HM21A	A	EK41A	B
HM31A	A	EZ33A	A
LA141A	B	HK31A	B+
M1A	A	HZ32A	C
ZE10A	A	K1A	A
ZK21A	B	QE22A	B
		ZE41A	C
		ZH62A	C–
		ZK51A	D
		ZK61A	D

Welding Ratings:
A-Excellent B-Good C-Fair D-Welding not recommended

The American Welding Society

CHART 14-2

DATA FOR TIG WELDING MAGNESIUM									
Material Thickness – inches	Number of Passes	Welding Current in amperes		Electrode diameter – inches			Welding Rod Diameter – inches	Gas Flow – Cu. Ft./Hr.	
		AZ31B	HK31A HM21A	HFAC	DCRP	DCSP		Argon	Helium
0.040	1	35	40	1/16	5/32	.040	3/32	12	24
0.063	1	50	55	1/16	3/16	.040	3/32	12	24
0.80	1	65	70	1/16	3/16	1/16	3/32	12	24
0.100	1	85	95	3/32	1/4	1/16	1/8	18	30
0.125	1	100	110	1/8	1/4	1/16	1/8	18	36
0.190	1	140	155	1/8	–	3/32	5/32	18	36
0.190	2	100	110	1/8	1/4	1/16	1/8	18	36
0.250	1	180	200	5/32	–	3/32	5/32	18	48
0.250	2	115	125	1/8	1/4	1/16	1/8	18	36
0.375	1	250	275	3/16	–	1/8	3/16	24	48
0.375	2	140	155	1/8	–	3/32	5/32	24	48
0.500	2	310	340	1/4	–	1/8	3/16	24	48
0.750 and over	2	350	385	1/4	–	5/32	3/16	36	48

Current values given for welding with a backing plate and for making fillet welds. Slightly lower current values used for welding without a backing plate and for making corner or edge joints.

The same types of joints and joint preparation used in welding aluminum apply here so the description is not repeated. In general, the joints are tightly butted together or, if a gap is left, it is usually less than 1/16 inch wide.

For manual TIG welding, chart 14-2 supplies the necessary data with the exception of the gas nozzles. The formula which requires that the nozzle be five to six times the diameter of the electrode applies equally in this case except for DCRP in which the nozzle should match the electrodes used for HFAC or be slightly larger. The small nozzles tend to increase the velocity of shielding gas and cause turbulence in the weld and, on occasion, draw air into the shielding zone.

Magnesium alloys are usually supplied with the surface etched to remove impurities and then oiled to preserve this surface. The oil must be removed and the surface either mechanically cleaned with abrasives or chemically cleaned prior to welding to eliminate any harmful inclusions.

SAFETY

All the standard protective equipment used in arc welding should also be used with magnesium welding. Magnesium is not particularly a fire hazard except in the form of chips, turnings, filings or other small particles. These should never be in the welding area.

One other safety precaution is necessary when welding magnesium-thorium alloys. Thorium is a radioactive element and can be toxic. A suggested limit of 0.1 mg of thorium per cubic meter of air (about 30 cubic feet of air) is a safe limit in the welder's breathing zone.

REVIEW QUESTIONS

1. What precautions are necessary when welding the magnesium-zinc alloys?

2. Why is the thermal conductivity rating of an alloy important to a welder?

3. What advantage is there in using DCRP when welding magnesium?

4. What can happen if magnesium grinding dust or particles come in contact with the welding arc?

5. On thick sections of magnesium, what might be necessary before welding?

UNIT 15 TIG WELDING STAINLESS AND MILD STEEL

Stainless steel is one of the most widely used alloys in our modern day living. It is found in the home in tableware, kitchenware and equipment as well as in some decorative pieces. It is equally important but not as obvious in the form of heating elements in toasters, grills, electric stoves, space and water heaters and many useful applications. Stainless steel is used in hotels, restaurants, transportation, communications, paper making, and chemical and food processing industries. It is supplied in sheets, strip, plate, structural shapes, tubing, pipe, wire extrusions and in a wide variety of alloys and finishes.

In general, a welder who has mastered the TIG process in the preceding units should have no difficulty in producing welds of excellent quality and appearance in stainless and mild steel. The absence of flux and slag, the high degree of visibility in the weld area, the concentration of heat in a very narrow zone and the elimination of splatter, all contribute to produce strong, sound, smooth welds at a high rate of travel. However, the net results are equal to the degree of attention given to all the factors involved in producing TIG welds in stainless steel. The techniques of TIG welding mild steel parallel those for the welding of stainless steel very closely.

PROPERTIES AND COMPOSITION OF STAINLESS STEEL

All stainless steels derive their stainless characteristics from the chromium content. The other chief alloying element is nickel, which is added to reduce the tendency of the metal to harden as well as to give other desirable characteristics. The major producers of stainless steel can furnish details and specifications for any type or grade they produce.

WELDING STAINLESS STEEL

The actual welding of stainless steel is not difficult. However, like other metals and alloys, it has certain characteristics which must be recognized and compensated for if satisfactory joints are to be produced by any fusion process.

The thermal conductivity of the alloys is about 40 percent to 50 percent less than for carbon steel. This means that the heat is retained in the weld zone much longer.

The thermal expansion is much greater than for carbon steel — about 50 percent more. This means that the shrinkage stresses are much greater and the resulting warpage becomes more of a problem if proper jigs and fixtures are not used.

In general, the metallurgical aspects of stainless steel welds are improved if the heat is carried away from the weld zone at a rapid rate. This can be accomplished by using jigs made of copper or by using copper inserts of adequate size. The electrical resistance of stainless steel varies from six times as great as carbon steel to twelve times as great, depending on the condition of the stainless steel.

JOINT PREPARATION

In the lighter gages (up to .040 inch), the best results are obtained in a butt joint by flanging both pieces. In the intermediate gages (16 through 10 gages), a square butt joint

CHART 15-1

DATA FOR MANUAL WELDING OF STAINLESS STEEL						
Thickness in Inches	Type of Weld	DCSP Welding Current Flat* Amperes	Tungsten or Thoriated Electrode	Welding Speed Inches per minute	Filler Rod Diameter	Recommended Argon Flow Cu. Ft. per Hr.
1/16	Butt	80-100	1/16	12	1/16	11
	Lap	100-120	1/16	10	1/16	11
	Corner	80-100	1/16	12	1/16	11
	Fillet	90-100	1/16	10	1/16	11
3/32	Butt	100-120	1/16	12	3/32	11
	Lap	110-130	1/16	10	3/32	11
	Corner	100-120	1/16	12	3/32	11
	Fillet	110-130	1/16	10	3/32	11
1/8	Butt	120-140	1/16	12	3/32	11
	Lap	130-150	1/16	10	3/32	11
	Corner	120-140	1/16	12	3/32	11
	Fillet	130-150	1/16	10	3/32	11
3/16	Butt	200-250	3/32	10	1/8	13
	Lap	225-275	1/8	8	1/8	13
	Corner	200-250	3/32	10	1/8	13
	Fillet	225-275	1/8	8	1/8	13
1/4	Butt	275-350**	1/8	5	3/16	13
	Lap	300-375**	1/8	5	3/16	13
	Corner	275-350**	1/8	5	3/16	13
	Fillet	300-375**	1/8	5	3/16	13
1/2	Butt	350-450**	3/16	3	1/4	15
	Lap	375-475**	3/16	3	1/4	15
	Corner	375-475**	3/16	3	1/4	15

* – Current values are for flat position only. Reduce the above figures by 10% – 20% for vertical and overhead welds.

** – For current values over 250 amps., use a torch with a water-cooled nozzle.

usually gives good results. In the lighter gages, a closed square butt joint is usually used. In the heavier gages, an open butt joint gives better penetration and fusion. An opening equal to one-third to one times the thickness of the material is recommended. The exact amount varies with the thickness of the material, the amount of current used, and the rate of travel. The operator, when preparing any joint for welding, must be very critical of any evidence of dirt, oil, grease, moisture or any material which might affect the finished weld.

CHOICE OF CURRENT

Generally DCSP is recommended when welding stainless steel or mild steel. In this case the cleaning action of HFAC is not necessary. It also results in a higher input of heat to the work, as discovered in the experiments in unit 5. The net result is greater penetration and higher welding speeds.

In welding the thinner sections of stainless and mild steel, the use of HFAC is recommended because of the lower heat input. The tendency to burn through or pierce the work is reduced. In the case of very light sections, the use of DCRP further reduces the tendency to burn through, but the operator must be aware of the hazards found in this type of TIG welding. Small electrodes are consumed at a rapid rate. The general rule for DCRP is that the electrode must be at least four times as great in diameter as for DCSP for a given current value. This must be rigidly followed in order to produce sound joints economically. For mild steel, use high frequency on start only.

CHOICE OF ELECTRODES, CURRENT VALUES AND SHIELDING GAS FLOW

Chart 15-1 gives the electrode size, current value and argon gas flow for a variety of sheet and plate thicknessess. For mild steel, use an argon gas flow of 20 cubic feet per hour. A 2 percent thoriated tungsten electrode is most desirable. For the smaller thicknesses of mild steel, the tungsten electrode should be sharpened to a point. The gas nozzle should be of sufficient size to insure complete shielding of the welding zone with a low velocity flow of the shielding gas.

JIGS AND FIXTURES

The use of jigs and fixtures to clamp and align the work is strongly recommended when welding stainless steel. This material has a relatively high coefficient of thermal expansion. A good jig helps to hold the work in good alignment and eliminate some of the warpage due to thermal stresses.

The thermal conductivity of stainless steel is relatively low and rapid removal of heat from the work results, in general, in a finished product with better physical and metallurgical properties. This rapid removal of heat can best be accomplished by using clamping jigs made of copper, which has a high rate of thermal conductivity, or by using jigs with copper inserts. Figures 15-1 and 15-2 show two types of steel jigs with copper inserts.

TECHNIQUES AND PRECAUTIONS

The gas nozzle should be as close as possible to the work consistent with good visibility. When welding with DCSP, the best results are obtained if thoriated tungsten electrodes are

Fig. 15-1 Copper insert jig for butt welds

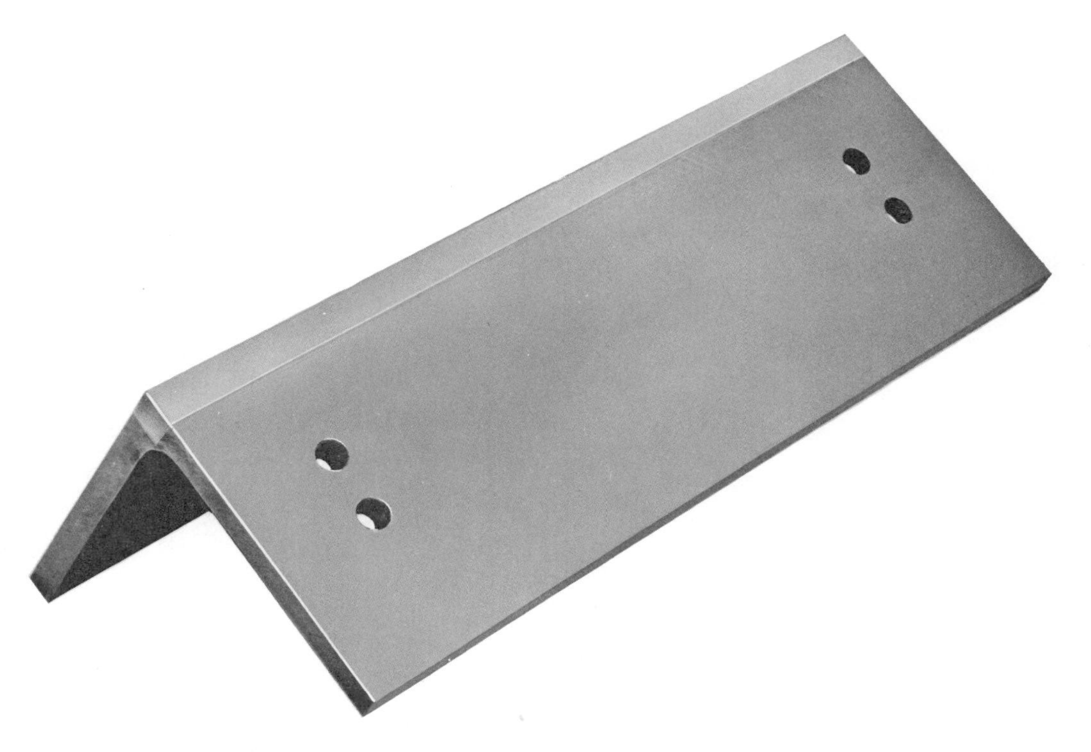

Fig. 15-2 Copper insert jig for corner welds

used. Welding in a still atmosphere free from drafts insures maximum effectiveness of the shielding gas used.

Argon gas shielding gives the smoothest arc action and a resulting smooth weld. Helium gas shielding produces a somewhat hotter arc and contributes to higher welding speeds. One of the large steel companies recommends experimenting with varying mixtures of these two gases to get the best results. In general, argon shielding gives good results when welding with all the alloys, especially in the thinner gages.

Electrode contamination must be guarded against to insure proper current flow. Accidental dipping of the electrode into the molten pool contaminates the electrode and greatly reduces the current-carrying capacity.

When filler rods are necessary, the use of a rod of the same composition as the base metal being welded gives good results. The TIG process does not reduce the percentage of the alloying elements to any great extent.

PRACTICE WELDS

At this point, the student should have developed enough skill and judgment to make a series of practice welds using various thicknesses of metal to make the standard types of joints. The coordinated motion of the TIG torch and filler wire remains the same as in figure 8-2.

It is found that 10-, 12-, and 14-gage stainless steel can be welded at a slow enough rate to permit the observation of the effects of varying arc length, electrode angles and filler-rod angles. While these thicknesses are welded with little difficulty, it is advisable to use workpieces from .040 inch to 1/16 inch thick and to practice until becoming proficient in producing welds in these light gages.

Some experimenting in dipping the electrode in the molten pool and observing the arc action and the weld produced helps to point out the undesirable characteristics of electrode contamination when welding in stainless steel. On both stainless and mild steel, the bead width should not be greater than 1/8 inch on light-gage sheet. Discoloration at each side of the weld should not extend more than twice the width of the bead.

If polishing and buffing equipment is available, some of the joints made should be polished, buffed and inspected. This is done to determine if the joints are of a quality

Fig. 15-3 Correctly made corner weld

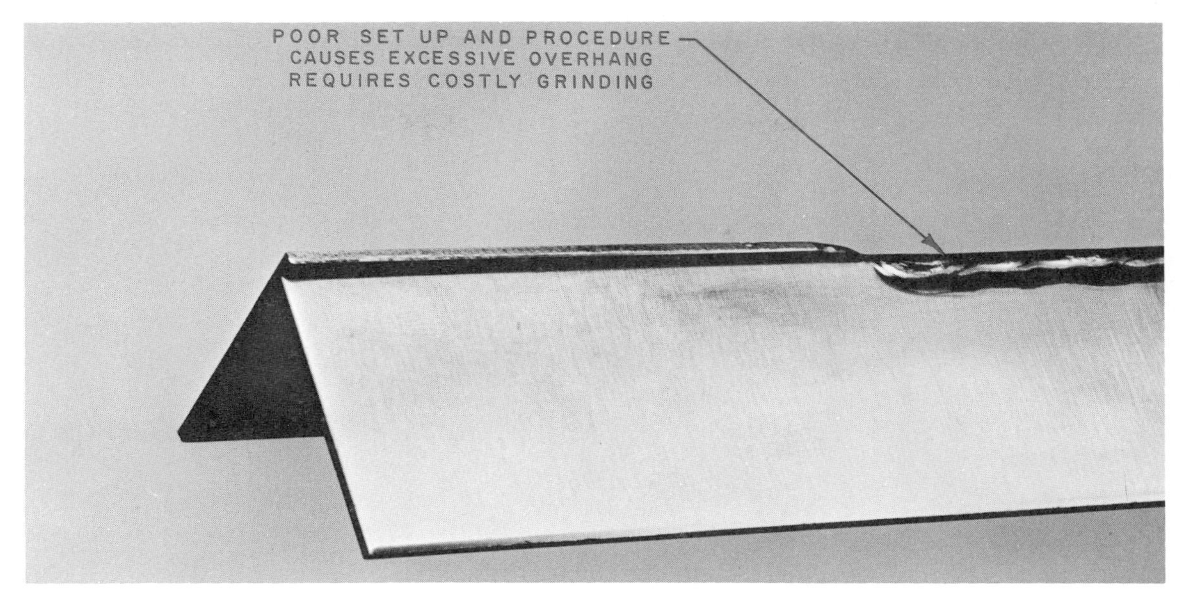

Fig. 15-4 Incorrectly made corner weld

which will yield the desired finish with a minimum of finishing time. Figure 15-3 shows a corner weld which has been buffed. Note the smooth contour and the line of fusion.

Figure 15-4 shows the same type of joint treated in the same manner, but made in such a way that the line of fusion would need much more grinding and polishing to give the smooth contour shown in figure 15-3.

REVIEW QUESTIONS

1. Why are jigs and fixtures more necessary when welding stainless steels?

2. How does the flow of shielding gas compare with that used for TIG welding aluminum and magnesium? Why?

3. If the finished bead is to be bright and shiny over its entire length, what procedure would be used?

4. How does the electrode extension and arc length compare with that used when welding other types of metals?

5. What is the best filler rod angle to use when TIG welding?

6. What are two reasons why DCSP is recommended for welding stainless steel?

UNIT 16 WELDING COPPER AND COPPER-BASE ALLOYS

Copper and its alloys are among the earliest metals used by man. Its low melting point, 1981 degrees F., plus the ease with which it could be refined from its ores made it readily available. Since copper could be worked and formed easily, it was a desirable material for many articles.

The use of copper and its hundreds of alloys is so extensive today that it would be hard to visualize present-day living without these valuable materials. From the welder's viewpoint, the ability to join copper and its alloys by one or several of the welding, brazing or soldering methods is an important factor.

DESIRABLE CHARACTERISTICS OF COPPER

- Copper and its alloys are highly resistant to many forms of corrosion.
- Copper and its alloys can be fabricated and formed by all standard methods.
- Copper is one of the best conductors of electricity and of heat energy.
- Many elements can be combined with copper to form a wide variety of alloys, each with specific characteristics. For example, some of these alloys are made to be highly ductile so that they can be drawn or spun. Other alloys may be highly wear-resistant and are used for many types of bearings.
- Some of the alloys are highly resistant to fatigue and are used to make corrosion-resistant springs. When alloyed with beryllium, the resulting material can be cold worked and heat treated to produce hard tools such as hammers and cold chisels. These are useful in areas where other materials might cause dangerous sparks.

UNDESIRABLE CHARACTERISTICS OF COPPER

- Copper and many of its alloys are susceptible to a condition known as hot shortness. This means that the material becomes brittle at high temperatures. Therefore, it can present many of the same problems as aluminum does in welding.
- Copper and many of its alloys owe much of their strength to the fact that they have been cold worked. Welding operations heat the metal in the weld zone to the point where it becomes annealed and loses much of its strength.
- While pure copper possesses excellent electrical and thermal conductivity, the addition of any elements, either deliberately to form alloys, or accidentally in the form of oxides, causes a sharp decrease in both electrical and thermal conductivity. It is usually found that pure metals possess better electrical and thermal conductivity than any of their alloys.
- Repeated stresses applied to copper, such as cold-forming operations or stress reversals caused by the normal operation, can cause the metal to become increasingly hard and brittle. This can lead to rupture or a fatigue type of failure.

CHART 16-1

SUGGESTED PROCEDURES FOR INERT-GAS TUNGSTEN-ARC WELDING COPPER AND EVERDUR®						
Base Metal Thickness Inch	Weld Groove	No. of Beads	Bead No.	Filler Rod Diameter Inch	DCSP Welding Current Ranges, Amperes	
			On Backing Bar		COPPER	EVERDUR®
1/16	Square	1		3/32	150-250	80-120
3/32	Square	1		1/8	180-300	100-150
1/8	Square	1		1/8	200-350	100-200
			Without Backing			
1/8	Square	2		1/8	150-300	100-150
5/32	Square	2		1/8	150-300	100-150
3/16	Square	2		1/8	180-350	100-200
3/16	Single-vee	2	1	3/16	150-300	100-200
			Root	1/8	200-300	150-200
1/4	Single-vee	3	1	1/8	200-350	100-150
			2	3/16	200-350	150-200
			Root	1/8	200-350	150-200
3/8	Single-vee	4	1	1/8	200-350	100-200
			2	3/16	250-350	150-200
			3	1/4	350-500	150-250
			Root	1/8	250-350	150-200
1/2	Single-vee	5	1	1/8	300-450	100-200
			2	3/16	300-450	150-250
			3	1/4	350-500	150-250
			4	1/4	350-500	200-300
			Root	1/8	300-450	150-200
3/4	Double-vee	6	1, 2	1/8	300-400	100-250
			3, 4	3/16	300-450	150-300
			5, 6	1/4	350-550	200-350
1	Double-vee	8	1, 2	1/8	300-400	100-250
			3, 4	3/16	300-450	150-300
			5, 6	1/4	350-550	200-350
			7, 8	1/4	350-600	200-400

CURRENT REQUIREMENTS FOR WELDING COPPER

A study of chart 16-1 and a comparison with charts 6-2, 9-1, 10-1 and 11-1 indicate that copper requires current settings of 50 percent to 75 percent higher than equal sections of aluminum. As an example: 1/8-inch aluminum requires currents up to 160 amperes; 1/8-inch copper requires 200 amperes to 350 amperes. For 1/8-inch tungsten electrodes using straight polarity, 250 amperes is the top current limit for TIG torches with ceramic nozzles.

WELDABILITY

Pure copper presents no problems. It is originally supplied in a clean state. However, the material should still be given a thorough cleaning just before welding. The presence of oxides on the surface is indicated by the color which ranges from light green to black. No welding should ever be attempted until these oxides have been removed either chemically or mechanically.

Atmospheric conditions can affect the finished joint in copper. With some types of copper welding, a tendency to porosity in the joint increases in direct proportion to the increase in relative humidity. TIG welding offers a decided advantage by providing an ideal atmosphere for the welding process.

The copper-silicon alloys are generally readily weldable. Some of these alloys are known by their trade names such as Everdur®, Herculoy®, and Olympic®. The copper-nickel alloys, such as super-nickel and cupro-nickel, are also readily welded using the TIG process if they are carefully cleaned before welding. The copper-aluminum alloys, known as aluminum-bronze, can be welded by the TIG process if the welder uses as much care as for welding aluminum. The copper-phosphorous alloys, usually referred to as phosphor-bronze, can also be welded economically with the TIG process.

While some of its alloys present no difficulties to the welder, many of the copper alloys are not as weldable. The group of copper-zinc, copper-tin, and copper-lead alloys and combinations of the four elements are either difficult or impossible to weld by the TIG process. The difficulty comes from the tendency of the zinc, tin or lead to vaporize under the intense heat of the arc. As the percentage of any of these three elements is increased in a copper-base alloy, the probability of making an acceptable TIG weld decreases. However, the low-temperature brazing process of oxyacetylene welding may be used to join many of these otherwise unweldable alloys. The many low-temperature brazing alloys available produce joints of high strength, excellent appearance and, in many cases, good color match when it is an important factor. Chart 16-2 provides an opportunity to compare copper and some of its alloys as to strength, ductility, melting point, and electrical and thermal conductivity.

JIGS AND FIXTURES

Copper and most of its alloys present unique problems when jigs and fixtures are necessary. Steel would make an excellent material for constructing jigs, but if it becomes too hot, copper and many of its alloys tend to bond or braze to it.

Carbon or carbon inserts eliminate this tendency but the material is brittle and wears rapidly. If the arc is allowed to strike the carbon and the tungsten electrode is negative, much carbon is transferred to the tungsten (just as it was transferred from positive to negative in the experiments in unit 5). This contaminates the tungsten and greatly lowers its current-carrying characteristics.

The best material for backing bars in jigs and fixtures appears to be copper itself, if it is thick enough so that it does not melt and fuse with the material being welded. Stainless steel is also satisfactory as a backing material if it has been heated until the surface is oxidized. This oxide is very hard to remove and prevents any bonding of the copper or copper-base alloys. It.has the advantage of low thermal conductivity, thus more heat can be used in

CHART 16-2

PHYSICAL CHARACTERISTICS OF COPPER AND SOME OF ITS ALLOYS							
Material	**Composition**	**Average Tensile Strength k.s.i.**		**% Elongation in 2 inches (annealed)**	**Melting Point Degrees F.**	**Conductivity**	
		Annealed	Hard			Electrical	Thermal
Copper (Tough pitch, Electrolytic Lake)	99.9 Copper .03-.07 Oxygen	30,000 to 40,000	40,000 to 67,000	35	1981	100+	.92
Deoxidized Copper	0-10 Phos. 0.25 Si Bal. Copper	30,000 to 35,000	40,000 to 50,000	35	1981	80	.80
Common Brasses Brazing, Spring, Cartridge, Yellow Brass	21-37 Zinc 0-4 Lead 0-2 Tin Bal. Copper	30,000 to 48,000	55,000 to 100,000	45 to 15	1823 to 1634	28 to 20	.31 to .22
Naval Brass, Tobin Bronze and Muntz Metal	37-43 Zinc 0-1½ Tin 0-2 Mn 0-1½ Iron 0-13 Ni 0-2 Lead	45,000 to 60,000	50,000 to 80,000	50 to 25	1742 to 1598	26 to 6	.28 to .08
Phosphor-Bronze Bearing Bronze and Gun Metal	1-30 Tin 0-4 Zinc 0-15 Lead 0-50 P Bal. Copper	30,000 to 60,000	60,000 to 150,000	50 to 15	1967 to 1418	45 to 8	.55 to .09
Copper-Silicon, Everdur® and Herculoy®	.25-5 Sil. 0-1½ Mn 0-5 Zinc 0-2.5 Iron Bal. Copper	40,000 to 60,000	65,000 to 145,000	75 to 20	1931 to 1832	12 to 4.5	.13 to .05
Super-Nickel and Cupro-Nickel	2-30 Nickel Trace Mn Bal. Copper	35,000 to 55,000	45,000 to 90,000	30 to 50	2012 to 2237	35 to 4.5	.40 to .06
Aluminum-Bronze	1-11 Al 0-4 Iron 0-5 Nickel 0-2 Tin Bal. Copper	40,000 to 100,000	50,000 to 125,000	15 to 60	1967 to 1886	35 to 7	.37 to .10
Beryllium-Copper	1-2.5 Be 0-1 Nickel Bal. Copper	50,000 to 70,000	70,000 to 190,000	55 to 45	1889 to 1742	45 to 17	.50 to .22

welding process. Since the thermal expansion of stainless steel is very close to that of copper, the possibility of excessive stresses being set up in the weldment is eliminated.

PRACTICE WELDS

The practice welding in this unit will depend on the availability of copper and copper-base alloys. The welder should gain as much experience as possible in welding the intermediate gages (1/16 inch to 1/8 inch thick) to make butt, lap, fillet and corner welds. It would be well to experiment with welding some of the copper-base alloys which contain zinc, tin and lead in varying amounts. Examine the finished welds and determine the effects these elements have on the finished joint.

At this point the student should have enough experience to be able to test finished welds and to be able to draw intelligent conclusions.

REVIEW QUESTIONS

1. Why is DCSP recommended for copper and copper-base alloy welding?

2. In terms of safety, can welding copper be hazardous?

3. From welding experiences, what is the best type of current for welding aluminum-bronze?

4. From a study of charts 16-1 and 16-2, what can be determined about current requirements when welding copper-base alloys as compared to welding pure copper?

5. What kind of backing bars are best suited for use with copper?

UNIT 17 TIG WELDING NICKEL AND NICKEL-BASE ALLOYS

Nickel and many of its alloys are widely used in the chemical industry and in food processing plants. Nickel is resistant to most of the alkalies and many acids. In chemical plants, nickel and its alloys are used to resist the highly corrosive effects of alkalies and the resultant contamination of the finished product. In food processing plants, nickel and its alloys also add to a high-purity product.

ADVANTAGES OF USING TIG WELDING FOR NICKEL

When welding nickel and its alloys, the inert-gas arc-welding process has some advantages. The flux used in electric arc welding of nickel and the flux used in oxyacetylene welding of nickel are no more of a problem than in the welding of steel. However, there is the possibility of flux entrapment in any metallic-arc welding process. TIG welding eliminates this possibility. The flux used for nickel and nickel alloys causes no difficulty at ordinary temperatures; but, if this flux is not thoroughly removed, it becomes a problem when the weldment is used at high temperatures. In this case it attacks the weld and adjacent metal and corrodes them rapidly. TIG welding avoids this difficulty. Splatter and coarse ripples, which are other defects caused by metallic arc welding, are not found when using the TIG process.

The TIG process does not have any advantage over metallic arc welding in the matter of grain growth or grain structure. However, both of these processes are carried on at a much more rapid rate than with oxyacetylene welding. This more rapid rate of heating and cooling generally results in a much finer grain structure in the weld and the heat-affected zone. In general, if all the factors that go into the cost of the finished product are considered, TIG welding is no more expensive than any of the other fusion processes used to join nickel and its alloys.

SURFACE PREPARATION

While no welding should be done without proper attention to surface preparation and cleanliness, nickel and its alloys are particularly sensitive to many chemicals. A good rule to follow is — unless proven to be safe, all foreign materials must be considered harmful.

Nickel and its alloys are usually supplied in a clean condition. However, many of the fabrication processes may leave the surface in a contaminated condition. Mechanical or chemical cleaning must be done before welding to avoid porosity and cracks in the finished joint. Lead, sulfur, phosphorus and some low-melting alloys are particularly harmful, as is the residue from alkaline cleaners. All traces of these materials should be removed from both sides of the joint before welding.

If the fabricating processes have required the work to be heated, nickel oxide forms. It is dark-colored, hard and highly refractory, melting at a temperature of 3794 degrees F. It can be removed by pickling or by one of the mechanical processes such as sand blasting, light grinding or the use of abrasive cloth. The quality of the finished joint depends greatly on how well the oxides and other contamination have been removed from both sides of the joint.

The amount of joint preparation also relates to the thickness of the material being welded. In general, a U-groove is recommended for the heavier sections to keep the price of the finished joint as low as possible.

JIGS AND FIXTURES

Jig design and tacking procedures which work well on carbon steel usually give equally good results for nickel. Copper is recommended as the best material for backing bars.

CURRENT REQUIREMENTS

DCSP is recommended for welding nickel and its alloys. However, on thin sections where piercing or burn-through is a problem, HFAC gives the advantage of lower heat input to the work. Amperage requirements and electrode sizes are equal to, or very close to those used to weld equal sections of carbon steel.

SHIELDING GAS

Helium is preferred for most TIG welding operations on nickel. It results in a hotter arc and increased welding speeds. Argon gas is recommended for the lighter sections where burn-through is a problem. A gas flow from 8 to 30 cubic feet per hour is usually enough, depending on the thickness of the material. The gas nozzle should always be of sufficient size to supply the shielding gas to the weld zone at a low velocity.

TECHNIQUES AND PRECAUTIONS

The nozzle and gas cap should be checked often to be sure they are tight. A loose joint can cause the flow of shielding gas to act as a venturi and draw oxygen from the air into the shield, causing weld contamination.

The torch should be held as nearly vertical as possible for good vision. The International Nickel Company recommends that, when welding on flat surfaces, the torch never be held at an angle greater than 35 degrees from vertical. Their investigation indicates that a sharper angle may draw air into the shielded zone, causing contamination.

In general, the amount of electrode extension does not vary from the normal procedure; that is, the electrode extension is as small as possible consistent with good vision and ease of manipulation. The tungsten electrode should be ground and maintained as a pencil point. Thoriated or zirconium alloyed tungsten electrodes maintain this desired point better than pure tungsten.

A superimposed high-frequency arc also helps to maintain this shape by eliminating the need to touch start the arc. In this case, if the machine is equipped with a switch for Start Only and Continuous, the switch should be set on Start Only.

The soundness of the welds produced depends on the arc length. In general, a long arc produces porosity in nickel and nickel alloys. Whenever possible an arc length of not over

CHART 17-1

RECOMMENDED COMBINATIONS FOR TIG WELDING

BASE METAL		FILLER ROD	
Nickel	200	Nickel	61
Monel	400	Monel	60
Monel K	500	Monel	64
Inconel	600	Inconel	62
Inconel	X-750	Inconel	69
Inconel	722	Inconel	69

1/16 inch is recommended. As with most other metals, the presence of hydrogen-producing moisture also results in porosity in the weld.

There seems to be a relation between the speed of welding and the porosity. When making welds in which no filler rod is used, such as outside corner welds and flanged butt welds, an increase in the rate of travel results in a more dense weld.

When filler metal is used, care should be taken to keep the hot end of the rod in the gas shield at all times to avoid oxidation and the resulting contamination. The electrode should not be permitted to come in contact with either the filler rod or the molten pool. Electrode contamination results in inferior joints in nickel just as in other metals.

The filler metals used for oxyacetylene welding of nickel and its alloys are not suitable for TIG welding. International Nickel Company supplies the filler rods shown in chart 17-1 in 36-inch lengths and diameters from 1/16 inch to 3/16 inch by thirty-seconds.

PRACTICE WELDS

The practice welding in this unit depends to some extent upon the availability of nickel and its alloys. Most plants which make these metals also make scrap material available to welding schools on a loan basis. If the material is available, thicknesses of about 1/8 inch give the best conditions for studying the action of the arc in and around the molten pool.

If possible, some experience should be gained in welding nickel and its alloys to steel. A large amount of this type of welding is used in lining steel tanks and containers with nickel. This is done to make corrosion-resistant containers that are strong and low in cost. Experience should also be gained in welding materials from .040 inch to .065 inch in thickness.

The operator should experiment by allowing the electrode to touch the molten pool and observing the arc action and bead in order to recognize the hazards in electrode contamination. Experimenting with various arc lengths and observing the finished beads gives the welder good experience in judging the source of defects caused by deviating from standard procedures. Testing of the finished joints by the methods used in previous units allows the operator to compare welds made in these materials with those made in other materials.

REVIEW QUESTIONS

1. When nickel is attached to steel as a liner material, what effect do variations in temperature have on the stresses set up between the two dissimilar metals?

2. How does the oxide of nickel compare with aluminum oxide?

3. What major point of dissimilarity is there between aluminum oxide and nickel oxide?

4. What is the major advantage of a U-groove over a V when welding thick sections?

5. How does the arc length that is used in this unit compare with that used in previous units?

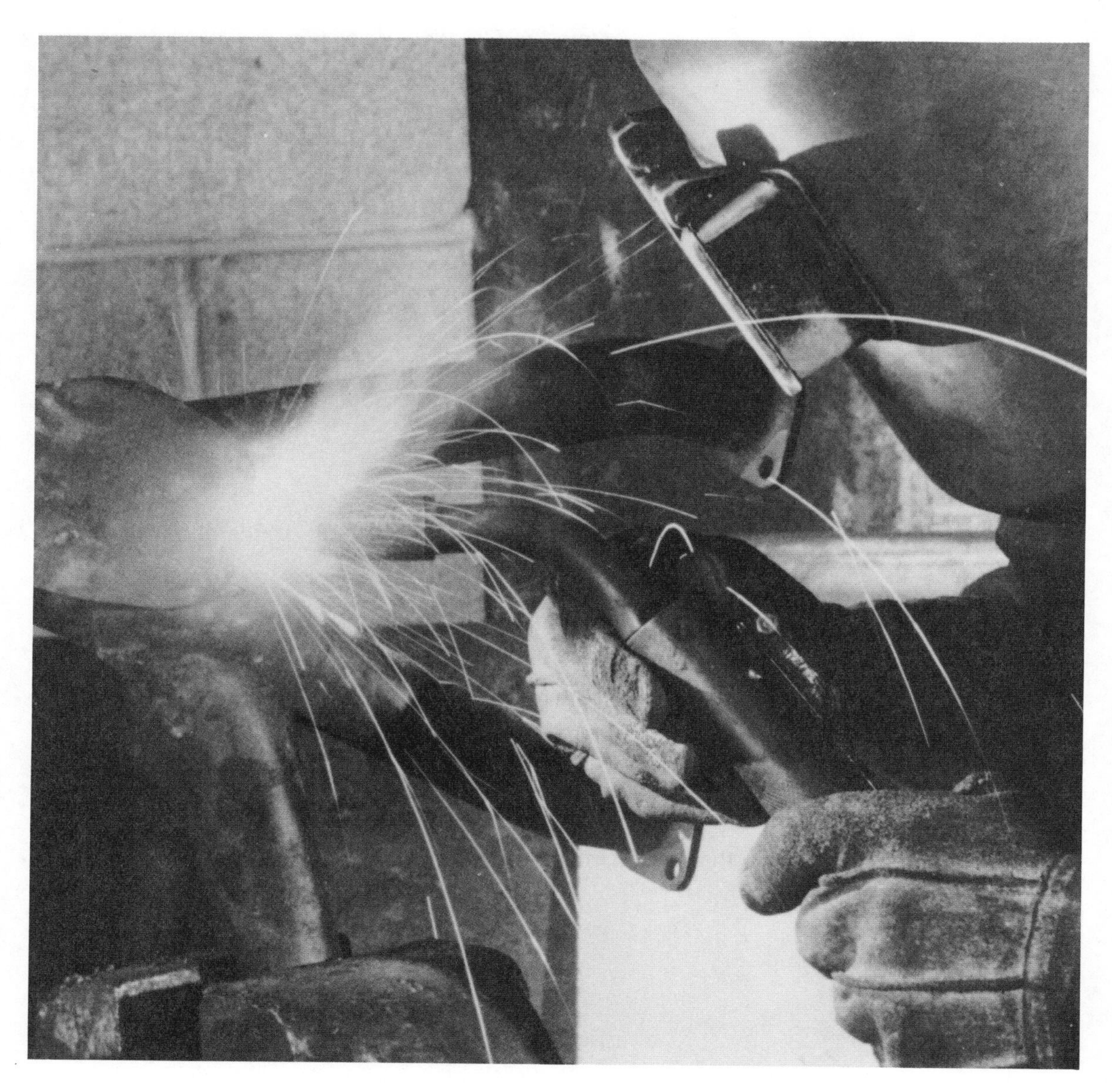

SECTION **2**

MIG WELDING

In the metallic inert-gas, shielded-arc process, a consumable electrode in the form of wire is fed from a spool through the torch, often referred to as a welding gun. As the wire passes through the contact tube in the gun, it picks up the welding current.

MIG welding differs from TIG in that it is a one-handed operation. Also, it does not require the same degree of skill as the two-handed process.

An important factor in the MIG process is the high rate at which metal can be deposited. This high rate of metal deposit and high speed of welding, which are characteristic of MIG, result in minimum distortion and a narrow heat-affected zone.

UNIT 18 THE METALLIC INERT-GAS WELDING PROCESS

From an operator's viewpoint, it is easier to gain skill in the MIG process than in the TIG process. The deposition rate is much faster with MIG than TIG although the same metals can be joined with both. The thickness of material to be joined is a factor in choosing the correct process.

MIG welding (often called metal inert-gas or gas-metal arc welding) is done by using a consumable *wire electrode* to maintain the arc and to provide filler metal. The wire electrode is fed through the torch or gun at a preset controlled speed. At the same time, an *inert gas* is fed through the gun into the weld zone to prevent contamination from the surrounding atmosphere.

ADVANTAGES OF MIG WELDING

- Arc visible to operator
- High welding speed
- No slag to remove
- Sound welds
- Weld in all positions

TYPES OF MIG WELDING

- *Spray-arc welding,* figure 18-1, is a high-current-range method which produces a rapid deposition of weld metal. It is effective in welding heavy-gage metals, producing deep weld penetration.

  At high currents, the arc stability improves and the arc becomes stiff. The transition point, when the current level causes the molten metal to spray, is governed by the wire type and size, and the type of inert gas used.

- *Short-arc welding,* figure 18-2, is a reduced-heat method with a pin arc for use on all common metals. It was developed for welding thin-gage metals to eliminate distortion, burn-through and spatter. This technique can be used in the welding of heavy thicknesses of metal.

- *MIG CO_2 (carbon-dioxide) welding* is a variation of the MIG process. Carbon dioxide is used as the shielding gas for

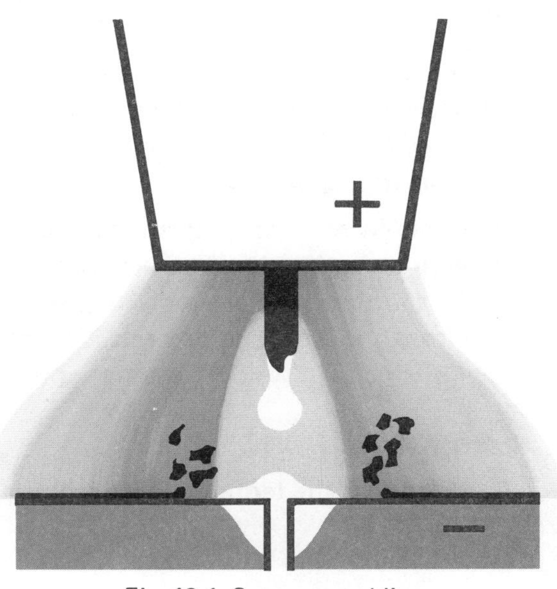

Fig. 18-1 Spray arc welding

Start of the short arc cycle — High temperature electric arc melts advancing wire electrode into a globule of liquid metal. Wire is fed mechanically through the torch. Arc heat is regulated by presetting the power supply.

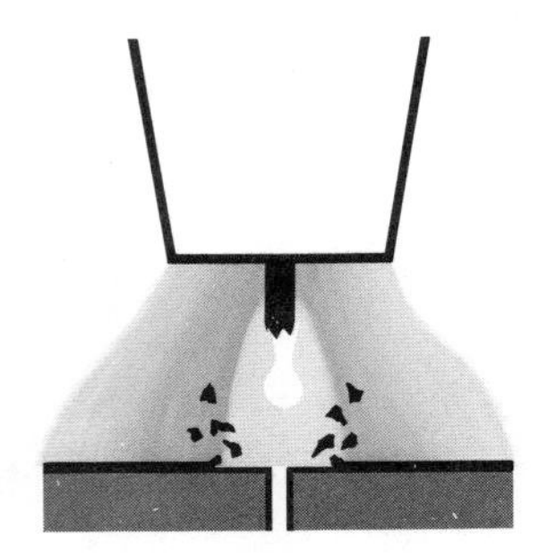

Molten electrode moves toward workpiece. Note cleaning action. Argon gas mixture, developed specifically for short arc, shields molten wire and weld seam, insuring regular arc ignition, controlling spatter and weld contamination.

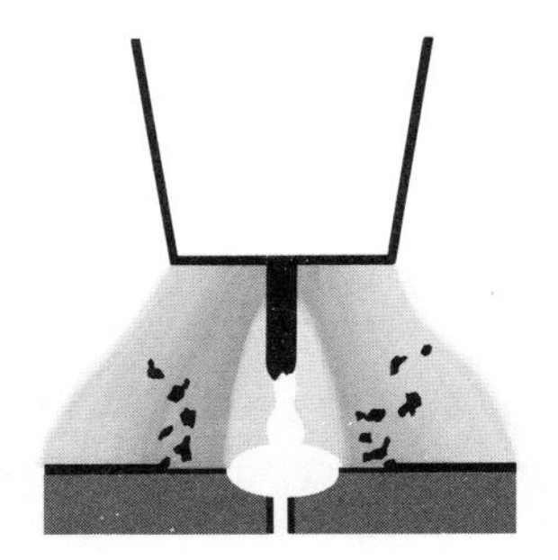

Electrode makes contact with workpiece, creating short circuit. Arc is extinguished. Metal transfer begins due to gravity and surface tension. Frequency of arc extinction in short arc varies from 20 to 200 times per second, according to "preset" conditions.

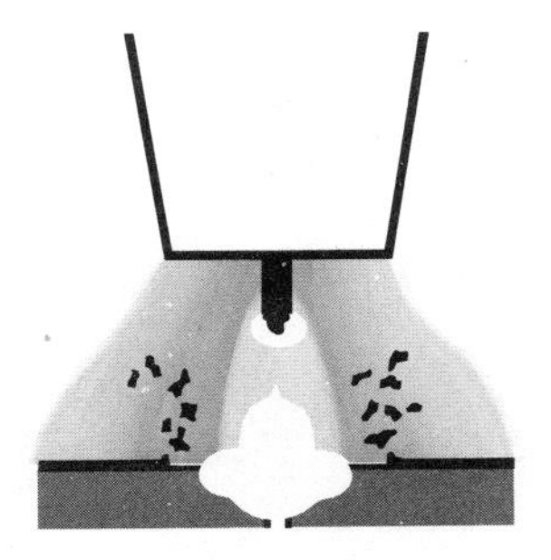

Molten metal bridge is broken by pinch force, the squeezing action common to all current carriers. Amount and suddenness of pinch is controlled by power supply. Electrical contact is broken, causing arc to reignite.

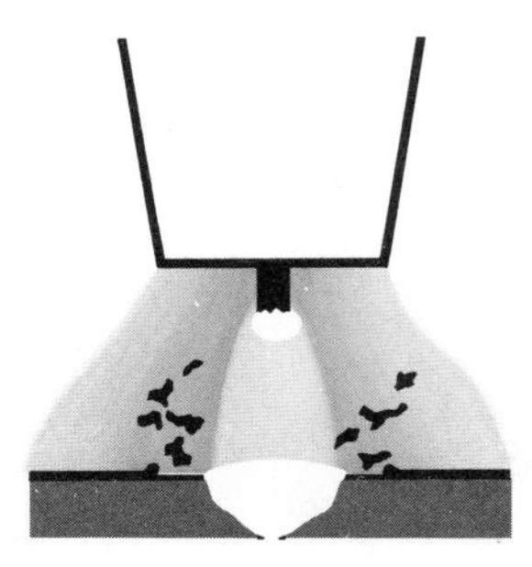

With arc renewed, short arc cycle begins again. Because of precise control of arc characteristics and relatively cool, uniform operation, short arc produces perfect welds on metals as thin as .030-in. with either manual or mechanized equipment.

Fig. 18-2 Short arc welding

the welding of carbon and low-alloy steel from 16 gage (.059 inch) to 1/4 inch or heavier. It produces deeper penetration than argon or argon mixtures with slightly more spatter. Carbon-dioxide MIG welding costs about the same as other processes on mild steel applications.

- *Cored-wire welding* is an intense-heat, high-deposition-rate process using flux-cored wire on carbon steel. Electrically, cored-wire welding is similar to spray-arc welding. In addition to inert-gas shielding, a flux contained inside the wire forms a slag that cleans the weld and protects it from contamination. In application, it is recommended for large fillet welds in the flat or horizontal position.

REVIEW QUESTIONS

1. What does the term MIG welding mean?

2. What is the principle of the MIG welding process?

3. What are four types of MIG welding?

4. What polarity is used for MIG welding?

5. What are the advantages of MIG welding?

UNIT 19 EQUIPMENT FOR MANUAL MIG WELDING

A specially designed welding machine is used for MIG welding. It is called a *constant-voltage (CV) type* power source. It can be a DC rectifier or a motor- or engine-driven generator. (See figure 19-3.)

The output welding power of a CV machine has about the same voltage regardless of the welding current. The output voltage is regulated by a rheostat on the welding machine, figure 19-1. Current selection is determined by wire-feed speed. There is no current control as such.

The wire-feeding mechanism and the CV welding machine make up the heart of the MIG welding process, figures 19-2, 19-4, and 19-5. There is a fixed relationship between the rate of electrode wire burn-off and the amount of welding current. The electrode wire-feed speed rate determines the welding current.

The gun is used to carry the electrode wire, the welding current, and the shielding gas from the wire feeder to the arc area, figure 19-6. The operator directs the arc and controls the weld with the welding gun.

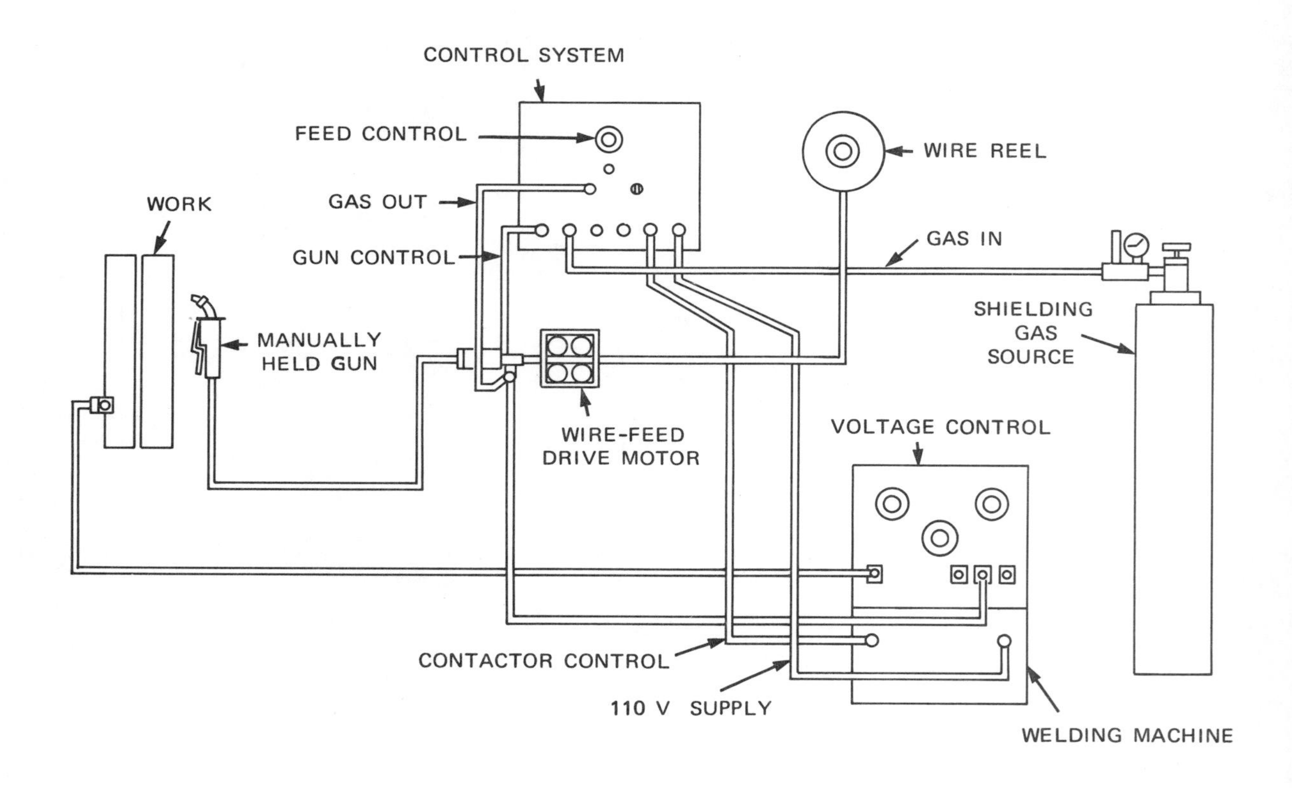

Fig. 19-1 MIG equipment

Fig. 19-2 Constant-voltage rectifier

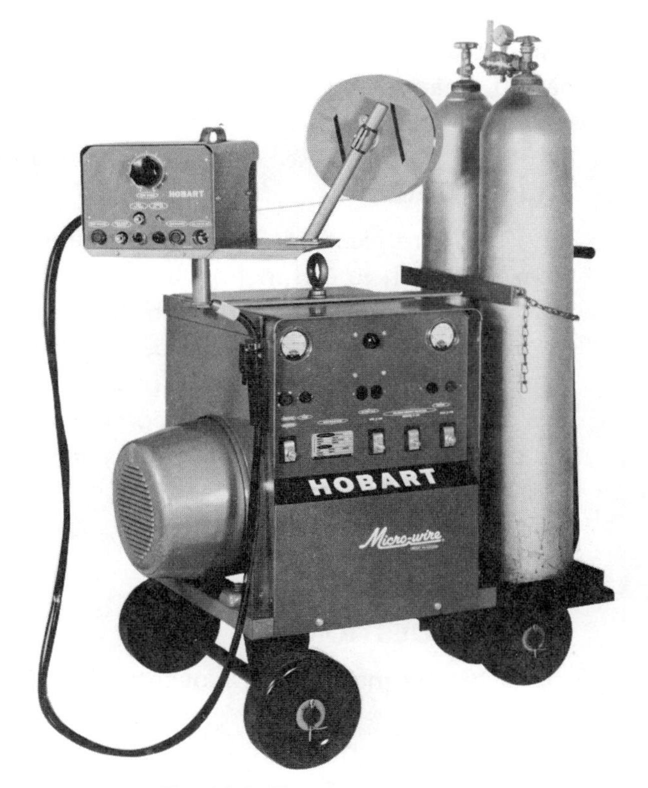

Fig. 19-3 Motor generator

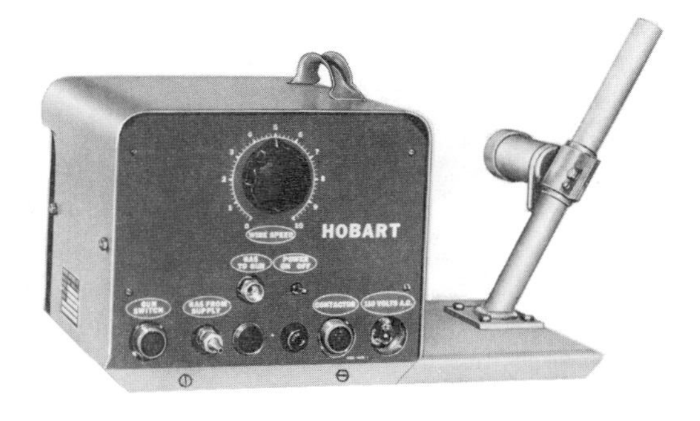

Fig. 19-4 Wire-feed control unit

Fig. 19-5 Wire-feed unit

SHIELDING GASES

The shielding gas can have a big effect upon the properties of a weld deposit. The welding is done in a controlled atmosphere.

Pure argon, argon-helium, argon-oxygen, argon-carbon dioxide, and carbon dioxide are commonly used with the MIG process. With each kind and thickness of metal, each gas and mixture affects the smoothness of operation, weld appearance, weld quality, and welding speed in a different way.

Gas-flow rate is very important. A pressure-reducing regulator and flowmeter are required on the gas cylinder. Flow rates vary, depending on types and thicknesses of the material and the design of the joint. At times two or more gas cylinders are connected (manifolded) together to maintain higher gas flow.

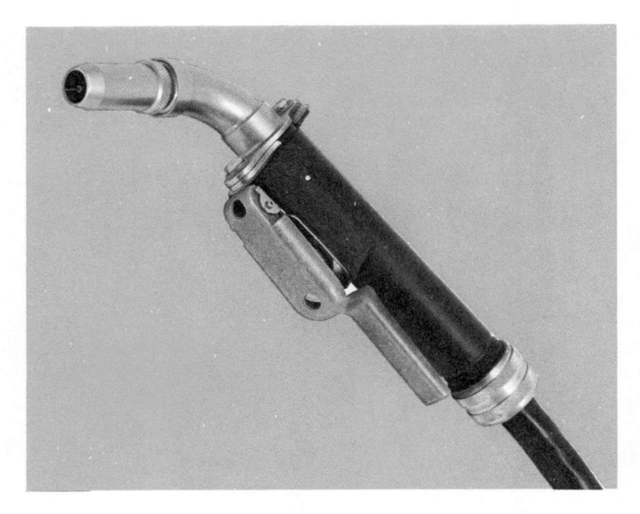

Fig. 19-6 Welding gun and cable assembly

FILLER WIRES

The wire electrode varies in diameter from .030 inch to 1/8 inch. The composition of the electrode wire must be matched to the base metal being welded. In the welding of carbon steel, the wire is solid and bare except for a very thin coating on the surface to prevent rusting. It must contain deoxidizers which help to clean the weld metal and to produce sound, solid welds.

REVIEW QUESTIONS

1. What are the main components of the MIG welding equipment?

2. What is considered to be the heart of the MIG welding process?

3. What does the wire-feed control determine?

4. What is the measurement of the flowmeter which registers gas flow to control the shielding atmosphere?

5. Is the voltage controlled by the wire feeder or the welding machine?

UNIT 20 MIG WELDING VARIABLES

Most of the welding done by all processes is on carbon steel. About 90 percent of all steel is plain carbon steel. This unit describes the welding variables in short-arc welding of 24-gage to 1/4-inch mild steel sheet or plate. The type of equipment usually found in training facilities lends itself well to these applications.

The applied techniques and end results in the MIG welding process are controlled by these variables and must be understood by the student. The variables are adjustments that are to be made to the equipment and also manipulations by the operator.

These variables can be divided into three areas.

- Preselected variables
- Primary adjustable variables
- Secondary adjustable variables

PRESELECTED VARIABLES

Preselected variables depend on the type of material being welded, the thickness of the material, the welding position, the deposition rate and the mechanical properties. These variables are

- Type of electrode wire
- Size of electrode wire
- Type of inert gas
- Inert-gas flow rate

Charts 20-1, 20-2, and 20-3 are references for the new MIG welding student. Manufacturers' recommendation also serve as a guide to be followed in these areas.

PRIMARY ADJUSTABLE VARIABLES

These control the process after preselected variables have been found. They control the penetration, bead width, bead height, arc stability, deposition rate and weld soundness. They are

- Arc voltage
- Welding current
- Travel speed

SECONDARY ADJUSTABLE VARIABLES

These variables cause changes in the primary adjustable variables which in turn cause the desired change in the bead formation. They are

- Stickout
- Nozzle angle
- Wire-feed speed

CHART 20-1

COMPARISON CHART MILD STEEL ELECTRODES FOR MIG WELDING									
MANUFACTURERS	American Welding Society Classification A5-18-6								
	E 70S-1	E 70S-2	E 70S-3	E 70S-4	E 70S-5	E 70S-6	E 70S-G	E 70S-1B	E 70S GB
Airco Welding Products Div. Air Reduction Co. Inc.	S-20		A 675		A 666	A 681	A 608	A 608	A 608
Alloy Rods Company Div. Chemetron Corporation			MINIARC 70					MINIARC 100	
Hobart Brothers Company	TYPE 20		TYPE 25		TYPE 30	TYPE 28		TYPE 18	
Linde Div. Union Carbide Corporation	LINDE 29S	LINDE 65	LINDE 82, 66	LINDE 85		LINDE 86	LINDE 83	LINDE 83	
Midstates Steel & Wire Co.			IMPERIAL 75			IMPERIAL 88	IMPERIAL 95		
Modern Engineering Co. Inc.			MECO 60S-3		MECO 70S-5		MECO 70S-G		
Murex Welding Products			MUREX 1316		MUREX 1315		MUREX 1313 MO	MUREX 1313 MO	
National Cylinder Gas Div. Chemetron Corporation			MINIARC 70				MINIARC 100		
National Standard Company	NS-106	NS-103	NS-101			NS-115	NS-116	NS-102	
P & H Welding Products Unit of Chemetron Corporation			P & H CO-85		P & H CO-86		P & H CO-87		
Page, Division of ACCO	PAGE AS-20		PAGE AS-25		PAGE AS-30	PAGE AS-28		PAGE AS-18	

CHART 20-2

ALL JOINTS ALL POSITIONS MILD STEEL							
MATERIAL THICKNESS		NUMBER OF PASSES	WIRE DIAMETER	WELDING CONDITIONS DCRP		GAS FLOW CFH	TRAVEL SPEED IPM
GAGE	INCH			ARC VOLTS	AMPERES		
24	.023	1	.030	15-17	30-50	15-20	15-20
22	.029	1	.030	15-17	40-60	15-20	18-22
20	.035	1	.035	15-17	65-85	15-20	35-40
18	.047	1	.035	17-19	80-100	15-20	35-40
16	.059	1	.035	17-19	90-110	20-25	30-35
14	.074	1	.035	18-20	110-130	20-25	25-30
12	.104	1	.035	19-21	115-135	20-25	20-25
11	.119	1	.035	19-22	120-140	20-25	20-25
10	.134	1	.045	19-23	140-180	20-25	27-32
	3/16 in.	1	.045	19-23	180-200	20-25	18-22
	1/4 in.	1	.045	20-23	180-200	20-25	12-18

CHART 20-3

SHIELDING GASES FOR MIG		
METAL	SHIELDING GAS	APPLICATION
CARBON STEEL	75% ARGON 25% CO_2	1/8 inch or less thickness: High welding speeds without burn-through; minimum distortion and spatter
	75% ARGON 25% CO_2	1/8 inch or more thickness: Minimum spatter, good control in vertical and overhead position
	CO_2	Deeper penetration, faster welding speeds
STAINLESS STEEL	90% HELIUM 7.5% ARGON 2.5% CO_2	No effect on corrosion resistance, small heat-affected zone, no undercutting, minimum distortion
LOW ALLOY STEEL	60-70% HELIUM 25-35% ARGON 4-5% CO_2	Minimum reactivity, excellent toughness, excellent arc stability and bead contour, little spatter
	75% ARGON 25% CO_2	Fair toughness, excellent arc stability, and bead contour, little spatter
ALUMINUM, COPPER, MAGNESIUM, NICKEL AND THEIR ALLOYS	ARGON AND ARGON-HELIUM	Argon satisfactory on lighter material, Argon-helium preferred on thicker material

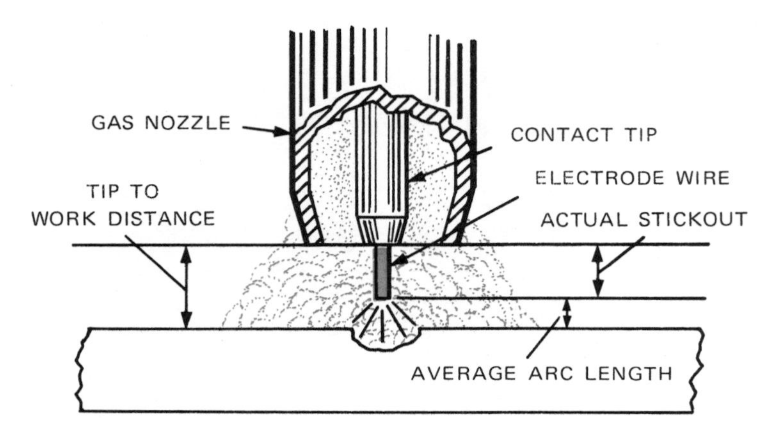

Fig. 20-1 Stickout

Stickout as shown in figure 20-1 is the distance between the end of the contact tip and the end of the electrode wire. From the operator's viewpoint, however, stickout is the distance between the end of the nozzle and the surface of the work.

Nozzle angle refers to the position of the welding gun in relation to the joint as shown in figure 20-2. The *transverse angle* is usually one-half of the included angle between plates forming the joints. The *longitudinal angle* is the angle between the centerline of the welding gun and a line perpendicular to the axis of the weld.

The longitudinal angle is generally called the nozzle angle and is shown in figure 20-3 as either trailing (pulling) or leading (pushing). Whether the operator is left-handed or right-handed has to be considered to realize the effects of each angle in relation to the direction of travel.

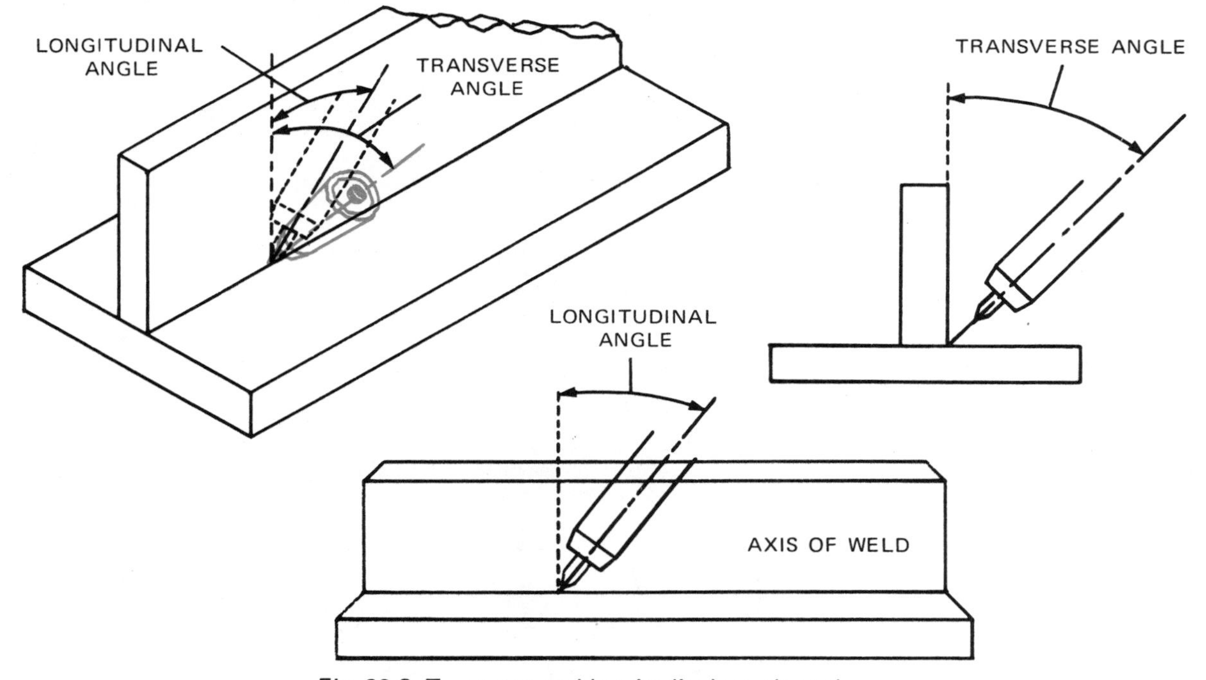

Fig. 20-2 Transverse and longitudinal nozzle angles

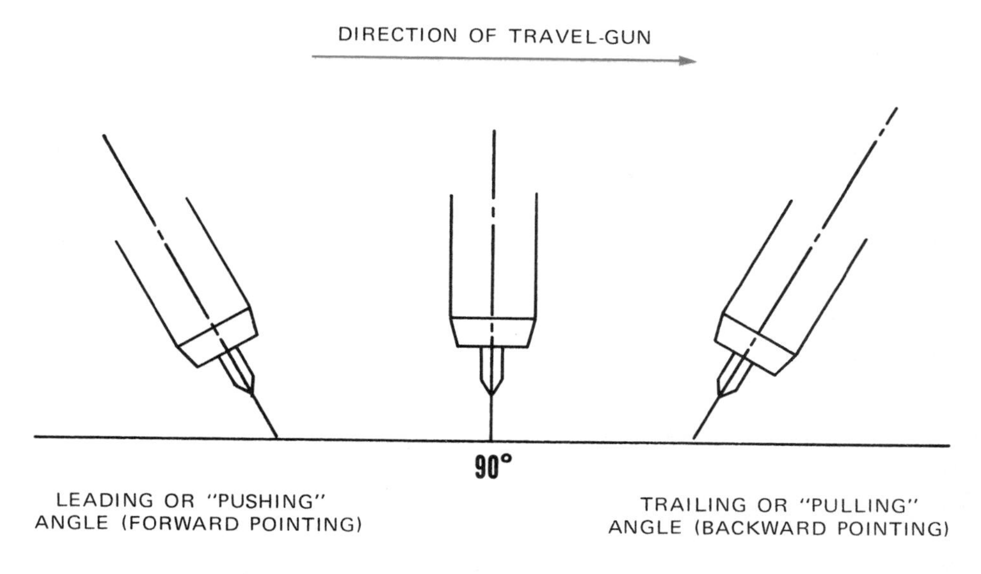

Fig. 20-3 Nozzle angle, right-handed operator

REVIEW QUESTIONS

1. What must be considered before selecting the type and size of electrode wire and type of inert gas?

2. What controls the penetration and the bead width and height?

3. What is the distance between the end of the nozzle and the work called?

4. What are the two nozzle angles called?

5. Where does the electrode wire pick up its electrical current?

UNIT 21 ESTABLISHING THE ARC AND MAKING WELD BEADS

It is assumed that the welding equipment has been set up according to procedures outlined in the appropriate manufacturers' instruction manuals. Students should know how to perform adjustments and maintenance on MIG welding equipment. As in TIG welding, the equipment is expensive, and the student must realize that the equipment can be destroyed if instructions are not followed.

Materials

10-, 11-, or 12-gage mild steel plate 6 in. x 6 in.
.035-inch E 70S-3 electrode wire
CO_2 shielding gas

Preweld Procedure

1. Check the operation manuals for manufacturer's recommendations.
2. Set the voltage at about 19 volts.
3. Set the wire-feed speed control to produce a welding current of 110 to 135 amperes.
4. Adjust the gas-flow rate to 20 cubic feet per hour.
5. Recess the contact tip from the front edge of the nozzle 0 to 1/8 inch.
6. Review standard safe practice procedures in ventilation, eye and face protection, fire, compressed gas and preventive maintenance. Safety precautions should always be part of the preweld procedure.

Welding Procedure

1. Maintain the tip-to-work distance of 3/8 inch (stickout) at all times. See figure 20-1.
2. Maintain the trailing gun transverse angle at 90 degrees and the longitudinal angle at 30 degrees from perpendicular. See figures 20-2 and 20-3.
3. Hold the gun 3/8 inch from the work, lower the helmet by shaking the head, and squeeze the trigger to start the controls and establish the arc.

   Note: Operators should not form the habit of lowering the helmet by hand since one hand must hold the gun and the other may be needed to hold pieces to be tacked or positioned.

4. Make a single downhand stringer weld.
5. Practice welding beads. Start at one edge and weld across the plate to the opposite edge.

   Note: When the equipment is properly adjusted, a rapidly crackling or hissing sound of the arc is a good indicator of correct arc length.

6. Practice stopping in the middle of the plate, restarting into the existing crater and continuing the weld bead across the plate.

Note: When the gun trigger is released after welding, the electrode forms a ball on the end. To the new operator, this may present a problem in obtaining the penetration needed at the start. This can be corrected by cutting the ball off with wire cutting pliers. The ball can cause whiskers to be deposited at the start. *Whiskers* are short lengths of electrode which have not been consumed into the weld bead.

This procedure should be practiced often by the new operator. A satisfactory performance in welding joints depends on the ability to do this basic manipulation.

Checking Application

1. Examine the base metal to be sure it is free from oil, scale, and rust.
2. Recheck the equipment settings according to the operation manual. This includes gas flow, stickout, gun angle, arc voltage, and amperage from wire-feed speed.
3. Keep equipment clean, specifically the gun nozzle, feeder rolls, wire guides, and liners. An anti-spatter spray should be used on the nozzle to help keep it clean.

REVIEW QUESTIONS

1. What is a good indication of correct arc length when the equipment is adjusted properly?

2. Why is it necessary to have the stickout distance correct, especially at the start and end of the weld?

3. Why is spatter buildup on the inside of the nozzle harmful to a good weld?

4. What is a whisker in MIG welding?

5. Why is the ball end cut off the wire?

UNIT 22 MIG WELDING THE BASIC JOINTS

The ability to manipulate the equipment and apply the single bead across a piece of sheet or plate is the basis for the welding of various joints. Being able to see and follow a joint helps to insure equal fusion on both pieces.

Note: Depending on the size of the gun, visibility may be a problem. The operator should be in the most comfortable position to see where the deposit is being made.

The welds that make up these basic joints can be applied in any position and can be single or multiple pass, depending on the thickness of the material being joined.

This unit describes the basics involved in welding the butt joint, lap joint, T joint, and corner joint in the flat position using 11-gage mild steel sheet. It is assumed that the operator can position and tack two pieces of this material to form these joints.

Materials

10-, 11-, or 12-gage mild steel sheet 1 1/2 in. x 6 in.
.035-inch E 70S-3 electrode wire
CO_2 shielding gas

Preweld Procedure

1. Check the operation manual for the manufacturers' recommendations.
2. Set the voltage at about 21 volts.
3. Set the wire-feed speed control to produce a welding current of 110 to 135 amperes.
4. Adjust the gas flow to 20 cubic feet per hour.
5. Adjust the voltage to get a smooth arc.

   Note: Before attempting to weld a joint, always adjust the machine, using a piece of scrap material.
6. Recess the contact tip from the edge of the nozzle 0 to 1/8 inch.
7. Make sure that the equipment is clean; specifically the gun nozzle and liner, feeder rolls and wire guides.
8. Remember that safety should always be part of the preweld procedure.

Welding Procedure (Butt Weld)

1. Place two pieces on the worktable in good alignment and with two 6-inch edges spaced 1/16 inch apart (root opening), as in figure 22-1.
2. Maintain a stickout of 3/8 inch. See figure 20-1.

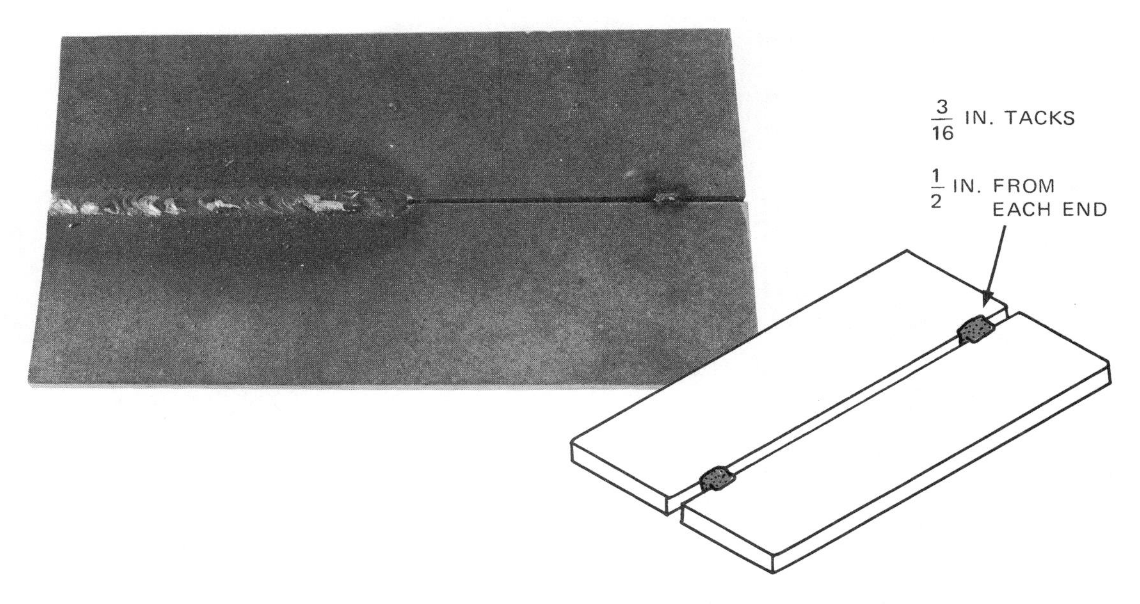

Fig. 22-1 Butt joint

3. Tack weld the two pieces together as shown in figure 22-1. The tacks should be placed about 1/2 inch from each end to avoid having too much metal and poor penetration at the start of the weld.

   Note: This method of tacking is used in all of the joints.

4. Use a transverse angle of 90 degrees or directly over and centered on the joint. Find the longitudinal angle by experimentation. The trailing gun angle is used first at 10 degrees perpendicular. The leading gun angle is used next at the same inclination. See figures 20-2 and 20-3.

   Note: Differences should be found and allowed for. The operator will prefer either the trailing gun angle or the leading gun angle, but the deciding factors are the degree of penetration, deposition rate, gas coverage of weld zone, and overall appearance.

5. Weld the joint on the tack side.

6. Cool the work and examine for uniformity.

7. Check the depth of penetration of the weld by first placing the assembly in a vise with the center of the weld slightly above and parallel to the jaws. Then bend the outstanding sheet toward the face of the weld. Penetration should be 100 percent with no faults.

8. Make more joints of this type and change the setting of the equipment slightly in different directions. Note what happens and what has to be done to compensate for it.

   Note: Thickness of material controls the root opening and whether or not the butt joint will need a bead on both sides.

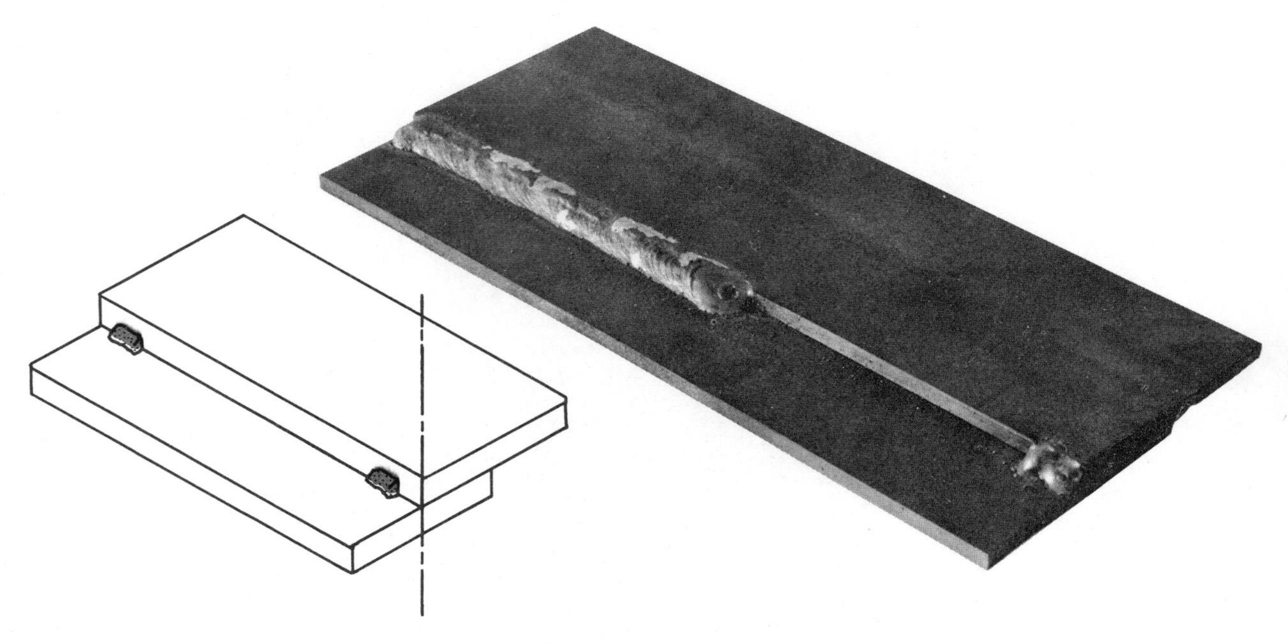

Fig. 22-2 Lap joint

Welding Procedure (Lap Joint)

1. Place two pieces on the worktable and tack as shown in figure 22-2.
2. Maintain a stickout of 3/8 inch. See figure 20-1.
3. The longitudinal angle is 10 degrees from perpendicular using a trailing gun angle. The transverse gun angle should be about 60 degrees from the lower sheet. The location of the nozzle in relation to the joint should be as shown in figure 22-3.

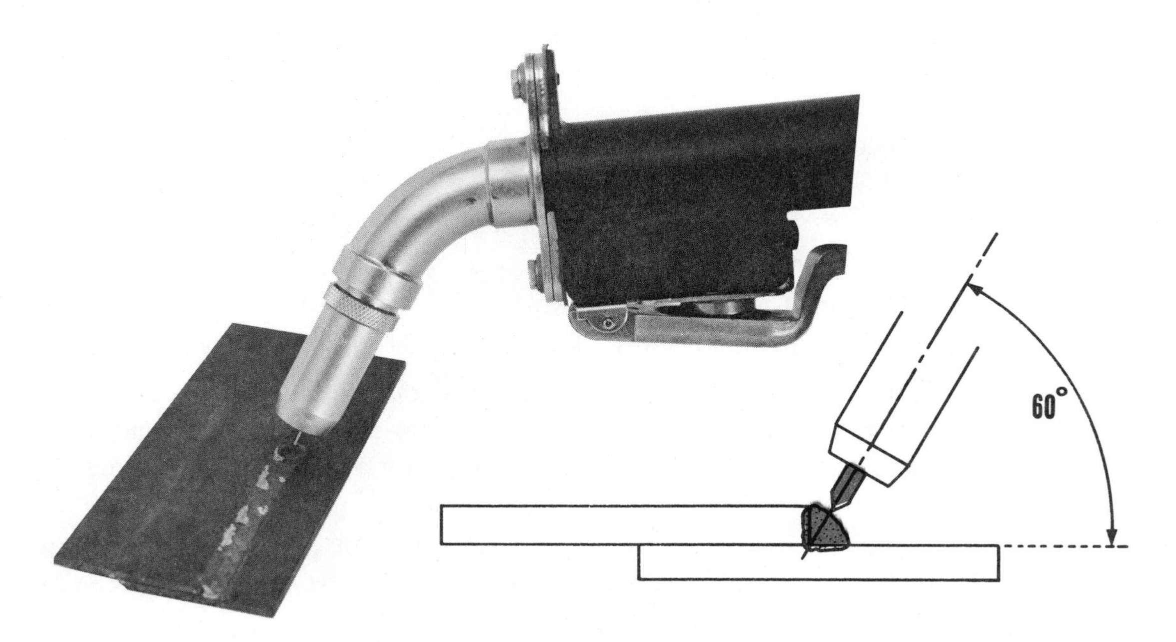

Fig. 22-3 Location of nozzle

Note: The operator's ability to compensate for the location of the gun in relation to the joint controls the uniformity of the bead and the desired amount of penetration.

4. Weld the joint. Allow for distortion by running the first bead on the opposite side from the tacks.
5. Cool and examine the bead for uniformity. Examine the line of fusion with the top and bottom sheets. This should be a straight line with no undercut.
6. Weld another lap joint on only one side (tack side). Place this piece in a vise in such a way that the top sheet can be bent from the bottom sheet 180 degrees, if possible, to check penetration and strength.
7. Make another test by sawing a lap-welded specimen in two and examining the cross section for penetration.
8. Make more joints of this type until they are uniform and consistent.

Welding Procedure (T Joint)

1. Place two pieces on the worktable and tack weld as shown in figure 22-4.

   Note: Hold one piece while tacking. This is good experience, as it is necessary to handle and manipulate the gun with one hand.

2. Maintain a stickout of 3/8 inch. See figure 20-1.

Fig. 22-4 T joint

3. The longitudinal angle is 10 degrees from perpendicular using a trailing gun angle. The transverse gun angle should be about 45 degrees from the lower piece. The location of the nozzle in relation to the joint should be as shown in figure 22-5.
4. Weld the joint. Allow for distortion by running the first bead on the opposite side from the tacks.
5. Cool and examine the bead for uniformity. The weld metal should be equally distributed between both pieces and show no signs of undercut.

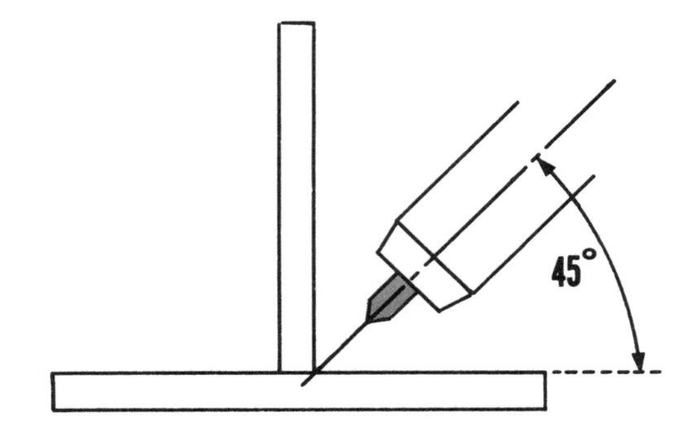

Fig. 22-5 Transverse angle

Note: A tack weld should be strong enough to resist cracking during the welding process but not large enough to affect the appearance of the finished weld. This is done in MIG welding by using slightly higher wire-feed speed.

6. Make another T joint welding only the tack side. Test this weld by bending the top piece against the joint a full 90 degrees. Examine the joint for root penetration and uniform fusion.
7. Continue to make the fillet weld until acceptable welds can be made each time. The fillet-type weld used on the lap and T joint is the most common weld.

Welding Procedure (Corner Joint)

Note: The technique used to set up and align the pieces to be joined for the downhand corner joint is more difficult. Figure 22-6 shows how this is done without using a fixture.

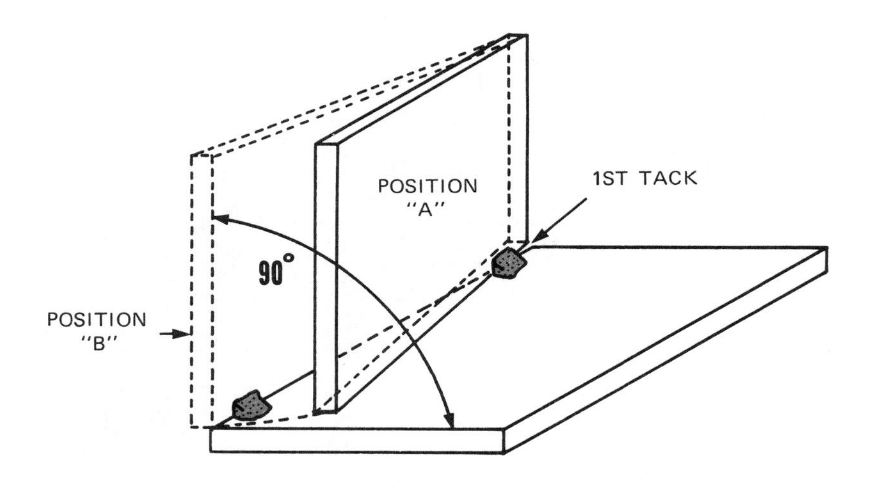

Fig. 22-6 Setting up for corner joint

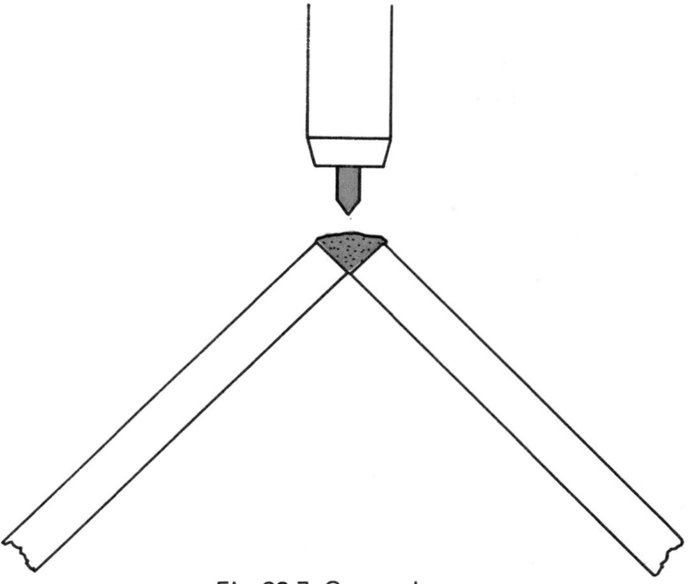

Fig. 22-7 Gun angle

1. Make the first tack while holding the piece as shown by position A, Figure 22-6. Then lift the hood and align the top piece to position B. Make a perfect open corner joint before placement of the second tack.
2. Maintain a stickout of 1/4 inch to 3/8 inch.
3. The longitudinal angle is 10 degrees from perpendicular using a trailing gun angle. The transverse gun angle should be perpendicular or bisect the included angle. See figure 22-7.
4. Weld the joint. Pay close attention to the start and end of the joint to avoid buildup or washout.
5. Cool and examine the bead for uniformity and penetration. The weld metal should be equally distributed between both pieces and show no signs of undercut or overlap. See figure 22-8.

Fig. 22-8 Examples of uniform welds

6. To test the corner joint, place the welded unit on an anvil and hammer it flat in order to examine root fusion and penetration.
7. Make more corner joints until they have uniform appearance and a good finish contour. The opposite side of this joint provides for good fillet weld practice.

Checking Application

1. Recheck the equipment settings according to the operation manual.
2. Keep the equipment clean.
3. Practice these four basic joints in the flat position using thicknesses of material up through 3/16 inch.

REVIEW QUESTIONS

1. When a tack is placed on two pieces of material being joined, what is the function of the tack?

2. What are the two types of nozzle angles as related to the longitudinal angle used in the operation of the gun?

3. What causes undercut and why is it harmful to the strength of the weld?

4. When welding any joint what is important concerning bead location?

5. Of the four joints, butt, lap, corner and T, which one might require more inert gas? Why?

UNIT 23 PROCEDURE VARIABLES

OUT-OF-POSITION WELDING

Upon satisfactory completion of the welds in the flat position, the student will be able to use the acquired skill and knowledge to weld out of position. This includes horizontal, vertical-up, vertical-down, and overhead welds. The basic procedures for each individual joint are no different out of position than in the flat position except a reduction in amperage of 10 percent is usually recommended. See chart 20-2.

MISALIGNED MATERIALS

The operator may, at times, have to weld pieces of material that are not in plane or aligned properly. There may be gaps or voids of various sizes which need a variation of stickout and/or wire-feed speed and voltage. The student should practice on the joints that require a deviation from standard procedures.

WELDING HEAVIER THICKNESSES

Heavier thicknesses of material can be welded with the MIG process using the multipass technique. This is done by overlapping single small beads or progressively making larger beads, using the weave technique, as in figure 23-1. The numbers refer to the order in which the passes are made. Individual job requirements govern the end result.

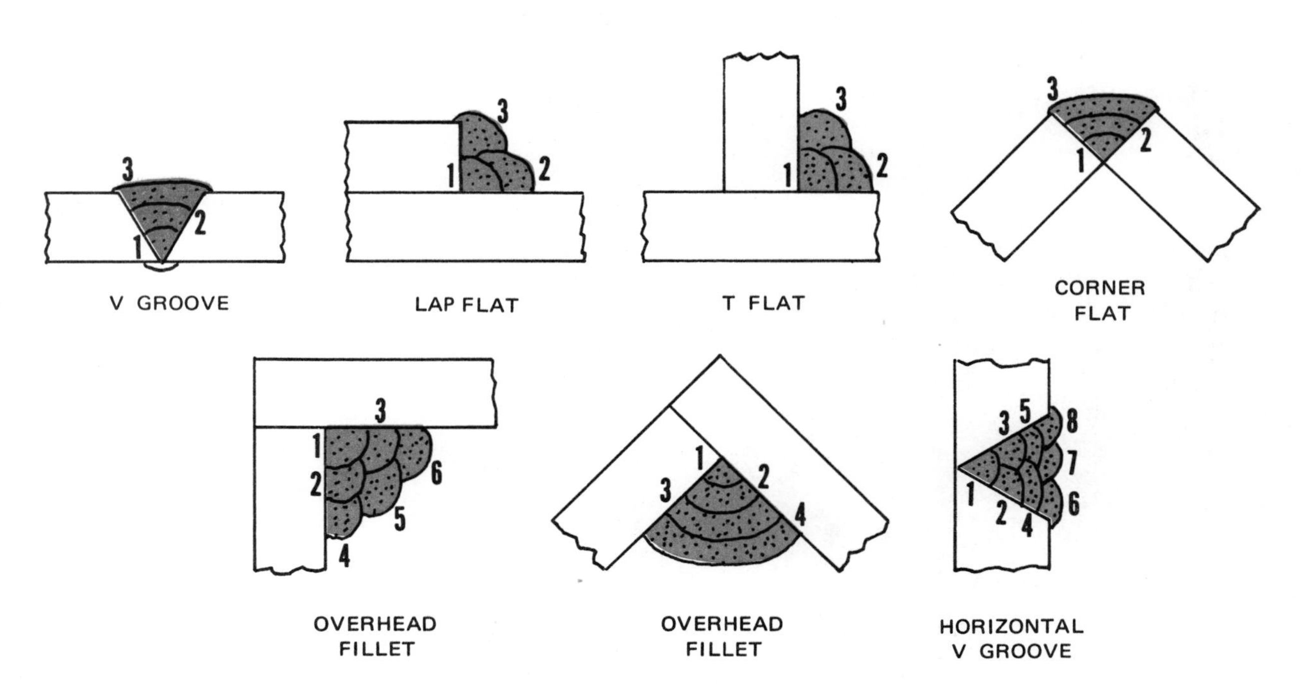

Fig. 23-1 Multipass welding

DEFECTS	PROBABLE CAUSE	CORRECTIVE ACTION
Fig. 23-2 Porosity 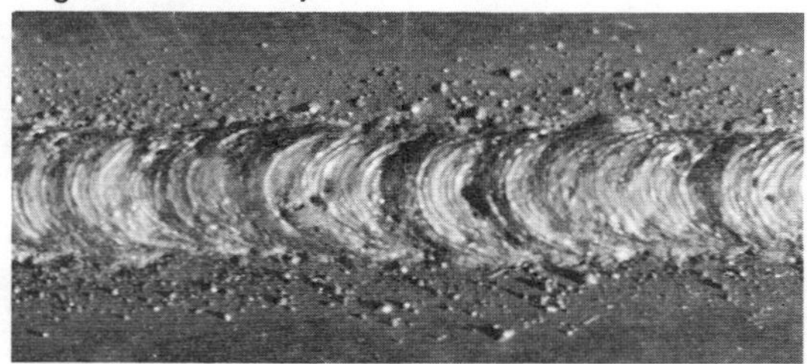	Gas flow does not displace air, clogged or defective system, frozen regulator	Set gas flow between 15 and 23 CFH. Clean spatter from nozzle often. Use a regulator heater when drawing over 25 CFH of CO_2
Fig. 23-3 Porosity in crater at end of weld 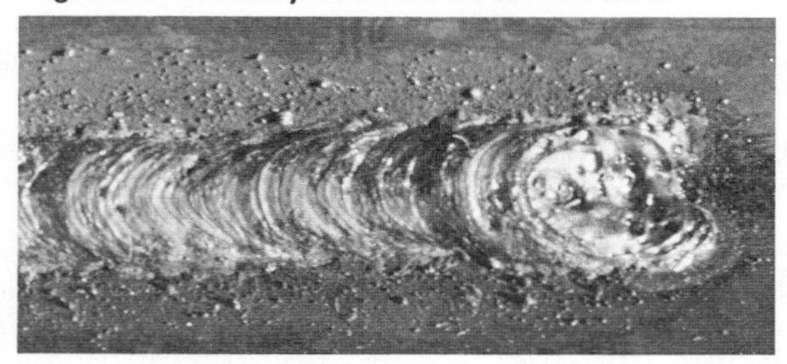	Pulling gun and gas shield away before crater has solidified	Reduce travel speed at end of joint
Fig. 23-4 Cold lap lack of fusion 	Improper technique preventing arc from melting base metal	Direct the welding arc so that it covers all areas of the joint. Do not allow the puddle to do the fusing. Use a slight whip motion.
Fig. 23-5 Burn-through and too much penetration 	Heat input too high in the weld area	Reduce wire-feed speed to obtain lower amperage. Increase travel speed. Oscillate gun slightly. Increase stickout to 1/2 inch maximum

Fig. 23-6 Lack of penetration

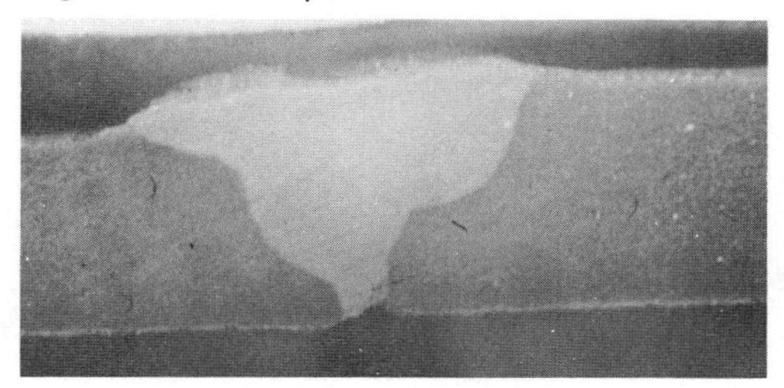

Heat input too low in the weld area

Increase wire-feed speed to obtain higher amperage. Reduce stickout to 1/4 inch.

Fig. 23-7 Whiskers

Electrode wire pushed past the front of the weld puddle leaving unmelted wire on the root side of the joint

Cut off ball on end of wire with pliers before pulling trigger. Reduce travel speed and, if necessary, use a whipping motion.

Fig. 23-8 Wagon tracks

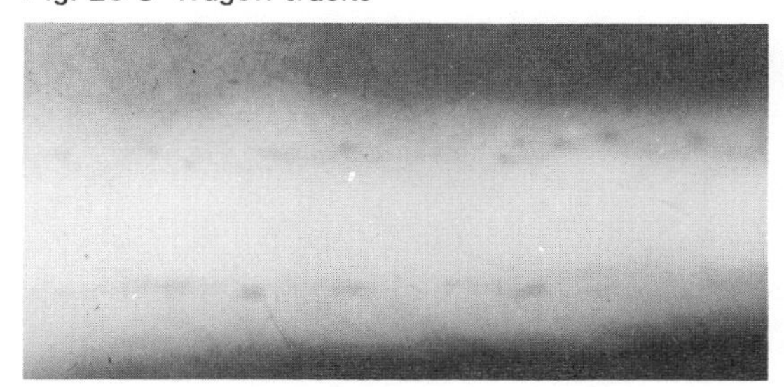

Too high bead contour or too high crown. Area where bead fuses to side of joint is depressed and next bead may not completely fill depressed area or void.

Arc voltage and travel speed should be high enough to prevent crown. When welding over these areas, be sure that the welding arc melts the underlying weld and base metal.

CAUSE AND CORRECTION OF DEFECTS

The operator has to be able to recognize and correct possible welding defects. MIG welding, like the other processes, must be properly applied and controlled to consistently give high-quality welds. The defects are shown in figures 23-2 through 23-8, accompanied by the causes and corrective actions to be taken.

REVIEW QUESTIONS

1. What positions does out-of-position welding refer to?

2. What is porosity in a MIG weld?

3. What quality will the weld probably lack if the current input is too low at the arc?

4. In MIG welding heavy plate, is more amperage and voltage required than welding light plate? Why?

5. Can the MIG gun be oscillated to improve bead conformity?

UNIT 24 MIG WELDING ALUMINUM

The welding of aluminum using the MIG process is advantageous over the TIG process because heavier sections of metal can be welded much faster. Aluminum and its alloys differ from mild steel in that there is no color change as the temperature from welding increases. Aluminum also develops a refractory oxide when exposed to air. Although material of .040 inch can be MIG welded, 3/16 inch is about the minumum for spray-arc welding.

Equipment

The constant-voltage machine should have a potential output of 500 amperes. Due to this increased output the standard gun is usually water cooled. There are guns which hold one pound of smaller diameter aluminum wire which do not require water for cooling.

Materials

1/4-inch aluminum plate
1/16-inch No. 4043 electrode wire
One cylinder of argon gas

Preweld Procedure

1. Set the voltage at 23 to 27 volts.
2. The wire-feed speed should produce a welding current of 225 to 300 amperes.

   Note: From these basic settings, a smooth transfer of metal across the arc can be obtained with slight adjustments. Voltage, amperage, and welding techniques vary to suit the joint and position conditions.
3. Adjust the gas flow to about 35 CFH.
4. Always clean the aluminum before welding. This can be done with suitable commercial solvents or by mechanically filing, scraping or brushing with a stainless-steel wire brush.
5. Always check the equipment manufacturers' recommendations before welding.

Welding Procedure

1. Maintain the stickout from 1/2 inch to 3/4 inch on all joints.
2. Use the same transverse angle for all joints as for mild steel.
3. Use the same longitudinal angle for all joints as for mild steel, except a leading gun angle should be used. See figure 20.3.
4. Hold the electrode wire toward the leading edge of the puddle. The forward motion can be steady or oscillating, depending on the application.
5. Check the joint preparation of aluminum since it is critical and the operator must be sure of good fitup. The butt joint, lap joint, and T joint are the best joints for fabrication.

6. Do the setup, tacking, welding and testing of the aluminum joints as for mild steel.

   Note: The spray-arc type of metal transfer can be used for out-of-position welding. Because of its fluidity, however, it is more difficult than short-arc welding.

Checking Application

1. The surfaces of the plate must be thoroughly clean for the best results.
2. Good joint fitup is necessary to provide weld puddle control and to prevent unnecessary distortion.

REVIEW QUESTIONS

1. What are two characteristics in welding aluminum that are different from those of mild steel?

2. What does the term oscillate mean in reference to the manipulation of the MIG gun and what is its function?

3. What type of gun angle is used in aluminum MIG welding?

4. What type of MIG welding is used on aluminum?

5. What is the minimum thickness of aluminum that can be MIG welded economically?

UNIT 25 MIG WELDING STAINLESS STEEL

The techniques involved in MIG welding stainless steel are similar to those used for mild steel. Spray-arc or short-arc welding can be used in the welding of stainless steel, depending on the thickness of the material being welded and the amount of current being produced. Refer to charts 25-1 and 25-2 for the welding conditions of spray-arc and short-arc welding, respectively.

Shielding gas should be argon with 1 percent oxygen, or argon with 2 percent oxygen depending on the thickness of the material being welded. In abbreviated form this is written O_2-1 and O_2-2.

CHART 25-1

GENERAL WELDING CONDITIONS, SPRAY ARC							
Plate Thickness (In.)	Joint & Edge Preparation	Wire Dia.	Gas Flow	Current (DCRP Amps)	Wire Feed (ipm)	Welding Speed	Passes
.125	Square Butt with Backing	1/16	35	200-250	110-150	20	1
.250	Single Vee Butt 60° Inc. Angle No Nose	1/16	35	250-300	150-200	15	2
.375	Single Vee Butt 60° Inc. Angle 1/16-in. nose	1/16	(O_2-1)	275-325	225-250	20	2
.500	Single Vee Butt 60° Inc. Angle 1/16-in. nose	3/32	(O_2-1)	300-350	75-85	5	3-4
.750	Single Vee Butt 90° Inc. Angle 1/16-in. nose	3/32	(O_2-1)	350-375	85-95	4	5-6
1.000	Single Vee Butt 90° Welded Angle 1/16-in. nose	3/32	(O_2-1)	350-375	85-95	2	7-8

CHART 25-2

GENERAL WELDING CONDITIONS, SHORT ARC								
Plate Thickness (In.)	Joint and Edge Preparation	Wire Dia. (In.)	Gas Flow (CFH)	Current DCRP (amps)	Voltage	Wire-Feed Speed (ipm)	Welding Speed (ipm)	Passes
.063	Nonpositioned fillet or lap	.035	15-20	85	15	184	18	1
.063	Butt (square edge)	.035	15-20	85	15	184	20	1
.078	Nonpositioned fillet or lap	.035	15-20	90	15	192	14	1
.078	Butt (square edge)	.035	15-20	90	15	192	12	1
.093	Nonpositioned fillet or lap	.035	15-20	105	17	232	15	1
.125	Nonpositioned fillet or lap	.035	15-20	125	17	280	16	1

A leading gun angle is used to give more visibility. An oscillating motion back and forth in the direction of the joint is desirable for fusion and uniformity.

Materials

16-gage stainless steel
.035 inch, type 308 electrode wire
75% argon and 25% CO_2 gas mixture

Preweld Procedure

1. Check the operation manual for the manufacturers' recommendations.
2. Copper backup bars are required for welding stainless steel, especially when welding only one side.

Welding Procedure

1. Arrange pieces of stainless steel to form joints as detailed in unit 22.
2. Stickout should be 1/4 inch to 3/8 inch.
3. Use either a leading or a trailing gun angle. The transverse gun angle always bisects the joint.

Checking Application

1. The workpiece must be thoroughly cleaned.
2. The wire-feed speed and the voltage settings are critical. A slight variation from the correct settings could produce unsatisfactory welds.

REVIEW QUESTIONS

1. What shielding inert gas is used in the MIG welding of stainless steel?

2. What kind of material is used for backup bars on stainless steel?

3. What does the word bisect mean?

4. What two types of MIG welding are used on stainless steel?

5. How is the thickness of stainless steel sheet specified?

UNIT 26 MIG WELDING COPPER

No other process is as good or as fast as MIG for welding copper or its alloys such as manganese-bronze, aluminum-bronze, silicon-bronze, phosphor-bronze, cupro-nickel, and some of the tin bronzes. To obtain high-quality welds in copper it is necessary to use deoxidized, non-oxygen-bearing forms of copper, copper-base material and filler material. Chart 26-1 gives the conditions for welding copper.

Argon is the preferred inert gas for welding 1/4 inch and thinner material. A flow of 50 cubic feet per hour gives a good shielding atmosphere. For greater thicknesses, a mixture of 65 percent helium and 35 percent argon offsets the high-heat conductivity.

Preweld Procedure

1. Preheat thicknesses of over 1/4 inch to 400 degrees F.
2. Use steel for backup when required on light thicknesses.

Welding Procedure

1. Arrange pieces of copper having thicknesses of at least 1/4 inch to form groove joints and T joints.
2. Use a stickout of 1/2 inch to 3/4 inch and an oscillating motion to weld the joints.
3. Use a leading gun angle for better visibility.

   Note: Good ventilation is important when welding copper and its alloys. The fumes that are produced are highly toxic and must be carried away from the operator.

Checking Application

1. Check the operation manual for manufacturers' recommendations.
2. The welding operator must be properly protected from radiation because of its intensity.

CHART 26-1

COPPER WELDING CONDITIONS						
Thickness (In.)	Amps DCRP	Volts	Travel (ipm)	Wire Dia. (In.)	Wire-Feed Speed (ipm)	Joint Design
1/8	310	27	30	1/16	200	Square butt, steel backup strip required
1/4 (1) 1/4 (2)	460 500	26	20	3/32	135 150	Square butt
3/8 (1) 3/8 (2)	500 550	27	14	3/32	150 170	Double bevel, 90° included angle, 3/16-in. nose
1/2 (1) 1/2 (2)	540 600	27	12 10	3/32	165 180	Double bevel, 90° included angle, 1/4-in. nose

REVIEW QUESTIONS

1. In terms of safety, what has to be foremost in the mind of the welder who is welding copper?

2. How many cubic feet per hour of inert gas is preferred for welding copper?

3. What kind of a backup is required on copper?

4. How much preheat is required on copper plate prior to welding?

5. What does heat conductivity mean?

UNIT 27 FLUX-CORED CO_2 SHIELDED MIG WELDING

The cored-wire welding process is a gas-shielded metal-arc welding process which uses the intense heat of an electric arc between a cored, consumable, continuously-fed electrode wire and the work.

SHIELDING GAS

The shielding gas (CO_2) (carbon dioxide, welding grade) displaces the air surrounding the arc and the weld puddle thus preventing contamination of the weld metal by atmospheric oxygen and nitrogen. This process requires a gas-flow rate of 35 to 60 cubic feet per hour. It may be necessary at times to have two or more cylinders manifolded together.

ELECTRODE WIRES

The flux-cored electrode wires, chart 27-1, recommended for use with this process, range in size from .045 inch to 1/8 inch diameter. The fluxes present within the core of these electrode wires add deoxidizers and strengthening elements to the deposited metal. The core also contains slag-forming and arc-modifying substances which protect the weld metal from atmospheric contamination, figure 27-1.

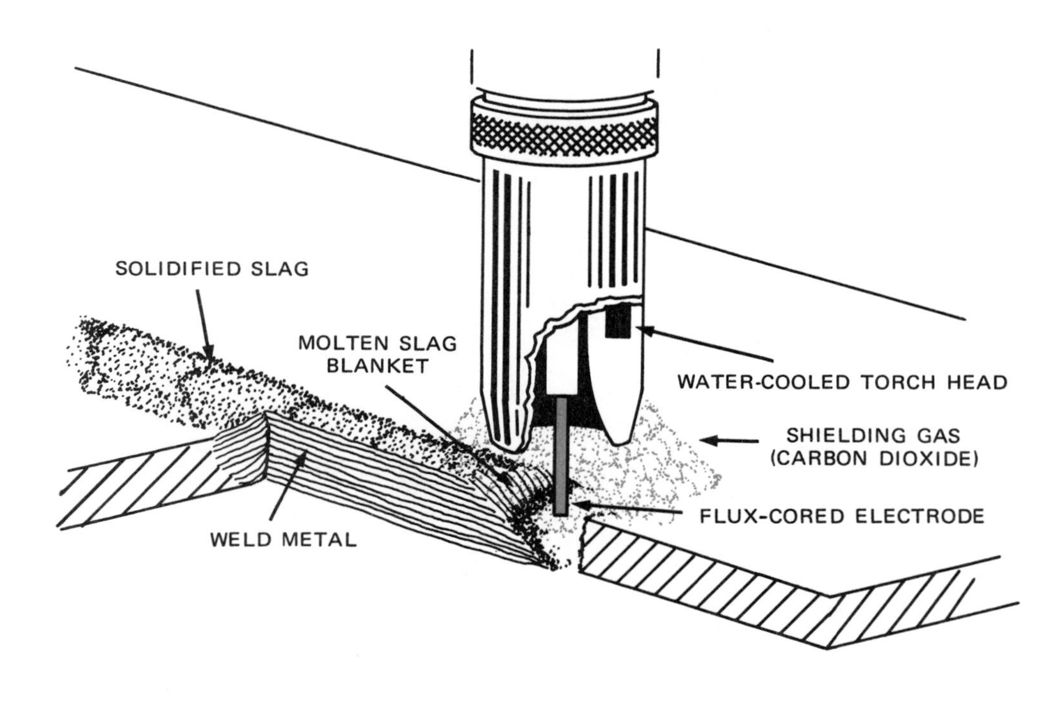

Fig. 27-1 Flux-cored gas-shielded welding

CHART 27-1

FLUX-CORED ELECTRODES									
MANUFACTURERS	American Welding Society Classification AWS–A5-20-69								
	E 60T-7	E 60T-8	E 70T-1	E 70T-2	E 70T-3	E 70T-4	E 70T-5	E 70T-6	E 70T-G
Airco Welding Products Div. Air Reduction Co. Inc.			Super-Cor Tuf-Cor	Flux-Cor No. 1					Tensil-Cor
Alloy Rods Company Div. Chemetron Corporation			Dual Shield 78, 111A, 111AC	Dual Shield SP, 110					
Arcos Corporation			Arcosarc 72	Arcosarc 70 Arcosarc 70X		Arcoshield 70			
Hardfacing Alloys, Ltd.				Har-CO_2-71		Har-CO_2-74	Har-CO_2-75		
P & H Welding Products Unit of Chemetron Corporation			P&HT-62 P&HT-64	P&HT-63	P&HT-73				P&HT-76 P&HT-100
Hobart Brothers Company			FABCO 71, 79, 76	FABCO 80		FABSHIELD 4			
Linde Div. Union Carbide Corporation			Linde FC-71	Linde FC-72					
McKay Company			Speed Alloy 71		Speed Alloy 73	Speed Alloy 74	Speed Alloy 75		
Murex Welding Products			Mure-Cor 2A-3	Mure-Cor 1					Mure-Cor MNM
National Cylinder Gas Div. of Chemetron Corporation			Dual Shield 78, 111A 75, 111AC	Dual Shield SP, 110					
Page, Division of ACCO			PFC-701	PFC-702					

APPLICATION

- Due to the very fluid weld puddle, satisfactory welds are best obtained by welding in the flat and horizontal positions.
- This process is used mainly for welding low- and medium-carbon steels and low-alloy, high-strength steels in thicknesses above 3/16 inch.
- Thicknesses up to about 1/2 inch are weldable with no edge preparation, and above 1/2 inch with edge preparation. Maximum thickness is unlimited.
- The electrode wires are designed for either single-pass or multipass procedure.

Welding Procedure

1. Set the stickout distance at 1 inch. It can be varied somewhat to control the weld.
2. Use a slightly leading gun angle. See figure 20-3.
3. Place the voltage setting between 25 and 36 volts. The wire-feed speed should produce from 250 to 700 amperes, depending on the application.
4. Using 1/2-inch plate, set up the lap joint and the T joint and weld according to preceding information.

Checking Application

1. Weld bead should have good conformity and penetration.

   Note: Penetration can be checked by cutting a specimen with a cutting torch through the weld and then grinding and filing the surface.

2. There should be no evidence of undercut.

REVIEW QUESTIONS

1. In the cored-wire welding process, how is the electrode wire different from standard MIG welding?
2. How does the deposit from cored wire compare to that of solid wire?
3. What kind of gun angle is used in cored welding?
4. What does the symbol CO_2 stand for?
5. How is the weld protected from atmospheric contamination?

UNIT 28 MIG SPOT WELDING

There are many welding applications that do not require 100 percent continuous welding of a particular joint. MIG spot welding is gaining wide acceptance today. It is competitive with riveting and resistance spot welding.

OPERATION

- Little welding skill is required.
- The operator places the gun nozzle against the metals to be joined and pulls the trigger.
- The welding control completes the welding cycle while the operator holds the torch in position.
- The preparation of the material depends on the strength required in the finished work.

APPLICATION

- Equipment should provide instantaneous and positive arc ignition, a constant rate of filler-wire feed, and precise timing of the weld cycle.
- An equipment setup includes constant voltage power supply, and automatic spot-welding control, wire-feed unit and a gun and lead assumbly. A different type of nozzle than that used with normal MIG welding is required.
- MIG spot welding can be used on carbon steel, stainless steel, copper-bearing alloys, and all weldable aluminum alloys.

REVIEW QUESTIONS

1. What controls are different for MIG spot welding?

2. What kind of welding machine is used for MIG spot welding?

3. Does MIG spot welding require a lot of practice?

4. What one part on the MIG gun is different when spot welding?

5. What governs the strength of a weld?

ACKNOWLEDGMENTS

The authors wish to express their appreciation to the following for their assistance in the development of this text:

- Air Reduction Sales Company, New York, NY 10017
- Allegheny Ludlum Steel Corporation, Pittsburgh, PA 15222
- Aluminum Company of America, Pittsburgh, PA 15219
- American Welding Society, Miami, FL 33125
- Bausch and Lomb, Rochester, NY 14602
- Detroit Testing Machine Company, Detroit, MI 48213
- Dow Metal Products Company, Midland, MI 48640
- Hobart Brothers Company, Troy, OH 45373
- Lincoln Electric Company, Cleveland, OH 44117
- Linde Company, Division of Union Carbide Corp., New York, NY 10017
- Los Angeles Trade-Technical College
- Miller Electric Company, Appleton, WI 54911
- Niagara Mohawk Power Corp., Syracuse, NY 13202
- Norton Company, Worcester, MA 01616
- Revere Copper and Brass, Inc., New York, NY 10017
- Sylvania Electric Products Company, Towanda, PA 18848
- Tempil Corporation, New York, NY 10011
- United States Steel Corporation, Pittsburgh, PA 15230
- Victor Equipment Company, Denton, TX 76203
- Welding Design and Fabrication, Cleveland, OH 44113
- Wilson Instrument, Division American Chain and Cable, New York, NY 10017

The following members of the staff at Delmar Publishers contributed to the preparation of this edition:

Director of Publications — Alan N. Knofla
Source Editor — Mark W. Huth
Associate Editor — Judith E. Barrow
Director of Manufacturing/Production — Frederick Sharer
Production Specialists — Debbie Monty, Patti Manuli, Sharon Lynch, Jean LeMorta, Betty Michelfelder, Lee St. Onge, Margaret Mutka
Illustrators — Tanya Harrell, George Dowse, Tony Canabush, John Orozco

This material has been used in the classroom by the Oswego County Board of Cooperative Educational Services, Mexico, New York.